Springer Series in Advanced Manufacturing

Series Editor

Duc Truong Pham, University of Birmingham, Birmingham, UK

The **Springer Series in Advanced Manufacturing** includes advanced textbooks, research monographs, edited works and conference proceedings covering all major subjects in the field of advanced manufacturing.
The following is a non-exclusive list of subjects relevant to the series:

1. Manufacturing processes and operations (material processing; assembly; test and inspection; packaging and shipping).
2. Manufacturing product and process design (product design; product data management; product development; manufacturing system planning).
3. Enterprise management (product life cycle management; production planning and control; quality management).

Emphasis will be placed on novel material of topical interest (for example, books on nanomanufacturing) as well as new treatments of more traditional areas.

As advanced manufacturing usually involves extensive use of information and communication technology (ICT), books dealing with advanced ICT tools for advanced manufacturing are also of interest to the Series.

Springer and Professor Pham welcome book ideas from authors. Potential authors who wish to submit a book proposal should contact Anthony Doyle, Executive Editor, Springer, e-mail: anthony.doyle@springer.com.

More information about this series at https://link.springer.com/bookseries/7113

Kunpeng Zhu

Smart Machining Systems

Modelling, Monitoring and Informatics

Springer

Kunpeng Zhu
Precision Manufacturing Laboratory
Institute of Advanced Manufacturing Technology
Chinese Academy of Sciences
Changzhou, Jiangsu, China

Institute of Precision Manufacturing
Wuhan University of Science and Technology
Wuhan, Hubei, China

ISSN 1860-5168 ISSN 2196-1735 (electronic)
Springer Series in Advanced Manufacturing
ISBN 978-3-030-87877-1 ISBN 978-3-030-87878-8 (eBook)
https://doi.org/10.1007/978-3-030-87878-8

This Springer imprint is published by the registered company Springer Nature Switzerland AG
The registered company address is: Gewerbestrasse 11, 6330 Cham, Switzerland

Dedicated To My Family

for their love and support

Preface

The Motivation of This Book

The Computerized Numerical Control (CNC) machining has been the backbone to the modern industry. The CNC machining, in particular the precision machining (i.e. micro-machining, high-speed machining) is a very complex thermo-mechanical material removal process, in which any changes in the process will have an effect on the machining quality. The conventional machining system does not take the process changes into account, but complete the process only according to the given process parameters and tool path. It cannot take adaptive actions under un-preceded machining state variations and hard to achieve the real-time process control and optimization. Therefore, the machinability of the system is fully utilized to ensure the high precision requirement and improve the productivity.

With the fast development of artificial intelligence (AI) and modern sensing technology, more efforts have been conducted on the application of advanced AI technologies to enhance the manufacturing intelligence and to improve the machining precision and productivity. In this aspect, there has been a booming of research on intelligent monitoring and control of CNC machining system, which would ultimately lead to smart machining to ensure the optimum performance of the machining systems. The concept of smart machining system is introduced in this context.

The Smart Machining System (SMS) is based on modern metal cutting theory and intelligent manufacturing technology, which can intelligently predict and optimize the cutting process based on the system perception and feedback. It adopts advanced online sensing and data processing theory and carries out real-time monitoring of the machining process. By applying the AI technology (i.e. machine learning approaches), the machining state is determined and the process parameters are optimized in real time, which enable the intelligent control of the machining system to obtain the ideal workpiece quality and machining efficiency.

With the arrival of industry 4.0 era, smart machining system has become increasingly important in modern precision machining technology and it shows a great prospect in the industrial applications. Therefore, it is very important to study the basic theory of smart machining system based on the process monitoring and intelligent control and explore its technical framework under the theory of big data and cyber-physical system. At present, there are many books on machining process mechanics and automation, but without books on data-oriented smart machining. This book aims to bridges this gaps and develops the general idea of design and implementation of the smart machining system, especially on the tool condition monitoring system.

The Content of This Book

Through reviewing the background knowledge and the latest development, this book introduce the state-of-the-art of smart machining theory and technology. It takes the tool condition monitoring as case studies and systematically discusses the modeling, simulation, monitoring, and intelligent control of machining process to provide synergic solutions for the establishment of smart machining system.

This book starts from the introduction of modern manufacturing system and the motivation of machining precision and intelligent requirement, and then it systematically elaborates the fundamental theory as well as the applications of smart machining system. It focuses on one of the machining process modeling and simulation methods, the state monitoring theory, and intelligent monitoring and control methods. These theoretical analyses are the basis for the establishment of smart machining system.

The book has an in-depth investigation of the latest development of machining theory and machine learning tools for intelligent process monitoring and control, describes the most common monitoring techniques used and the signal processing methods when applied to real problems. At the same time, the book introduces new methods and develops a cyber-physical production system (CPPS) framework of smart machining monitoring system. Based on the latest development of cyber-physics system theory, it integrates the machine tool system dynamics, metal cutting mechanics, tool wear mechanism, big data analytics and the machine learning theory for heterogeneous data fusion. Under this framework, case studies are elaborated on deep fusion of physical process and cyber process to realize online monitoring, tool wear compensation, and smart control of machining process.

There are 12 chapters in the book, which are divided into four parts: the basic theory of machining technology, the fundamental theory of smart system modeling and monitoring, the application topics, and the development trend. Specifically,

- Chapter 1–3 covers the state-of-the-art approaches applied to metal cutting theory.
- Chapter 4–6 addresses up-to-date signal processing techniques and machine learning methods for intelligent process modelling and monitoring.
- Chapter 7–10 integrates theory and practical implementations of smart machining process modelling and monitoring system.
- Chapter 11–12 develops new smart machining monitoring system via big data and cyber-physical production system framework.

The content of this book extends the theory and the application of high performance cutting technology in the field of precision and intelligent manufacturing. It is the integration of the author's research and teaching experience in the field, and would be a valuable reference for the researchers, graduate students, engineers as well as a complements to the higher level undergraduate students.

Acknowledgement

This book summarizes the author's research in this field in the past 20 years, and it represent the contributions of many collaborators. The author would like to give his special thanks to Prof. Wong York San and Prof. Hong Geok Soon of National University of Singapore, who lead the author to this important and interesting research filed and continuously encourage and inspire him ever since the PhD study.

The author thanks to Dr. Liu Tongshun, Dr. Duan Xianyin, Dr. Li Guochao, Dr. Li Si, and PhD students Zhang Yu and Guo Hao from the Precision Manufacturing laboratory, Institute of Advanced Manufacturing Technology, the Chinese Academy of Sciences. In the process of writing this book, the author would like to thank a group of excellent and dedicated postgraduates. This book was completed with their support, especially, Zhang Yu, Guo Hao, Imran Hussian, Yuan Dezhi, Ling Zhihao, Huang Chengyi, Li Xin, Sun Jian, and research engineer Li Gang and Wang Qisheng.

The author would also like to thank Prof. Jerry Fuh Yin Hsi and Prof. Lu Wen-Feng for their long-term support and guidance.

Thanks to the Institute of Advanced Manufacturing Technology (IAMT), Hefei Institute of Physical Science, and the Chinese Academy of Sciences, for providing an amenable scientific research environment for this work being successfully completed.

Thanks to the Alexander von Humboldt Foundation and Professor Birgit Vogel-Heuser of Institute of Information and Automation of Technical University of Munich (TUM) for their support during my stay in Germany. Part of the lecture notes in the course of Signal Processing in Mechatronics (2011-2013) in TUM forms the basis of Chap. 6.

Thanks to the Ministry of science and technology, the National Science Foundation of China, and the Chinese Academy of Sciences for their support.

Changzhou, Jiangsu, China/Wuhan, Hubei, China Kunpeng Zhu
April 2021

Contents

Chapter 1
Introduction to the Smart Machining System

1.1 The Development of Modern Manufacturing System

We are now living in the new era of the fourth industrial revolution (Industry 4.0). The important sign of this era is the fast developing of 3C (computation, communication, control) and AI (artificial intelligence) technologies, as well as their pervasive applications uses in the manufacturing industry. Due to these benefits, the machinery manufacturing system is undergoing tremendous growth in automation, optimization, flexibility, integration, intelligence, and precision. The increasing impact of system science and methodology on the machinery manufacturing system is resulting in the new idea of modern machinery manufacturing systems.

Because of the rapid advancement of science and technology since the beginning of the century, machinery manufacturing has made tremendous progress. The use of computer technology and automation in the design and management of machinery manufacturing industries is one of the most prominent features. Thanks to computer technology and automation, with which the labor productivity and integrated work process have been significantly improved [1–4]. In the meantime, they have achieved significant technical and economic benefits by shortening the product design and manufacturing cycle, ensuring product quality, improving working conditions, operations, and management, and reducing energy consumption. As a result, the automation of the manufacturing system is highly valued and developed rapidly. It can be said that the development process of modern machinery manufacturing systems is equivalent to the development of automation and integration of mechanical manufacturing.

Looking back on the developmental stages of manufacturing systems, five distinct stages can be identified. These stages are listed with more descriptions in Table 1.1.

Stage 1: The development period for stand-alone automation, which is marked by a rigid automation production line. From the beginning of the twentieth century to the 1950s and 1960s, automation technology was widely used in various types of manufacturing processes. Mechanization, semi-mechanization, and assembly lines have all been replaced by automatic production lines in industry. This automation

K. Zhu, *Smart Machining Systems*, Springer Series in Advanced Manufacturing,
https://doi.org/10.1007/978-3-030-87878-8_1

Table 1.1 The five stages of the development of manufacturing system

Stage	Name	Application	New technology	Feature	Period
Stage 1	Automatic production line	Mass production	• Relay program control • Machine tool combination	Rigid and high productivity	1940–1950
Stage 2	CNC machine tool machining centre	Multi-species production of single or more	• Numerical control (NC) • Computer numerical control (CNC)	Flexible process focus	1950–1980
Stage 3	Flexible manufacturing systems	Multi-species production of small batch, Mass production	• CAD • Robot • Group technology (GT) • DNC • Automation	The ideal combination of flexibility and efficiency	1970–1980
Stage 4	Computer integrated manufacturing system	Factory automation of design, manufacturing and economic management	• CAD/CAM/CAPP • Production management and scheduling • Information technology • Simulation technology	Fully automated, optimized, intelligent and distributed communication network	1980–present
Stage 5	Cyber-physical production system	Smart factory, digital workshop	• Digital twin • Dynamic scheduling • Cloud computing • Big data analytics	Adaption to changing environment and smart decision	2014–present

is compatible with the mass production model, which resolves the mass production and process automation problem.

Stage 2: The development period for special equipment automation, which is defined by numerical control (NC) system and CNC machine tools. The machinery industry's production scale and productivity expanded significantly from the early 1950s to the 1970s [1]. Meanwhile, the manufacturing system's complexity and automation are increasing. The in-depth application of modern control theory and computer technology gives the manufacturing industry a huge boost. The CNC technology is developed to meet machining automation and control requirements, which employs digital technology to achieve a programmable tool movement trajectory. The core equipment in this stage is the CNC machining center.

Stage 3: The period of system automation development, which is defined by a flexible manufacturing system (FMS) [3, 4]. The automation of individual device equipment into the entire production line, as well as the automation and flexibility of entire sections and workshops, transforms from the 1970s to the 1980s. The FMS is being developed to adapt small batch production automation. It is usually composed of three components: a CNC machine processing center, a storage system, and a computer system for central control. The flexibility of the manufacturing system refers to its ability to perform a variety of processing tasks. Flexibility evaluates not only the system's ability to be rearranged and adjusted but also evaluates the cost to complete these production tasks, as well as the cost of rearranging and adjusting the system. Flexible manufacturing systems are divided into three basic types based on the size and flexibility of the system scale: flexible manufacturing cell (FMC), flexible manufacturing system (FMS), and flexible automatic line (FAL). The flexible manufacturing cell not only operates independently but also as part of a flexible manufacturing system that includes several flexible manufacturing cells. It can be said that "flexibility" is also one of the important features of the modern manufacturing system.

Stage 4: The integrated automation development period, marked by the Computer Integrated Manufacturing System (CIMS) [1]. The science and technology of 3C have been developing rapidly in the last 40 years. They provide an important system theory and technology for manufacturing automation by combining it with a machine tool system. They enable the machining system to use a unified data format and protocol to ensure the transfer of information and the sharing of resources. In addition, they also replace a blueprint for product engineering or other technical documents. As a whole, the various subsystems can constitute an organic link through automated factory communication, i.e. automated factories. Computer Integrated Manufacturing (CIM) reflects this new development of the machinery manufacturing system. The CIMS is being developed with the development of computer-aided design and manufacturing. Based on information technology, automation, and manufacturing, it gradually integrates various isolated automation subsystems scattered throughout the process of product design and manufacturing through computer technology to form an integrated and smart manufacturing system that is suitable for multi-variety and small batch production and achieves overall benefits. It is the strategic goal of the development of machinery and advanced manufacturing technology of the machinery industry in the twenty-first century. CIMS is pursuing not only modernization but also optimization, flexibility, intelligence, and integration [5].

Stage 5: The Cyber-physical system (CPS) in manufacturing, which is a multi-dimensional complex system integrating computing units, network, and physical environment [6]. Through the holistic integration and deep cooperation of 3C technology, it realizes the real-time perception, dynamic control, and information service of large-scale manufacturing systems. The CPS realizes the integrated design of computing, communication, and physical system, which can make the system more reliable, efficient, and real-time collaborative, and has an important and wide application prospect. In the CPS, the integrated network, the information processing, sensing, and driving technologies make cyberspace and physical equipment deeply integrated, so that the system can achieve enough intelligence and meet the action needs in the changing environment [7].

The cyber-physical production systems (CPPS) is the specific form of CPS in manufacturing field. The development of CPPS is dependent on the advancement of 3C technology, as well as the breakthrough of manufacturing equipment and advanced processes.

The framework of CPPS, which is a systematic concept and fundamental to Industry 4.0, is relatively new. The CPPS is still in its early stages of development, and many core issues and technical details have yet to be effectively defined or addressed. Different manufacturing focus on different processes, and their cognition of CPPS differs as well. Intelligent Manufacturing Systems (IMS), Digital Factory (DF), and Reconfigurable Manufacturing Systems (RMS) [7] are three well-known models that have developed closely related the CPPS rationale. The CPPS requires more research and development, and there is still a long way to go before it becomes a mature system for applications.

1.2 Modern Machining Technology

The machining technology is the foundation and core of manufacturing technology, which is determined by various methods of process. The machining technology refers to the use of a tool (including grinding wheels) or energy flow through removal, modification, addition or other means of making a workpiece materials that satisfy certain design requirements for finished products. The purpose of the machining is to obtain a certain geometry of the surface and to have a certain geometric accuracy. There are many types of machining processes, such as turning, milling, grinding, polishing, drilling, EDM, lithography, ultrasonic machining, laser machining, and so on [8, 9]. The main content of this book will cover the conventional CNC machining technology, in particular the high-speed milling in modern precision manufacturing. Facing the increasing demand of technology, the development trend of modern machining technology has shown the following important characteristics.

1.2.1 High Precision Machining

Obtaining higher machining accuracy has always been the aim of machining technology. Over the last 100 years, the precision of ordinary machining has been improved by about three orders of magnitude, while the precision of precision machining has reached a level of 10 nm and has even improved by about five orders of magnitude. In 1983, Professor Taniguchi [10] summarized the current state of the art of precision and ultra-precision machining technology and predicted its development trend, as shown in Fig. 1.1. Today, 40 years later, the development of precision and ultra-precision machining technology continues to follow the trend shown by some

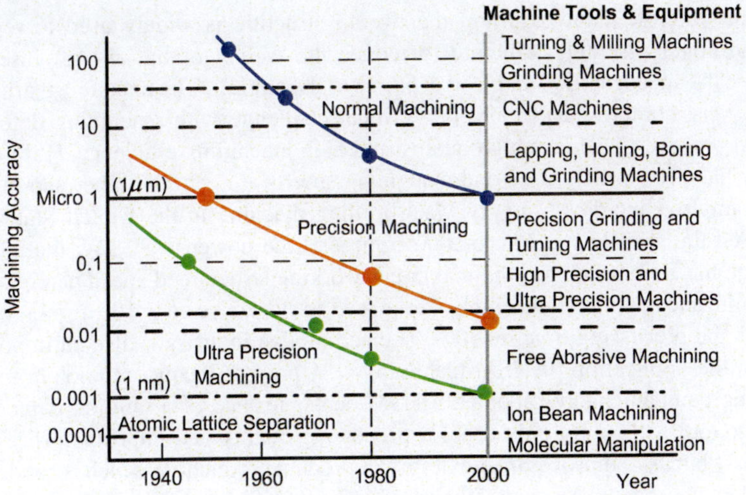

Fig. 1.1 The machining accuracy improvement in the twenty-first century [10]

curves in the figure. Although the figure shows that the precision of ultra-precision machining will soon reach the limit of the atomic lattice distance, there is still a huge requirement for improvement in precision for conventional machining, and a further improvement in machining precision is still an important trend in the development of machining technology [11].

In addition, the advancement of micro machining technology promotes the implementation of precision and micro machining, which leads to the birth of micro electro mechanical system (MEMS) and micro mechanical cutting technology [12]. Micro cutting has the characteristics of micromachining and ultra-precision machining and is gradually integrated with nanomachining due to its object size becoming small to micron level, its dimensional tolerance and geometric tolerance becoming as small as tens of nanometers, and its surface roughness is as low as nanometers. Today, scientist have completed the relocation and arrangement of individual atoms in the laboratory, and the width of the large scale integrated circuits has reached 10 nm. Micro-machining and nano-machining have broad prospects with the development of these technologies.

1.2.2 *High Speed Machining*

With the development of high-speed spindle, linear motor, and high-speed control unit, the high-speed machining (HSM) has become an important machining process to achieve high product quality and high machining efficiency [13]. In recent years, HSM has been widely applied in aviation, aerospace, automotive, and mold manufacturing.

In the aircraft manufacturing, the overall structure is widely used to replace the traditional assembly structure to reduce the weight of the aircraft fuselage, improve the speed, agility, and load capacity of the aircraft. The machining structure is also very complex, and there is a critical challenge with machining deformation. At present, to achieve high quality and high machining efficiency, HSM technology has been widely used and machining speeds are getting higher and higher. For example, Cincinnati's hyper Mach milling machine in the United States has increased the spindle speed to 60,000 r/min and the power to 80 kW; the milling machine has a linear motor, the maximum working stroke feed speed can be up to 60 m/min, and the maximum empty stroke feed speed can be up to 100 m/min, which reduces the processing time by 50%. The automotive industry is also an important field in the application of HSM technology. At present, many automobile manufacturing companies have adopted a HSM center instead of a multi-axis modular machine tool, which not only ensures processing quality and improves processing efficiency but also improves the flexibility of product production, which is conducive to product upgrades. The successful introduction of HSM technology in other fields, such as mold production and electronic manufacturing, has greatly encouraged the development of the industry.

The HSM center is still under fast development. Its spindle speed has reached up to 300,000 rpm and its linear feed speed has reached 200 m/min. With the development of HSM process, machine tools, the understanding of cutting mechanics, the applications of HSM will be increasingly widespread.

1.2.3 Green Machining

The green machining has emerged from the concept of green manufacturing. It aims to adopt less advanced and pollution-free machining technology in the production and to save as many resources as possible. Its main characteristics are energy savings, low consumption, and free of pollution [13, 14].

Energy savings are aimed to decrease energy losses as much as possible in the machining process. For example, cutting power consumption can be reduced by optimizing cutting performance. In general, the energy required to remove the unit volume of material should be reduced as much as possible, i.e. the specific energy required to remove the material. Low consumption refers to the consumption of raw materials by simplifying the composition of the machining system during the production process. Material consumption can be reduced by optimizing machining process, and adopting less chip removal process, dry machining technology, remanufacturing technology, and other methods.

In addition, in the green machining efforts are made to ensure that 'no waste' is produced, that is, to adopt advanced processing methods or to take special measures to reduce or eliminate waste liquids, gas, residues, noise, and any other substances that may harm the environment and operators.

Modern machining technologies are paying attention to green environmental protection to achieve sustainable development and finally achieve real harmony between man and nature. With the development of science and process and the advancement of human society, green machining has become an inevitable requirement and trend.

1.2.4 Smart Machining

In the machining process, any changes of state will affect the process and the final machining efficiency. The conventional CNC machining technology does not take into account the change of tool state, but only the geometry of the workpiece, the process parameters, and the tool path. It cannot handle un-preset situations in the real-time machining process. Corresponding measures cannot be taken based on the change of state in the processing process and the real-time optimization of the processing state cannot be realized. Therefore, the machining capability is not fully utilized to ensure an optimum final quality. These problems can be well resolved through the use of smart (or, intelligent) machining technology [15]. Smart machining is a technological innovation of the existing CNC machining technology, in which the state monitoring, intelligent optimization, and adaptive control of the machining process can be achieved through simulation analysis and optimization. Data processing and sharing throughout the entire machining process enables the various changes in the cutting process to be intelligently "predicted", "sensed", "controlled" and "optimized" for smart machining [16–19].

Smart machining technology is the main technical characteristic of state-of-the-art manufacturing process and equipment. It has been explored widely and has been the core of many national advanced manufacturing strategies, such as the "Advanced Manufacturing Partnership Program (AMP)" of United States, "Industry 4.0" of Germany, and "Made in China 2025" of China [18, 19]. It is not only the fundamental technology of intelligent manufacturing systems but also the key machining process for high quality, high efficiency, and excellent control.

1.3 The Smart Machining System

Smart machining is a process that based on process monitoring and intelligent control technology, designed to adaptively solve many uncertain problems in the process that require manual intervention. Its ultimate goal is to realize the intelligent decision-making, monitoring, and control of the machining process. The fundamentals of smart machining is the modern cutting theory and digital manufacturing technology [20]. In the machining process, advanced monitoring and information analysis are used to monitor and extract the state-of-the-art features of machine tools, work-piece, and cutting tools in real-time. Artificial intelligence, combined with theoretical

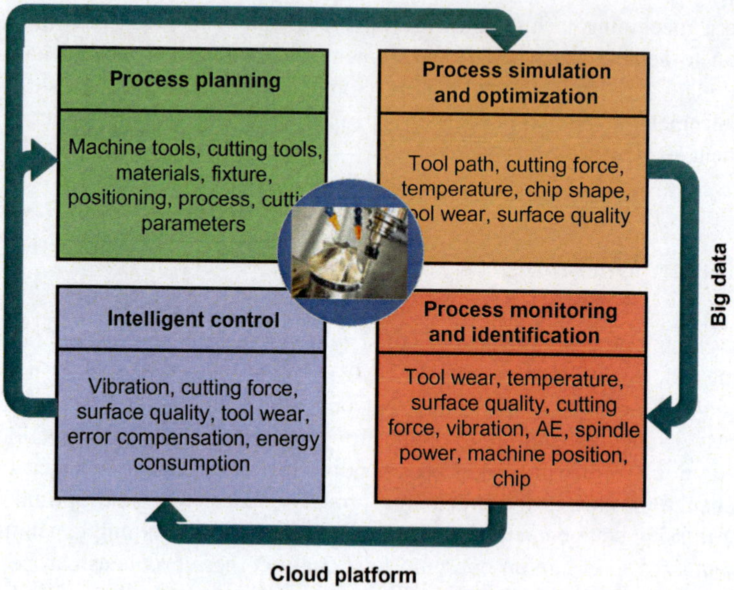

Fig. 1.2 The content of smart machining system

knowledge and processing experience, is used to assess the machining state, with data analysis, reasoning, decision-making, and real-time optimization to achieve the intelligent control of the machining process. In this way, the optimal machining is completed and the ideal workpiece quality and machining efficiency are achieved. Figure 1.2 shows the main factors involved in smart machining.

The online monitoring and optimization of the machining process lies in the core of smart machining system, which mainly includes online monitoring, decision-making optimization, and real-time control modules. The on-line monitoring module can monitor and extract state signal processing features that can 'sense' the specific working conditions of the machine tool, tool, and workpiece and assess the machining state.

The process parameters can be adjusted online via the real-time control module. Intelligent control of the cutting process is achieved by adjusting the cutting parameters (speed, cutting depth, feed rate), tool position and posture, and fixture compensation position, so that the machining process is always in the ideal state.

Information processing is carried out through the whole process of smart machining. The data involved in machining system include basic machine tool parameters, fixture, tool and workpiece, cutting parameters, optimization strategies, measurement and inspection data, and data control parameters, etc. Based on data and cloud computing technology, combined with artificial intelligence algorithms, data is analyzed and optimized for intelligent machining. It achieves multi-terminal data extraction through data communication and enables data cloud platform sharing by uploading data to the cloud.

1.3.1 Intelligent Process Planning

For the traditional machining process planning, the analysis of the process is mainly based on the personal experience of the process engineer, the selection of machine tools, tooling fixtures and tools, and, finally, the selection of cutting parameters according to the requirements of the process and the mechanical machining process. The main problem with this process planning is that human factors have a significant impact on the final processing quality of the parts. Due to the difference of the personal knowledge and experience, the process parameters chosen by different process engineers vary a lot for the same part and the quality of the processed parts is not consistent.

The key feature of smart machining process planning is the introduction of data processing technologies such as database, knowledge base, big data, and cloud platforms into machine tool selection, tooling fixtures, tools, and process parameters [21–23]. The simulation method is introduced to simulate and optimize the planning of the process. It is very important to select the new parts process parameters by referring to the machining parameters accumulated by the same type of parts. By analyzing and extracting a large number of reference, the appropriate process parameters for the current parts are selected. The process planning process is not only based on the personal knowledge and experience of process engineers but is also on the processing data of many engineers. The obtained process parameters are more reasonable than the process parameters of a single process engineer, and significantly avoids the influence of human factors on the processing quality. Meanwhile, process information and processing quality parameters will also be stored, and data sharing will be done through cloud data, which will provide a reference for further process planning.

Simulation of process planning could predict the process, identify potential problems in the process as earlier as possible, and provide suggestions for system improvement and optimization plan. For example, it is decided whether the selected machine tool, fixture, and tool are reasonable, whether the tool path is impaired, and whether the selected processing parameters are optimized.

1.3.2 The Process Simulation and Optimization

The cutting process can be predicted and optimized by the machining process modeling using geometric, physical and numerical simulations before actual cutting. Through simulation analysis, the physical quantity changes in the machining process are obtained to predict the actual machining process, the cutting parameters are then optimized to guide the selection of the actual processing parameters, and potential problems in the processing are identified.

The tool path is constantly changing in the CNC machining process. Through the geometric simulation, we can find out whether the tool path is correct, identify whether there is overcutting and undercutting, possible collision between the tool and the workpiece, and the associated problems such as low machining efficiency and poor surface quality due to the unreasonable tool path. At the same time, the simulation of the machining process can predict and optimize the machining process, shorten the machining time, extend the tool life, improve the surface quality, and improve the machining quality and efficiency as a result.

The main factor to be considered in the tool path generation process is the geometry of the workpiece while ignoring the influence of the position of the tool axis on the machine tool process and the motion of the machine tool. Tunc et al. [24, 25] analyzed the impact of the cutter shaft position on the cutting process by simulation and proposed a new method of cutting shaft vector optimization. Using the simulation method, the tool axis vector in the machining process is controlled along the generated tool path to improve the machining process in terms of cutting strength, stability, and machine motion. The method is effective under certain constraints.

The simulation of the cutting process allows us to predict cutting strength, cutting temperature, tool wear, chip shape, and other stages of the cutting process [26, 27]. By analyzing the results obtained from the simulation, we can identify the problems that may arise in the cutting process, such as excessive loading, high temperature, serious wear, and so on, to propose solutions and change the processing parameters. Through physical simulation and optimization of the cutting process, the cutting parameters and angle of the tool can be optimized to obtain the best performance. At the same time, the results obtained by simulation can be used as a reference for monitoring the actual cutting data, which enable to optimize the cutting parameters and the tool angle.

The variation of the cutting force is the most sensitive to the dynamic changes of the cutting process and forms the basis for understanding the cutting mechanism. At the same time, force modeling is an important part of process optimization and intelligent process monitoring. In recent years, from macro to micro, scholars have developed many models, which focuses on the accurate solution of instantaneous cutting thickness considering the size effect, cycloid trajectory, tool runout, and other factors [28]. Sun et al. [29] calculated by vector method the instantaneous undeformed chip thickness and the cutting contact position under the influence of tool eccentricity and established a cutting force model of five-axis machining. Altintas et al. [30] summarized the empirical model, maximum shear stress and minimum energy principle respectively, resolved cutting parameters such as shear stress, shear angle, and friction angle by an iterative algorithm, and established a cutting force prediction model by oblique cutting mode. In the microscale, the blunt radius and the size effect of the tool have a significant effect on the cutting force.

The simulation can predict the machining quality of parts, such as geometry size, dimensional precision, surface roughness, and residual stress, which can further guide the selection and optimization of machining parameters and tools.

1.3.3 The Machining Process Monitoring

The most direct means of "sensing" the machining state is by monitoring the machine tool, tool, and workpiece in the machining process. Metal cutting is a very complex process involving the dynamic changes of these components, for example, the chatter and vibration of machine tools, wear, and damage of tools. The monitoring status is many and complex, including machine tool and tool position, cutting force, tool temperature, tool wear, vibration, acoustic emission, workpiece surface quality, chip shape, etc. For example, monitoring the changes in tools and workpiece can identify the interaction between the tool and the workpiece, and the changes real-time to determine whether there are anomaly of cutting forces and tool temperature, serious wear, and severe vibration. It can therefore judge whether the machining state is correct and whether the stable cutting process is carried out and whether the "sensing" of the cutting process can be carried out.

How to monitor cutting state and predict tool life intelligently is one of the technical foundation for the realization of precision and smart machining. Shi et al. [31] model and classify the state of the tool by means of the time–frequency and dynamic characteristics of the cutting force and obtain good results. Nouri et al. [32] have proposed a new method for monitoring the wear of tools in real-time, which is not related to processing conditions. During the machining process, the model coefficients of the milling force are tracked. The results show that these coefficients can be used to describe the tool's wear status. Studies have found that acoustic emission signals and spectral characteristics can meet the requirements of tool wear monitoring for different application needs [33, 34]. Szydlowski et al. [35] and Zhu et al. [36] detected a change in the diameter of the tool surface to identify the changes of tool wear status. Other studies established prediction models by monitoring the surface quality of the workpiece, using neural networks or other machine learning approaches [37, 38], indirectly obtained the tool surface roughness.

The fusion of multi-sensor signals reflects more information on the changes in tool status and improves the state recognition capability [39]. Malekian et al. [40] used acoustic emission, force, vibration, and developed an adaptive, fuzzy neural network model to predict tool wear that had a better predictive effect. Due to the inherent flaws of traditional neural network methods in training and learning, more scholars turn to model methods that are easier to learn and expand, such as the hidden Markov model [41], the deep learning network [42, 43], etc., which have achieved ideal results in milling tool monitoring.

Based on these researches, many manufacturers have developed commercial machining process monitoring software [44–47]. Based on the adaptive monitoring system developed by OMATIVE (Israel), cutting power is kept to the maximum by changing feed rate to maximize machining efficiency without damaging the tool. Other system such as ARTIS (Germany), MARPOSS (Italian) and RENISHAW (UK) have systems for monitoring tool status and processing intermittent tool shapes. The types of sensors used in these commercial software are summarized in Table 1.2. These commercial systems are mainly based on the condition monitoring of

Table 1.2 Typical machining process monitoring modules

Product	Omative	Artis	Brankamp	Montronix	Nordmann	Renishaw	Marposs
Power	✓	✓	✓	✓	✓		✓
Torque		✓		✓			
Strain	✓		✓	✓			
Displacement	✓				✓		
Force	✓	✓	✓	✓			✓
Acoustic emission	✓	✓	✓	✓	✓		
Vibration	✓			✓			✓
Image		✓	✓				
Laser		✓			✓	✓	✓

cutting tools in the cutting process and on prior experimental database. They do not make full use of real-time monitoring data, system model, process parameters, and other data generated in the processing process to optimize the process intelligently. Self-learning and adaptive capabilities are therefore limited. On the other side, the important function of tool monitoring module is simple, which limits the use of these systems.

1.3.4 The Intelligent Control

The process optimization module feeds the information extracted from the on-line monitoring module and optimizes the single or multi-objective machining process according to the optimization requirements. The optimization objectives include force, tool life, workpiece surface quality, and machining efficiency. The optimization results are then sent to the Real-Time Control Module to complete the adjustment of the machine tool parameters [48]. It includes changing the cutting parameters, real-time position error compensation, path optimization, tool state adjustment (tool change, tool physical state change), to complete the real-time adjustment of the cutting state and achieve optimum machining efficiency.

Afazov et al. [49] used the variation features of the milling force to identify the chatter and avoid entering the chatter area by adjusting the process parameters. Quintana et al. [50] have used Fourier transform to detect milling chatter based on the frequency amplitude of the milling signal. Kuljanic et al. [51] identified the tool chatter through multi-axial force and vibration features. Zuperl et al. [52] developed an adaptive neural control strategy for force control in high-speed milling. Liu et al. [53] studied the dynamic characteristics in the time–frequency domain and the chatter monitoring and developed control strategy of the micro-milling process when the tool state changed. Results have showed that the state of tool wear has a significant impact on the dynamic characteristics of the milling system. Cao et al. [54] extracted the signal features of the high-speed spindle to establish an intelligent monitoring and diagnostic system, and realized fault self-diagnosis and adaptive adjustment of the high-speed spindle process parameters. However, the signal collected by the current sensor needs to be processed for evaluation and decision-making. The decision-making capacity is limited, and the reasoning and self-learning skills are inadequate, which makes it hard to achieve smart machining.

1.3.5 The Database and Big Data Analytics

Data processing and analytics is carried out through the entire machining system, including machine tool, workpiece, tool, fixture, process data processing information, and post-process data information. Intelligent processing of data needs a number of operations, such as data collection, classification, management, storage, extraction, optimization, and sharing. The development of the database, network, big data, and cloud computing technology will provide more space for the development of smart machining technology in the cutting process [55–59]. Due to the complexity of the cutting process, there is a large amount of data from process planning, simulation optimization, cutting process optimization and control, and quality inspection to completion of machining. By setting up a database and a knowledge base, data on the machining process can be well managed and inherited, data can be mined, accessed, and analyzed quickly and process parameters can be effectively determined through big data analytics. Through networking and cloud platforms for data communication and sharing, the circulation of data in the cutting process is increased and the machining experience is well shared and utilized.

1.3.6 Smart Machine Tool

Smart machine tools not only support the complex operations of machine tools, but also achieve intelligent machining to ensure high precision and high machining efficiency [60, 61]. A smart machine tool has the ability to perform intelligent perception, decision-making, and execution. The specific functions of the smart machine tool include human–computer interaction, machining simulation, self-monitoring, intelligent anti-collision, vibration control, adaptive technology (load adaptive, position adaptive, spindle power adaptive, motion-adaptive), error measurement, and compensation (geometric error, temperature error), intelligent spindle [62, 63], and intelligent tool system [64, 65], equipment maintenance, etc. The "brain" of the smart machine tool is a numeric control system. With the development of the numerical control system, the open numerical control system, i.e., Step-NC, has become the development direction of the machine tool control system, which enables the numerical control system to develop into modularization, platform, standardization, and serialization. The open architecture makes the CNC system more universal, flexible,

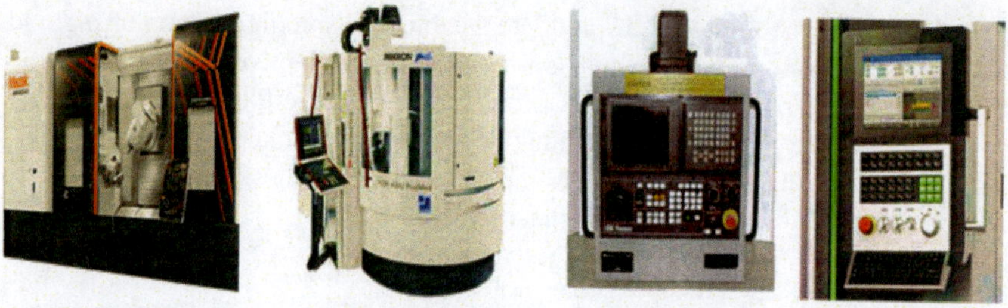

Mazak smart machine tool | The MIKRON center | The GE-Fanuc platform | Shenyang i5 system

Fig. 1.3 Intelligent machine tool and the intelligent machining platform

adaptable, and expandable. Users are allowed to develop the system for the second time, and it is convenient to reconstruct and modify the system according to their needs, to make multiple uses of the system.

Mazak, Okuma, Mikron, and other machine tool manufacturers have been engaged in the research and development of smart machine tools. Mazak's E-Series [66] smart machine tool can monitor itself, analyze a lot of machine tool-related information, processing status, environment, and other factors, and take appropriate action to ensure optimal processing. At present, Mazak's smart machine tool (Fig. 1.3) is capable of performing four intelligent functions: active vibration control function that can minimize vibration during machine tool processing; intelligent thermal barrier technology that can control the thermal displacement of the machine tool during processing; an intelligent safety barrier that is used to prevent collisions between machines. Mikron Smart Machine Tool [67] (Fig. 1.3) has launched four functional modules, and the users can receive information on the operation of the Mikron machining center via this system. Using the short message form of mobile phones, users can learn the status of the machine tool's operation and program execution status.

In the smart machine tool, the software and hardware systems are combined with advanced sensing, measurement, control technology, and information technology. This makes CNC machine tools not only have a real-time monitoring function, but also the ability to remotely monitor, communicate with people, manage production tasks, and even communicate with each other. It can be seen that advanced monitoring technology not only improves the processing capability of CNC machine tools but also makes CNC machine tools more intelligent in combination with information technology and monitoring technology.

Research on smart machine tools is still at the exploration and improvement stage. At present, the intelligent level of the smart machine tool cannot meet the requirements of intelligent processing of certain high-quality components. Smart machine tools need to be thoroughly and comprehensively studied in the areas of process planning, organization and process knowledge management, simulation and tool path optimization, smart online optimization control, etc.

1.4 The Trends of Smart Machining System

Smart machining system has been developed in many aspects and has achieved good results, but the existing smart machining technology is far from being able to solve problems in the machining process and therefore needs further research and applications, which are mainly described in the following aspects.

(1) **The Theoretical System of Smart Machining System**

The key to smart machining is to study and develop intelligent information interactions between quality-cutting tool-system. Due to the complexity of the cutting process, many factors have an impact on the cutting process, and the relationship between different factors is complex and coupled. With the continuous development of related technologies and the introduction of new processes and theories, the theoretical system of smart machining [17, 68], i.e., the CPPS, needs to be enhanced and improved.

(2) **Intelligent On-line Monitoring**

Heterogeneous sensor based online monitoring technologies needs to develop to monitor vibration, temperature cutting, tool wear, and the operation of the equipment in the NC machining process. Depending on the pre-established system control model, the parameters can be intelligently adjusted in real-time and errors in the machining process can be compensated. Intelligent monitoring and optimization systems are the premises of smart machining technology. Only by constantly making technological breakthroughs in intelligent monitoring methods and theories and by making optimization algorithms faster, more accurate and more intelligent can we ensure the rapid development of smart machining technology.

(3) **Joint Modeling and Optimization**

Process simulation plays an important role in the control of the machining process. Many situations in the machining process can be predicted by simulation. Most existing simulation software is only used as a separate module, and not in good synergy with other modeling and optimization software, which largely limits the process control capabilities. Joint simulation by combining modeling and optimization is the development direction for simulation cutting in the future. In addition, the existing simulation software pays more attention to the simulation of the cutting process while the optimization function of the cutting process is relatively weak. Further development of the cutting process optimization function is an important topic and research direction for process simulation.

(4) **Intelligent Machining Strategy and CNC System Integration**

There are two key technologies for the implementation of intelligent CNC systems, one is an intelligent machining strategy, the other is the integration of an intelligent machining strategy and a CNC system. To make the machining system as flexible as possible with a view to maintaining the required accuracy and high productivity, the smart machining system must be able to change its control strategy in line with the actual situation and finally achieve an optimum combination of quality, efficiency, and cost. How key technologies are coordinated, how the response time is shortened, and how the intelligence and accuracy of the control system are improved are important challenges to investigate.

References

1. Groover MP (2015) Automation, production systems, and computer-integrated manufacturing, 4th edn. Pearson
2. Kusiak A (2018) Smart manufacturing. Int J Prod Res 56(1–2):508–517
3. ElMaraghy HA (2005) Flexible and reconfigurable manufacturing systems paradigms. Int J Flex Manuf Syst 17:261–276
4. Koren Y, Gu X, Guo W (2017) Reconfigurable manufacturing systems: principles, design, and future trends. Front Mech Eng 13(2):1–16
5. Arinez JF, Chang Q, Gao RX et al (2020) Artificial intelligence in advanced manufacturing: current status and future outlook. J Manuf Sci Eng 142(11):110804
6. Foundations for innovation in cyber-physical systems. Workshop Report National Institute of Standards and Technology (2013)
7. Monostori L, Kádár B, Bauernhansl T, Kondoh S et al (2016) Cyber-physical systems in manufacturing. CIRP Ann 65(2):621–641
8. Byrne G, Dornfeld D, Denkena B (2003) Advancing cutting technology. Ann CIRP 52(2):483–507
9. Herzog D, Seyda V, Wycisk E et al (2017) Additive manufacturing of metals. Acta Mater 117:371–392
10. Taniguchi N (1983) Current status in, and future trends of, ultraprecision machining and ultrafine materials processing. CIRP Ann 32(2):573–582
11. M'Saoubi R, Axinte D, Soo SL et al (2015) High performance cutting of advanced aerospace alloys and composite materials. CIRP Ann Manuf Technol 64:557–580
12. Qin Y (2010) Micro-manufacturing engineering and technology. Elsevier
13. Dornfeld DA, Lee D-E (2008) Precision manufacturing. Springer
14. Haapala KR, Zhao F, Camelio J et al (2013) A review of engineering research in sustainable manufacturing. J Manuf Sci Eng Trans ASME 135(4):041013
15. Wegener K, Gittler T, Weiss L (2018) Dawn of new machining concepts: compensated, intelligent, bioinspired. Procedia CIRP 77:1–17
16. Deshayes L, Welsch L, Donmez A et al (2005) Smart machining systems: issues and research trends. National Institute of Standards and Technology (NIST) report
17. Zhu KP, Zhang Y (2018) A cyber-physical production system framework of smart CNC machining monitoring system. IEEE/ASME Trans Mechatron 23(6):2579–2586
18. Monostori L et al (2016) Cyber-physical systems in manufacturing. CIRP Ann 65(2):621–641
19. Park HS, Tran NH (2014) Development of a smart machining system using self-optimizing control. Int J Adv Manuf Technol 74(9):1365–1380
20. Zhou Z, Xie S, Chen D (2012) Fundamentals of digital manufacturing science. Springer
21. Lihui W (2015) An overview of function block enabled adaptive process planning for machining. J Manuf Syst 35:10–25
22. Lihui W (2013) Machine availability monitoring and machining process planning towards Cloud manufacturing. CIRP J Manuf Sci Technol 6(4):263–273
23. Zhao W, Li J, Pei Y et al (2016) Research and development of intelligent cutting database cloud platform system. Int J Adv Manuf Technol 2016:1–13
24. Tunc LT, Budak E, Bilgen S et al (2016) Process simulation integrated tool axis selection for 5-axis tool path generation. CIRP Ann Manuf Technol 64(1):381–384

25. Tunc LT (2019) Smart tool path generation for 5-axis ball-end milling of sculptured surfaces using process models. Robot Comput Integr Manuf 56:212–221
26. Calamaz M, Coupard D, Girot F (2008) A new material model for 2D numerical simulation of serrated chip formation when machining titanium alloy Ti-6Al-4V. Int J Mach Tools Manuf 48(3–4):275–288
27. Attanasio A, Ceretti E, Rizzuti S et al (2008) 3D finite element analysis of tool wear in machining. CIRP Ann Manuf Technol 57(1):61–64
28. Afazov SM, Ratchev SM, Segal J (2010) Modelling and simulation of micro-milling cutting forces. J Mater Mach Technol 210(15):2154–2162
29. Sun YW, Guo Q (2011) Numerical simulation and prediction of cutting forces in five-axis milling processes with cutter run-out. Int J Mach Tools Manuf 51(10–11):806–815
30. Samuel J, Jun MBG, Ozdoganlar OB et al (2020) Micro/meso-scale mechanical machining 2020: a two-decade state-of-the-field review. J Manuf Sci Eng 142(11):110809
31. Shi X, Wang X, Jiao L, Wang Z, Yan P, Gao S (2018) A real-time tool failure monitoring system based on cutting force analysis. Int J Adv Manuf Technol 95:2567–2583
32. Nouri M, Fussell BK, Ziniti BL, Linder E (2015) Real-time tool wear monitoring in milling using a cutting condition independent method. Int J Mach Tools Manuf 89:1–13
33. Jemielniak K, Arrazola PJ (2008) Application of AE and cutting force signals in tool condition monitoring in micro-milling. CIRP J Manuf Sci Technol 1(2):97–102
34. Hung CW, Lu MC (2013) Model development for tool wear effect on AE signal generation in micro-milling. Int J Adv Manuf Technol 66:1845–1858
35. Szydłowski M, Powałka B, Matuszak M, Kochmański P (2016) Machine vision micro-milling tool wear inspection by image reconstruction and light reflectance. Precis Eng 44:236–244
36. Zhu KP, Yu XL (2017) The monitoring of micro milling tool wear conditions by wear area estimation. Mech Syst Signal Process 93:80–91
37. Hung PB (2016) An intelligent neural-fuzzy model for an in-process surface roughness monitoring system in end milling operations. J Intell Manuf 27(3):689–700
38. Teti R, Jemielniak K, O'Donnell G et al (2010) Advanced monitoring of machining operations. CIRP Ann Manuf Technol 59(2):717–739
39. Duro JA, Padget JA, Bowen CR (2016) Multi-sensor data fusion framework for CNC machining monitoring. Mech Syst Signal Process 66–67:505–520
40. Malekian M, Park SS, Jun MBG (2009) Tool wear monitoring of micro-milling operations. J Mater Mach Technol 209(10):4903–4914
41. Zhu KP, Liu T (2018) On-line tool wear monitoring via hidden semi-Markov model with dependent durations. IEEE Trans Ind Inf 14(1):69–78
42. Terrazas G, Martínez-Arellano G, Benardos P, Ratchev S (2018) Online tool wear classification during dry machining using real time cutting force measurements and a CNN approach. J Manuf Mater Process 2(72):1–18
43. Luo B, Wang H, Liu H, Li B, Peng F (2018) Early fault detection of machine tools based on deep learning and dynamic identification. IEEE Trans Ind Electron 66:509–518
44. OMATIVE. http://www.omative.com/cmpm
45. ARTIS. https://artis.de/eng/
46. BRANKAMP. https://brankamp.com/eng/
47. MONTRONIX. http://www.montronix.com/en/
48. Landers RG, Barton K, Devasia S, Kurfess T, Pagilla P, Tomizuka M (2020) A review of manufacturing process control. J Manuf Sci Eng 142(11):110814
49. Afazov SM, Zdebski D, Ratchev SM, Segal J, Liu S (2013) Effects of micro-milling conditions on the cutting forces and process stability. J Mater Mach Technol 213:671–684
50. Quintana G, Ciurana J, Ferrer I, Rodríguez CA (2009) Sound mapping for identification of stability lobe diagrams in milling processes. Int J Mach Tools Manuf 49(3–4):203–211
51. Kuljanic E, Sortino M, Totis G (2008) Multisensor approaches for chatter detection in milling. J Sound Vib 312(4):672–693
52. Zuperl U, Cus F, Reibenschuh M (2012) Modeling and adaptive force control of milling by using artificial techniques. J Intell Manuf 23(5):1805–1815

53. Liu MK, Halfmann EB, Suh CS (2014) Multi-dimensional time-frequency control of micro-milling instability. J Vib Control 20(3):643–660
54. Cao H, Zhang X, Chen X (2017) The concept and progress of intelligent spindles-a review. Int J Mach Tools Manuf 112:21–52
55. Sang Z, Xu X (2017) The framework of a cloud-based CNC system. Procedia CIRP 63:82–88. In: The 50th CIRP conference on manufacturing systems
56. Liu XF, Rakib Shahriar Md, Nahian Al Sunny SM, Leu MC, Hu L (2017) Cyber-physical manufacturing cloud: architecture, virtualization, communication, and testbed. J Manuf Syst 43:352–364
57. Wang L (2013) Machine availability monitoring and machining process planning towards Cloud manufacturing. CIRP J Manuf Sci Technol 6:263–273
58. Tapoglou N, Mehnen J, Vlachou A, Doukas M, Milas N, Mourtzis D (2015) Cloud-based platform for optimal machining parameter selection based on function blocks and real-time monitoring. J Manuf Sci Eng 137:040909–040911
59. Kurfess TR, Saldana C, Saleeby K, Dezfouli MP (2020) A review of modern communication technologies for digital manufacturing processes in industry 4.0. J Manuf Sci Eng 142(11):110815
60. Gao RX, Wang L, Helu M, Teti R (2020) Big data analytics for smart factories of the future. CIRP Ann Manuf Technol 69:668–692
61. Mohring H-C, Wiederkehr P, Erkorkmaz K, Kakinuma Y (2020) Self-optimizing machining systems. CIRP Ann Manuf Technol 69:740–763
62. Rizal M, Ghani JA, Nuawi MZ, Haron CHC (2015) Development and testing of an integrated rotating dynamometer on tool holder for milling process. Mech Syst Signal Process 52–53:559–576
63. Albrecht A, Park SS, Altintas Y, Pritschow G (2005) High frequency bandwidth cutting force measurement in milling using capacitance displacement sensors. Int J Mach Tools Manuf 45(9):993–1008
64. Zhang G, Ehmann KF (2015) Dynamic design methodology of high speed micro-spindles for micro/meso-scale machine tools. Int J Adv Manuf Technol 76(1–4):229–246
65. Cheng K et al (2017) Smart cutting tools and smart machining: development approaches, and their implementation and application perspectives. Chin J Mech Eng
66. Mazak e. https://www.mazakusa.com/machines/series/integrex-e-h
67. Mikron. https://www.mikron.com/machining-solutions
68. Gao R, Wang L, Teti R, Dornfeld D et al (2018) Cloud-enabled prognosis for manufacturing. CIRP Ann

Chapter 2
Modeling of the Machining Process

2.1 The Machining Process Modeling Methods

The primary goal of machining process modeling is to improve machining performance prediction. The most studied predictive methods are analytical, numerical, and artificial intelligence (AI) modeling, which are commonly validated with experimental data [1, 2]. There are also many studies attempt to develop hybrid modeling techniques to integrate the benefits of the different approaches.

The practical applications of the machining process prediction modeling are composed of the following two main stages.

Stage 1: modeling of machining parameters, such as cutting force, cutting torque and cutting power, tool wear, tool life, part precision, and surface quality, etc.

Stage 2: modeling of machining performance, in which sometimes optimization conditions need to be determined.

From 1995 to 1997, the International Academy for Production Engineering (CIRP) organized a modeling research group to conduct a special study on the modeling of machining processes [3]. It was divided into six groups to model the machining process of different types of cutting. According to the research results, more than 43% of the research groups adopt experimental/empirical modeling; 32% used analytical modeling; and 18% used numerical modeling, including finite element method (FEM) modeling accounts for the vast majority, and only a small number of research groups adopt the boundary element method (BEM) and finite difference method (FDM). The number of research groups using artificial intelligence technology is gradually increasing. CIRP annals updated the recent research results in 2013 [4]. Generally, there are several representative directions for new developments of machining process modeling strategies.

K. Zhu, *Smart Machining Systems*, Springer Series in Advanced Manufacturing,
https://doi.org/10.1007/978-3-030-87878-8_2

2.1.1 Modeling Based on Cutting Mechanics

The modeling of cutting force is an important basis for understanding the cutting mechanisms and dynamic processes. The instantaneous change of force can not only reflect tool deformation, wear, and energy loss but also provide an important reference for fixture design, tool wear monitoring, and process control.

The selection and determination of the cutting force model is the basis of the dynamic simulation of the milling process. There are two sources of cutting force: one is the resistance caused by the elastic deformation and plastic deformation of the cutting layer metal and the surface layer metal of the workpiece; the other is the friction resistance between the cutting tool, the chip, and the workpiece surface. Since there is a cutting force in the cutting process, it will consume energy. The mechanical energy is transformed into heat energy. The cutting force directly affects the cutting heat, and further affects the speed of tool wear and the quality of machining [5]. It is helpful to analyze the cutting process and solve the technological problems of metal cutting to understand the factors influencing the cutting force and master the changing law of the cutting force.

The main contents include:

(1) Chip formation and deformation zone mechanics.
(2) Cutting force modeling.
(3) The phenomenon and mechanism of tool wear and its influence on cutting force.
(4) The thermal–mechanical coupling relation of the cutting process is analyzed.
(5) Thermal error tracing and compensation mechanism in the cutting process.

2.1.2 Modeling Based on Machine Tool Vibration

Vibration has been a major concern for the productivity in high speed machining [6]. The forced vibration is studied in the design of machine tool, in which the rotation speed of workpiece, natural frequency of the machine tool, and excitation frequency (such as noise of gear chain, electromagnetic frequency of the driving motor, etc.) are considered. In the self-excited vibration, important topics such as the mechanism of self-excited vibration, the phase shift, and the existence of regeneration effects are widely studied. It has been found that the regeneration effect occurs not only before one rotation of the workpiece but also before two or more revolutions. This is called the multiple regeneration effect.

2.1.3 Modeling Based on Numerical Simulation

In the cutting process, the un-deformed chip thickness, the magnitude and direction of the instantaneous cutting force, the elastic deformation, and the instantaneous cutting temperature distribution change significantly. The off-line physical simulation of the cutting process is carried out employing the finite element method. Considering the material deformation and heat transfer from the perspective of dynamics, the chip shape, cutting force, cutting temperature, surface residual stress, and cutting deformation can be predicted [5]. Therefore, modeling the cutting process with numerical simulation has become one of the key research directions to uncover the mechanics and predict the machining performance.

In the past 20 years, scholars have made many remarkable achievements in finite element method (FEM) simulation of the metal cutting process, especially in high-speed machining [7]. In the modeling with FEM, the key of simulation is to build the property change law of materials in the machining and establish the dynamic constitutive model suitable for cutting materials.

2.1.4 Modeling Based on Measurement Information

In order to analyze or control the machining process, a large amount of experimental data is needed. Under the condition of a single piece and small batch production, it is often impossible to obtain so much data, even from public publications or well-designed experiments. It is also difficult to obtain data about various processing procedures, cutting tool materials, and workpiece materials. Accurate and reliable data can only be obtained through experiments in the workshop. However, the workshop is different from the laboratory, and the workshop staff is not trained to do a lot of measurements on the experimental results. The only solution is to install sensors and other information input devices on the machine tool to collect the required information [6]. The improved model can guarantee the optimal performance of the machining process. This method can be understood as the model of the machining process learning from experience, and the sensor is the provider of the experience.

2.1.5 Modeling Based on Artificial Intelligence (AI)

The modeling method based on artificial intelligence is developing fast recently. A large number of researches focus on monitoring or predicting the processing process by fusing sensor information [8]. Some researchers use this method to predict machinability or to design cutting tools. The most dominant AI method is to use an artificial neural network (NN). It does not need to understand the physical nature of the machining process to get an explicit model. Although it can't be used as an analytical model of the process, it helps to coordinate the functions of modeling, sensing, and learning. The NN provides a new method for empirical modeling, especially to model the nonlinear relationship between the machining state and the system measurements by learning the nonlinear mapping with the networks.

2.1.6 *Modeling Method Combining Data and Cutting Mechanics*

The current research only considers the CNC machine tool itself or the monitoring system itself but lacks the deep integration of processing physical process, process data, and monitoring information [9]. The separation of data and mechanisms greatly limits the function of machining process modeling and intelligent optimization. It is necessary to establish a new CPS theory and architecture for the complex machining process modeling. Taking CNC machine tools and intelligent monitoring function modules as objects, it is necessary to deeply integrate the description method of cutting physical model, tool status, process parameter information, and monitoring data information to improve the accuracy and reliability of the model. The modeling method can be realized by establishing the dynamic model of the machining process and the deep learning model and algorithm of monitoring data.

Researchers have long been trying to establish a general model of machining process, but it has been demonstrated that this does not produce good results. Therefore, reliable feasible solutions, although temporary and not necessarily perfect, always meet the production requirements faster than ideal solutions. By combining theory with application and promoting each other, the modeling of the machining process can be improved and made more practical.

2.2 Principles of Chip Formation

2.2.1 *Chip Formation*

Excess metal is removed from the raw material to form a real mechanical part with the required shape and accuracy. The excess metal has been removed in the form of chips and recycled [10]. As shown in Fig. 2.1, the undeformed cutting layer *AGHD* can be regarded as composed of many parallelograms, such as ▭ *ABCD*, ▭ *BEFC*, and ▭ *EGHF*. When these parallelograms are pushed by the rake face, they slide upward along the *BC* direction, forming other parallelograms, namely, ▭ *ABCD* → ▭ *A′B′C′D′*, ▭ *BEFC* → ▭ *B′E′F′C′*, ▭ *EGHF* → ▭ *E′G′H′F′*. It can be seen that the process of metal cutting is a process in which the metal in the cutting layer is pushed by the rake face of the cutting tool, and the plastic deformation is mainly shear slip to form chips.

2.2.2 *Mechanical Model of Chip Formation*

To study the effect of rake face friction on plastic metal chip deformation. First, analyze the force acting on the chip. Under right angle free cutting, the forces acting

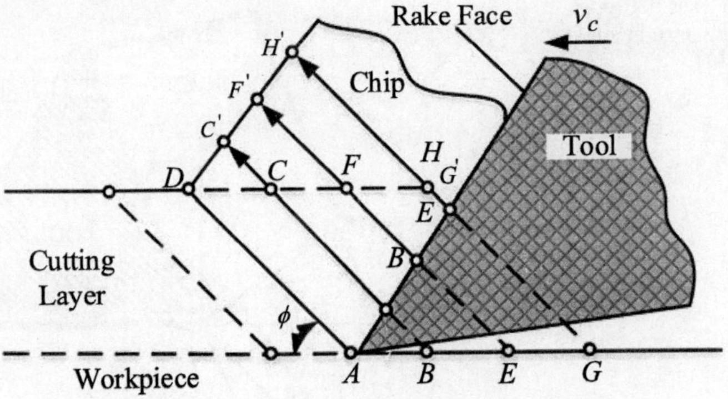

Fig. 2.1 Schematic diagram of a cutting process

on the chip include normal force F_n and friction force F_f on the rake face, and there is also a positive pressure F_{ns} and shear force F_s on the shear surface, as shown in Fig. 2.2. The resultant force of these two pairs of forces should be balanced with each other. If all the forces are drawn in front of the cutting edge, the relationship of the forces as shown in Fig. 2.2 can be obtained. In Fig. 2.3, F_r is the resultant force of F_n and F_f, which is called chip forming force [11, 12]. The shear stress on the shear plane is represented by τ, and the chip forming force F_r is:

$$\left.\begin{array}{l} F_s = \tau A_s = \dfrac{\tau A_D}{\sin\phi} \\ F_s = F_r\cos(\phi+\beta-\gamma_o) \end{array}\right\} \Rightarrow F_r = \frac{F_s}{\cos(\phi+\beta-\gamma_o)} = \frac{\tau A_D}{\sin\phi\cos(\phi+\beta-\gamma_o)} \tag{2.1}$$

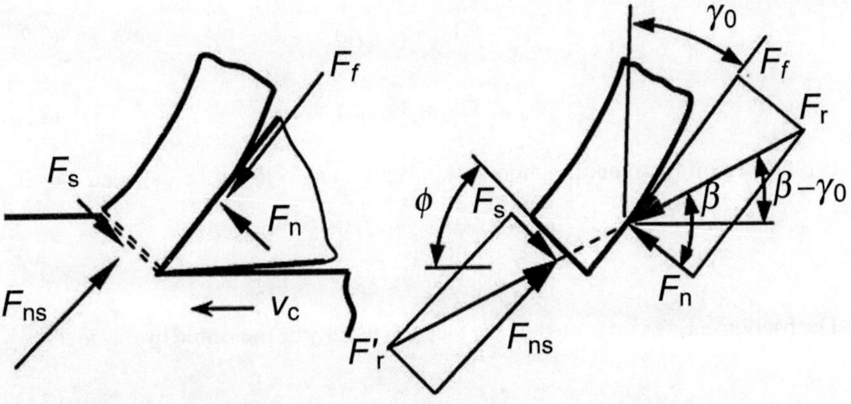

Fig. 2.2 Forces on chips

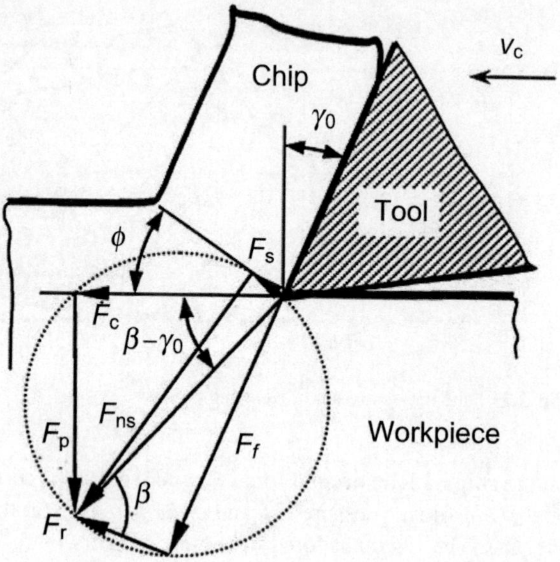

Fig. 2.3 Relations between force and angle in right angle free cutting [12]

where A_s is the sectional area of the shear plane. A_D is the cross-section area of the cutting layer which can be expressed as $A_D = h_D \cdot a_w$, where h_D is the cutting thickness and a_w is the cutting width. Moreover, shear angle ϕ refers to the angle between the surface where shear slip occurs and the direction of cutting speed. β is the angle between F_r and F_n, which is called friction angle. γ_o is the tool rake angle.

The cutting component force in the direction of cutting motion is F_c, and the cutting component force perpendicular to the cutting direction is F_p. The cutting forces F_c and F_p of the cutting edge in two directions in the machining plane can be expressed as follows:

$$F_c = F_r \cos(\beta - \gamma_o) \tag{2.2}$$

$$F_p = F_r \sin(\beta - \gamma_o) \tag{2.3}$$

If the force on the flank is ignored, the friction angle β can be calculated as in Eq. 2.4.

$$\beta = \gamma_o + \arctan\left(F_p / F_c\right) \tag{2.4}$$

The friction coefficient μ on the rake face can usually be measured by this method.

$$\mu = \tan \beta \tag{2.5}$$

2.2.3 Divisions of Deformation Zones

The metal will shear and slip in the process of machining. According to the streamline in the metal cutting process, as illustrated in Fig. 2.4, the flow path of a certain point of the metal to be cut in the cutting process, the plastic deformation of the metal in the cutting process can be roughly divided into three deformation regions.

(1) The first deformation zone (I) occurs from the plastic deformation of the *OA* line to the completion of the shear slip of the *OM* line.
(2) When the chip flows out along the rake face, it is further squeezed and rubbed by the rake face, which makes the metal microstructure fibrosis near the rake face, and its direction is parallel to the rake face. This area is called the second deformation area (II).
(3) The machined surface is extruded, rubbed, and rebounded by the blunt part of the cutting edge and the flank, resulting in fibrosis and work-hardening. The deformation in this region is also relatively dense, which is called the third deformation region (III).

A typical slip-line field model for orthogonal micro-cutting process presented by Jin and Altintas [13, 14], as shown in Fig. 2.5, which assumes that a surface layer of work material defined by the undeformed chip thickness flows into the primary shear deformation zone, and separates at a stagnant point *B*. The velocities of the workpiece, shear, and chip are indicated as V_w, V_s, and V_c, respectively. From point *B*, some of the material is trapped in the dead zone created by the chamfer, where the material flows over towards the regular rake face and forms the chip. Chamfer geometry is defined by the length of the chamfered edge (b_{cf}) and chamfer angle (second rake angle α_1), as shown in Fig. 2.5a. The slip line field is used to form energy equations for the primary shear zone (AB), chamfered zone (OH), and secondary deformation

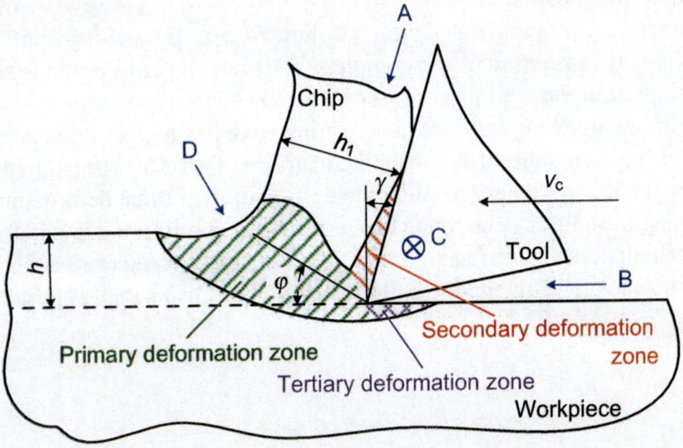

Fig. 2.4 Flow paths and three deformation zones in metal cutting [2]

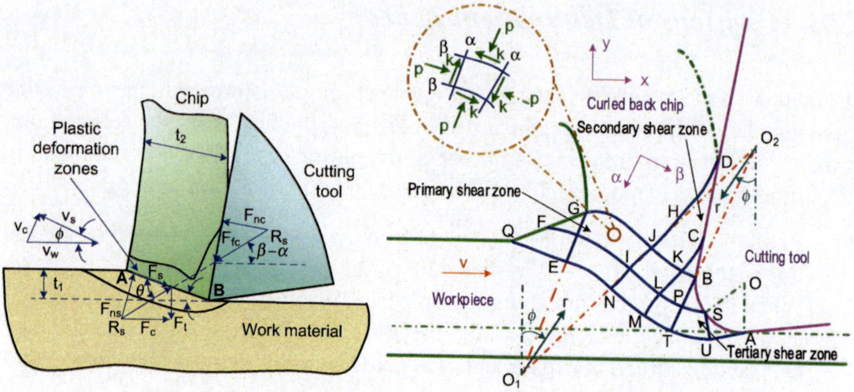

Fig. 2.5 Chip formation models for **a** chamfered tool [13] and **b** microcutting [14]

zone where the chip flows over the regular rake face (HG). The energy equations are used to predict the shear angle, forces, and temperature while considering the flow stress, strain, strain rate, and temperature effects. The slip-line field model of the orthogonal micro-cutting process with a round edge tool is shown in Fig. 2.5b. Plane strain deformation and steady-state cutting conditions are assumed. The primary shear zone consists of three regions: triangular region QEG is due to the pre-flow effect since line QG is a stress-free boundary, all of the slip-lines in QEG intersect with QG at a 45° angle. Region GJNE is convex upward because of the curling-back effect of the chip; region JBTN is concave downward due to the friction between the chip and the tool rake face. Both region GJNE and region JBTN is composed of circular arcs and straight radial lines; O_1 is the center of the circular arcs in region GJNE, and O_2 is the center of the circular arcs in region JBTN. The shear flow stress and hydrostatic stress are modeled by including the effects of strain, strain-rate, and temperature of the material [14].

These three deformation zones are concentrated near the cutting edge, and the stress is relatively concentrated and complex. The metal layer to be cut is separated from the material of the workpiece, part of which is turned into chips, and the other part is left on the machined surface. The cutting edge has a great influence on chip removal and the formation of the machined surface. Therefore, to study the metal cutting process, it is necessary to study the deformation of three deformation areas. Further in-depth studies can be found on shear deformations in the first deformation zone, the calculation of shear angle [15–18], extrusion deformations in the second deformation zone [19, 20], and the built-up edge mechanisms [21, 22].

2.3 Cutting Forces

In the process of cutting, the deformation of the cutting layer metal is mainly due to the force given by the cutter. The force acting on the workpiece or tool is called the cutting force. The cutting force not only deforms the cutting layer metal, consumes work, but also generates cutting heat, which makes the tool wear blunt and affects the surface quality and production efficiency. At the same time, the cutting force is also the main factor for the selection of machine tool motor power, the design of machine tool main motion, and feed motion mechanism. The size of cutting force can be used as one of the indexes to measure the machinability of workpiece and tool materials, and also can be used as a controllable factor to adapt to the control of the cutting process.

2.3.1 Sources of Cutting Forces

According to the research on the process of cutting metal by cutting tool in Sect. 2.2, it can be known that the cutting force comes from three aspects, as shown in Fig. 2.6.

(1) To overcome the resistance of elastic deformation of processed materials.
(2) To overcome the resistance of plastic deformation of processed materials.
(3) It can overcome the friction between chip and tool rake face, workpiece transition surface, and machined surface on tool flank. They are represented by $F_{f\gamma}$ and $F_{f\alpha}$, respectively.

Elastic deformation resistance and plastic deformation resistance exist in the three deformation zones in cutting, but the resistance in the first deformation zone is the largest.

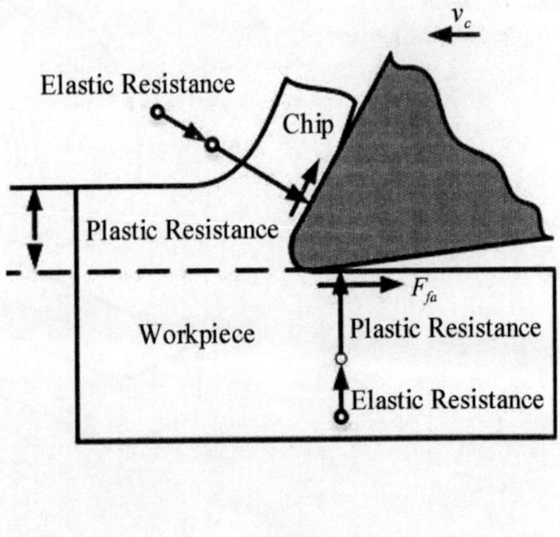

Fig. 2.6 The source of cutting forces

2.3.2 *Joint and Component Cutting Forces and Cutting Powers*

(1) **Joint and component cutting forces**

The above forces form a resultant force F acting on the turning tool. As shown in Fig. 2.7, for measuring and applying convenience, F can be decomposed into three vertical awesome forces, namely, feed force F_f, back-cutting force F_p, and cutting force F_c [12, 23]. It can be seen from Fig. 2.7.

$$F = \sqrt{F_c^2 + F_p^2 + F_f^2} = \sqrt{F_c^2 + F_D^2} \tag{2.6}$$

where F_c is the main cutting force, which is perpendicular to the base plane P_r and consistent with the cutting speed v_c. It consumes the main power of the machine tool and is the basis for calculating the cutting power, selecting the motor power of the machine tool, and designing the main transmission mechanism of the machine tool. F_f is the feed force, also known as radial force or awesome force. It acts on the base plane P_r and parallel to the feed direction. It is the basis for designing the feed mechanism of the machine tool. F_p is the back-cutting force, also known as axial force or cutting force. It acts on the base surface P_r and is perpendicular to the feed direction. It can deform the workpiece. It is the basis for verifying the stiffness of the machine tool spindle in the horizontal plane and the strength of the corresponding parts. F_D is the resultant force acting on the base plane, which is awesome for the projection of the total force on the plane of the cutting layer, the feed force, and the back-cutting force.

According to the experiment, for example, when $\kappa_r = 45°$, $\lambda_s = 0°$, and $\gamma_o \approx 15°$, there is an approximate relationship between F_c, F_f, and F_p:

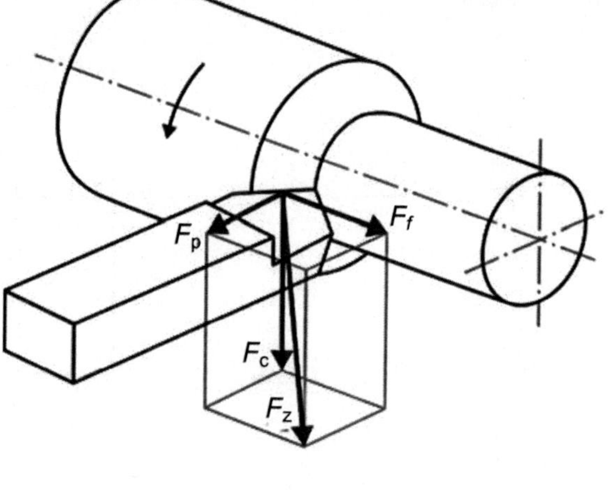

Fig. 2.7 Joint and component cutting forces

$$F_p = (0.4 \sim 0.5)F_c$$
$$F_f = (0.3 \sim 0.4)F_c$$

Substituting these equations into Eq. 2.10 givens:

$$F = (1.12 \sim 1.18)F_c$$

With the different cutting conditions such as tool material, geometric parameters and cutting parameters, workpiece material, and tool wear, the proportion of each component force can be changed in a large range.

(2) **Cutting powers**

The work consumed in the cutting process per unit time is called cutting power P_c. The cutting power is the sum of the collective power of cutting force F_c and feed force F_f, so the back-cutting force does not consume work because the displacement of F_p in the back-cutting force does not occur. Hence, the cutting power is given as:

$$P_c = \left(F_c v_c + F_f \frac{n_w f}{1000}\right) \times 10^{-3} \text{kW} \tag{2.7}$$

where n_w is the workpiece speed (r/s); v_c is the cutting speed (m/s); f is the workpiece feed rate (mm/r).

Since the work consumed by the feed motion relative to the main motion is very small (less than 1–2%), it can be ignored. The cutting power is simplified as:

$$P_c = F_c v_c \times 10^{-3} \text{kW} \tag{2.8}$$

The power P_E of the machine tool motor can be calculated as:

$$P_E \geq \frac{P_c}{\eta_c} \tag{2.9}$$

where η_c is the total transmission efficiency of the machine tool, generally taken as 0.75–0.85. The larger value can be used for new machine tools, and the smaller value can be used for old machine tools.

2.3.3 Empirical Models of Cutting Forces

The cutting force is measured by a force measuring instrument. Based on a large number of experiments, the data obtained are processed by mathematical method, and the empirical formula for calculating the cutting force can also be obtained. At present, there are two kinds of empirical formulas for calculating cutting force in practice: one is the exponential formula, and the other is the unit cutting force.

(1) **Exponential model**

It is widely used in metal cutting to calculate cutting force empirically by using the exponential model:

$$\begin{cases} F_c = C_{F_c} a_p^{x_{F_c}} f^{y_{F_c}} v_c^{n_{F_c}} K_{F_c} \\ F_p = C_{F_p} a_p^{x_{F_p}} f^{y_{F_p}} v_c^{n_{F_p}} K_{F_p} \\ F_f = C_{F_f} a_p^{x_{F_f}} f^{y_{F_f}} v_c^{n_{F_f}} K_{F_f} \end{cases} \tag{2.10}$$

where C_{F_c}, C_{F_p}, and C_{F_f} are the influence coefficients of workpiece material and cutting condition on the three components, which are determined by the workpiece materials and cutting conditions. x_{F_c}, x_{F_p}, and x_{F_f} are the influence indexes of back-cutting depth a_p on the three components. y_{F_c}, y_{F_p}, and y_{F_f} are influence indexes of feed rate f on three components. n_{F_c}, n_{F_p}, and n_{F_f} are influence indexes of cutting speed v_c on the three components. K_{F_c}, K_{F_p}, and K_{F_f} are products of correction coefficients affected by various factors on the three components. The cutting power P_c can be calculated by Eq. 2.11 with the cutting force F_c.

(2) **The unit cutting force**

The unit cutting force is the specific cutting force per unit area, denoted by k_c. If the unit cutting force is known, the cutting force F_c can be calculated using Eq. 2.14 as

$$F_c = k_c A_D = k_c a_p f = k_c h_D b_D \tag{2.11}$$

where k_c is the unit cutting force (N/mm^2). A_D is the nominal cross-sectional area of the cutting layer (mm^2). h_D is the nominal thickness of the cutting layer (mm). b_D is the nominal width of the cutting layer (mm). a_p is the back-cutting depth (mm). f is the feed rate (mm/r).

The power consumed per unit volume of metal material is called unit cutting power, denoted by p_c. If the unit cutting power is known, the cutting power P_c can be calculated as:

$$P_c = p_c Q_z \tag{2.12}$$

where p_c is the unit cutting power [kW/(mm^3·s^{-1})]. Q_z is the metal removal amount per unit time (mm^3/s).

For turning, the metal removal amount Q_z is given as:

$$Q_z = 1000 v_c A_D = 1000 v_c a_p f$$

Substituting Eqs. 2.14 into 2.11 yields:

$$P_c = F_c v_c \times 10^{-3} = k_c a_p f v_c \times 10^{-3}$$

Substituting these equations into Eq. 2.16 can obtain the relationship between unit cutting power and unit cutting force in turning machining:

$$p_c = k_c \times 10^{-6} \tag{2.13}$$

(3) **The exponential model**

The empirical formula of cutting force refers to the exponential formula of cutting force, which is established by the cutting experiment. There are many methods of cutting experiments, such as the single factor method and multi-factor method. Data processing methods include the graphic method, linear regression method, and computer data acquisition and processing method. The following only introduces the graphical method based on the single factor experiment to illustrate the establishment process of the index formula.

When the cutting force experiment is carried out, other cutting conditions are kept unchanged, only the back-cutting depth a_p is changed, and the cutting component force data of different AP are measured by a dynamometer. The obtained data is drawn on logarithmic coordinate paper, which is approximately a straight line, as shown in Fig. 2.8 [24]. The mathematical expression can be expressed as:

$$Y = kX + b$$

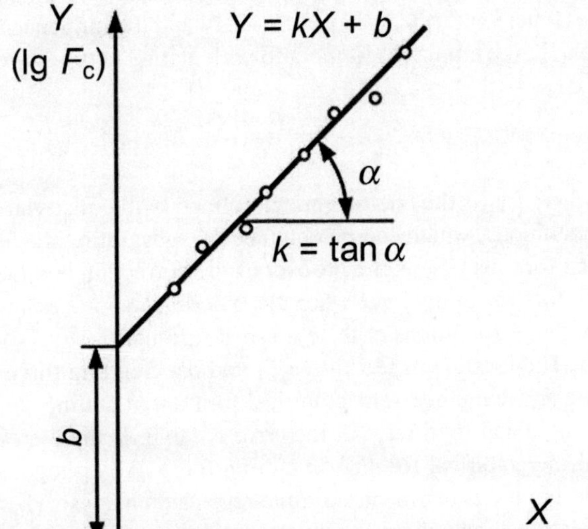

Fig. 2.8 Exponential model with logarithmic coordinates

where Y is the logarithm of cutting force F_c, i.e., $Y = \lg F_c$. X is the logarithm of back-cutting depth, i.e., $X = \lg a_p$. b is the longitudinal intercept of $F_c - a_p$ straight line in the logarithmic coordinate system (its value is $\lg C_{a_p}$). k is the slope of the $F_c - a_p$ line in the logarithmic coordinate system (its value is $\tan\alpha = x_{F_c}$).

The values of a and b can be measured directly from the graph, so the above equation can be written as:

$$\lg F_c = \lg C_{a_p} + x_{F_c} \lg a_p$$

i.e.,

$$F_c = C_{a_p} a_p^{x_{Fc}} \tag{2.14}$$

Similarly, the relationship between cutting force F_c and feed rate f and cutting speed v_c can be obtained

$$F_c = C_f f^{y_{Fc}} \tag{2.15}$$

$$F_c = C_{v_c} v_c^{y_{Fc}} \tag{2.16}$$

where C_f is the longitudinal intercept of $F_c - f$ line in logarithmic coordinate system; y_{F_c} is the slope of $F_c - f$ line in logarithmic coordinate system; C_{v_c} is its logarithm value is the longitudinal intercept of $F_c - v_c$ line in logarithmic coordinate system; n_{F_c} is the slope of $F_c - v_c$ line in logarithmic coordinate system.

After synthesizing Eqs. 2.18–2.20 and the influence of various secondary factors on F_c, the empirical formula for calculating cutting force can be obtained:

$$F_c = C_{F_c} a_p^{x_{Fc}} f^{y_{Fc}} v_c^{n_{Fc}} K_{F_c} \tag{2.17}$$

where C_{F_c} is the coefficient determined by the material to be processed and cutting conditions, which can be obtained by substituting the actual experimental data into the formula. K_{F_c} is the product of the correction coefficients of various influencing factors on cutting force when the experimental processing conditions are inconsistent with the conditions of the empirical formula.

The formula of feed force F_f and back-cutting force F_p can also be obtained by the above method. The empirical formula of cutting force by the graphic method is simple and intuitive, but the error is large. If there are three or more main factors in the empirical formula of cutting force, the orthogonal design should be used to design the experiment, multiple regression analysis should be used to deal with the relationship between variables, and the multivariate formula should be listed. More advanced and comprehensive methods are introduced in the later sections.

2.3.4 *Affecting Factors of Cutting Forces*

In the process of cutting, many factors have different degrees of influence on the cutting force. In addition to the three aspects of workpiece material, cutting parameters, and tool parameters, there are also the effects of tool material, flank wear, tool grinding quality, and cutting fluid. The influence degree and law of these factors are fully reflected in the theoretical formula and empirical formula of cutting force. For example, in the slip-line field theory, the theoretical cutting force model is [1]:

$$F_c = \tau_s hb \frac{1}{\sin\phi_c \cos(\phi_c + \beta_a - \alpha_r)} \tag{2.18}$$

where τ_s is shear stress. ϕ_c is the shear angle between the direction of cutting speed and the shear plane. β_a is the average friction angle between the rake face of the tool and the moving chip. α_r is the rake angle of the tool. b is the width of the cut (or the depth of cut in turning). h is the uncut chip thickness.

(1) **Workpiece materials**

As shown in Eq. 2.22, the higher the strength and hardness of workpiece material the greater τ_s, and the cutting force F_c also increases. However, with the increase of strength, the friction coefficient μ decreases and the friction angle β also decreases, which makes the shear angle ϕ increase (because $\phi = \pi/4 - \beta + \gamma_o$) [15]. Hence, the deformation coefficient Λ_h decreases, and the cutting force F_c also decreases. Combining the above two effects, it can be seen that the cutting force F_c still increases, but it is not directly proportional to the increase of strength.

When the strength and hardness of the workpiece material are similar, the greater the plasticity of the material, the greater the plastic deformation, and thus the greater the cutting force. When cutting brittle materials, the plastic deformation of the cutting layer is very small, and the friction between the chipping chip and the rake face is very small, so the cutting force of brittle materials is generally less than that of plastic materials.

It can be seen that the cutting force is not only affected by the original strength and hardness of the material but also affected by the work hardening ability of the material. For example, the strength and hardness of austenitic stainless steel are low, but the ability to work hardening is large. Small deformation will lead to an increase in hardness and cutting force.

If the stress concentration between the structural components is caused in steels containing sulfur (S) and lead (Pb), it is easy to form crack chips, and the cutting force will be reduced by 20–30% compared with the normal steel, so this kind of steel is also called free cutting steel. Different heat treatment states and metallographic structures of the same material will also affect the cutting force.

(2) **Cutting parameters**

The cutting depth a_p and feed rate f determine the cutting area. Therefore, the increase of a_p and f will increase F_c, but the influence degree of them is different. When a_p

increases, F_c also increases linearly. When f increases, F_c increases linearly and nonlinearly. This is due to the doubling of cutting depth a_p, cutting width b_D, and cutting force F_c. When f is doubled, the cutting thickness h_D also doubled, and F_c should be doubled. However, with the increase of h_D, the deformation coefficient Λ_h will decrease, which leads to the decrease of F_c.

As shown in Fig. 2.9, when v_c <100 m/min, the actual rake angle of the turning tool decreases due to chip accretion and wear of front and rear tool surfaces, which increases cutting force F_c. When v_c >100 m/min, with the increase of v_c, the friction coefficient μ decreases, the shear angle ϕ increases, and the cutting force F_c decreases. On the other hand, with the increase of v_c, the cutting temperature θ also increases, and the strength and hardness of the processed metal decrease, which also leads to the decrease of cutting force F_c [25].

Figure 2.10 shows the relationship between cutting speed v_c and cutting force F_c when machining gray cast iron. Because of the small plastic deformation and the small friction of the chip on the rake face when machining brittle materials, the cutting speed has little effect on the cutting force F_c [26].

(3) **Cutting tools**

The influence of rake angle γ_o on cutting force is the greatest. The results show that the change of γ_o and F_c is about 1.5% when cutting plastic metal. The greater the plasticity, the greater the change range. This is due to the increase of front angle γ_o, the increase of shear angle ϕ, the decrease of deformation coefficient Λ_h, and the decrease of shear force. Figure 2.11 shows the law of influence of rake angle γ_o on three-dimensional cutting force [2]. It can be seen that the influence of rake angle γ_o on F_f and F_p is greater than that on F_c.

It has been found that the negative chamfering on the rake face affects the cutting force significantly [27, 28]. When the rake angle has negative chamfering, the cutting force increases due to the increase of chip deformation. The negative chamfering affects the cutting force through the ratio of its width to feed rate. When the ratio

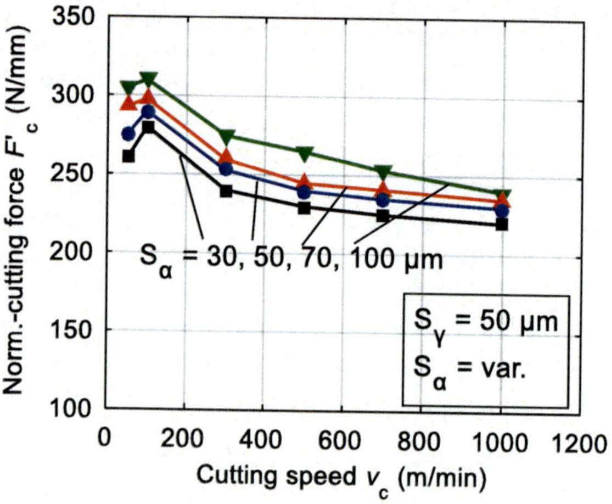

Fig. 2.9 Cutting forces F_c with cutting speeds v_c in different tool flank face wear with orthogonal turning operations of AISI1045 with coated WC–Co inserts. v_c = var. $f = 0.1$ mm. S_γ is tool rake face wear. S_α is tool flank face wear [25]

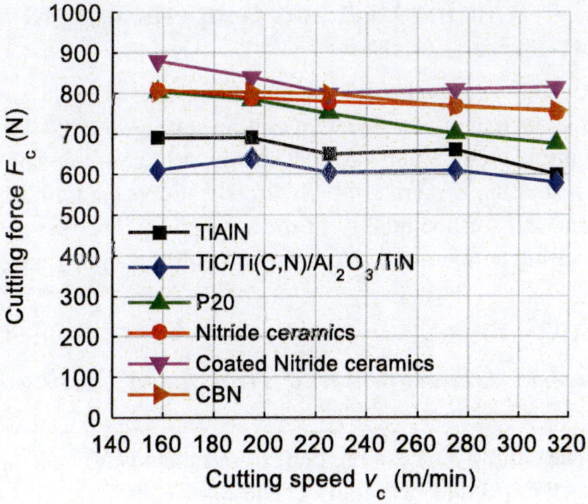

Fig. 2.10 Influence of cutting speed v_c on cutting forces F_c of different workpiece materials. Cutting parameters: nitride ceramics—$f = 0.16$ mm/rev, $a_p = 2$ mm; coated nitride ceramics and CBN—$f = 0.12$ mm/rev, $a_p = 3.3$ mm [26]

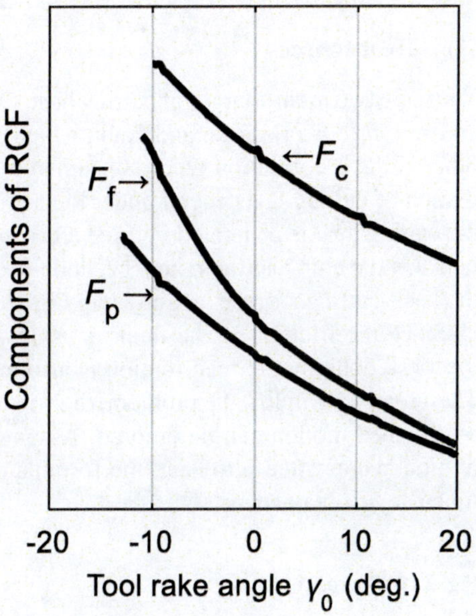

Fig. 2.11 Influence curves of the tool rake angle γ_o on three direction cutting forces [2]

reaches a certain value (steel ≥5, cast iron ≥3), the cutting force is no longer increased but tends to be constant [28].

2.4 Cutting Heat and Temperatures

Cutting heat and cutting temperature caused by it is one of the important physical factors that affect the metal cutting state. About 97–99% of the energy consumed in cutting is converted into heat energy. A large amount of heat makes the temperature of the cutting zone rise, which directly affects the tool life and the machining accuracy, and the surface quality of the workpiece. Therefore, the study of cutting heat and cutting temperature has important guiding significance for production practice.

2.4.1 Generation and Transfer of Cutting Heat

The cutting temperature depends on the position and amount of cutting heat, as well as the speed of heat transfer and loss. Therefore, to control the cutting temperature, it's very necessary to find a way to reduce the production of cutting heat, but also design a reasonable and effective way of heat loss.

(1) **Heat sources**

There are two main sources of cutting heat. One is due to the friction between the chip and rake face, workpiece, and flank, which are the main source of cutting heat. The other is the deformation work consumed by elastic deformation and plastic deformation of cutting layer metal under the action of the cutting tool. Correspondingly, the cutting heat is generated in three areas, namely, the shear plane, the contact area between the chip and the rake face, and the contact area between the workpiece and the back-cutting surface, as shown in Fig. 2.12.

Since the friction on the flank is usually much less than that on the rake face, the work done by the feed motion is also much less than that by the main motion. Therefore, to simplify the problem, the friction work on the flank and the work done by the feed motion can be ignored. Assuming that all the work done by the main motion is converted into heat, the formula of the cutting heat P_c generated in unit time can be obtained as.

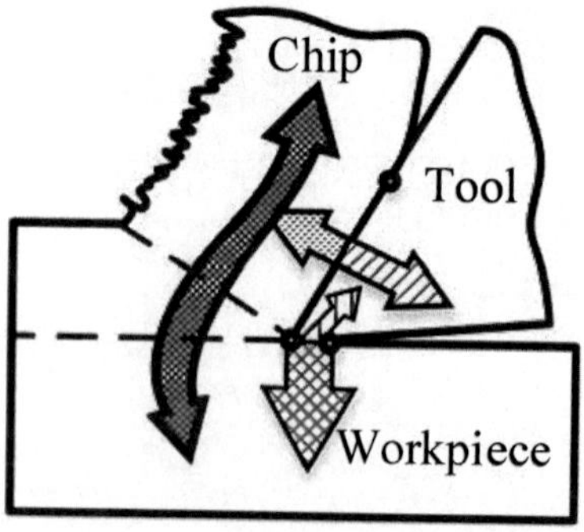

Fig. 2.12 Sources of cutting heat [29]

$$P_c = F_c v_c \tag{2.19}$$

where P_c is the cutting heat produced per second (J/s); F_c is the cutting force (N); v_c is the cutting speed (m/s).

(2) **Transfer of cutting heat**

The main ways of heat transfer are chip, workpiece, cutting tool, and surrounding media (such as air, cutting fluid, etc.), but the main factors affecting heat conduction are the thermal conductivity of workpiece, tool materials, and the condition of surrounding media. When the thermal conductivity of the workpiece material is high, the heat transferred from the cutting zone to the chip, and the workpiece increases. As a result of the chip and chip off the cutting process will not be adversely affected. The heat transferred to the workpiece will cause thermal deformation of the workpiece, which will affect the machining accuracy and surface quality of the workpiece. When the thermal conductivity of the tool material is high, the heat transferred from the cutting area to the tool is increasing, which will cause the temperature rise of the tool, thermal deformation, affect the machining accuracy, and also lead to the tool wear aggravation. To improve the conditions of the surrounding medium and more cutting heat can be dissipated from the surrounding medium by using cutting fluid and spray cooling methods. This part of heat has no adverse effect on the cutting process.

The contact time between the chip and the tool also affects the cutting temperature, which is due to the deformation work consumed by the shear slip on the shear surface and the friction work consumed by the internal friction formed by the relative motion between the rake face and the chip at the bottom of the chip. The heat transformed by these work needs time to be transferred from the chip. For example, in high-speed milling, because of the high cutting speed, the heat in the chip has been removed before a large amount of heat could transfer to the tool. Therefore, the cutting speed increases and the cutting temperature decreases. Another example is: when drilling, the chip is still in contact with the tool and the workpiece, the heat carried away by the chip is transferred to the tool and workpiece again, which increases the cutting temperature. For different cutting methods, the distribution of cutting heat along the non-conduction path is also different, as shown in Table 2.1. The higher the cutting speed and the larger the cutting thickness, the more heat the chip will take away.

Table 2.1 Heat conduction ratios of different machining methods [30]

Conduction pathways	Drilling (%)	Turning (%)	Milling (%)
Chips	55–75	74.6–96.3	65–74.6
Workpieces	10–35	**1.1–20**	1.3–25
Tools	5–15	2.1–18	5.3–10

2.4.2 Cutting Temperatures and Their Distributions

Cutting temperature is the average temperature of the cutting area, expressed by θ. Generally, the average temperature of the contact area between the rake face and the chip is approximately replaced. Temperature field refers to the temperature distribution of the workpiece, chip, and tool. The temperature field can be calculated by theoretical calculation, but the mathematical method is complex and the amount of calculation is large.

Figure 2.13 shows the temperature distribution on the front and back tool surfaces in the main section when turning different materials [31]. Through the study of these graphs, we can conclude some laws for temperature distribution.

(1) The temperature of each point on the shear plane is the same, so it can be inferred that the stress–strain law of each point on the shear plane changes little.
(2) The highest temperature on the rake face and the flank is at a certain distance from the cutting edge, which is due to the increasing friction heat along the cutting edge. The highest temperature appears on the rake face.
(3) In the shear region, there is a higher temperature gradient in the vertical shear direction, which is due to the high speed of the shear slip, so the heat can't be transmitted out on time, thus forming a larger temperature gradient.
(4) The temperature gradient of the chip bottom layer on the vertical rake face is large, and the temperature may drop by half when the distance from the rake face is 0.1–0.2 mm. This shows that the friction on the rake face is concentrated at the bottom of the chip, so the cutting temperature has a great influence on the friction coefficient of the rake face.

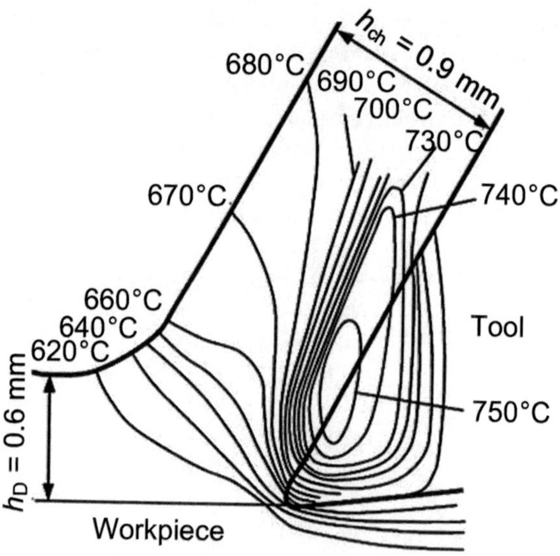

Fig. 2.13 Temperature distributions in two-dimensional cutting [31]. Workpiece: low carbon steel. Cutting tool parameters: $\gamma_o = 30°$ and $\alpha_o = 7°$; Cutting thickness: $h_D = 0.6$ mm; Cutting speed: $v_c = 100$ m/min; Dry cutting; Preheating temperature: 611 °C

(5) The contact length of the flank is very small, so rising and falling of temperatures are completed in a very short time, resulting in a thermal shock on the machined surface.
(6) The greater the plasticity of the workpiece material, the greater the contact length on the rake face, and the distribution of cutting temperature is more uniform. The greater the brittleness of the workpiece material, the closer the highest temperature to the cutting edge.
(7) The lower the thermal conductivity of the workpiece material, the higher the temperature of the front and back cutter surface.

2.4.3 Modeling of Temperature Fields

The milling process belongs to the category of cold working, but from the chip formation process, it has the characteristics of large deformation, high strain rate, and a lot of heat generation. In the process of milling, the distribution of cutting temperature is mainly determined by the plastic deformation of the workpiece and the heat generated by the friction between chip and tool interface. At present, the simulation of the cutting temperature field is mostly based on the mathematical model, using the finite element method (FEM) which can be conducted in ANSYS, ABAQUS, and other related simulation software [32]. The simulation and modeling of milling temperature field generally follow the process:

(i) determine the distribution of cutting heat;
(ii) determine the heat conduction equation and boundary conditions;
(iii) finite element software simulation and results analysis.

The main characteristic of the heat conduction problem is that the heat source has a certain shape and size, a certain dynamic state, and heat output, but the boundary conditions are mostly unknown values. Many heat conduction problems are solved by traditional analytical method or numerical method, but it is very difficult to implement. For the infinite heat conduction range and the heat source concentrated in a very small element volume, the simplest solution can be obtained by using the heat source method, and the calculation results are very close to the actual results. It is based on the solution of the temperature field at any time after the instantaneous point heat source emits a certain amount of heat in the infinite medium. According to the law of conservation of energy and Fourier law, the three-dimensional heat conduction equation of instantaneous point heat source can be established:

$$k\left(\frac{\partial^2 T}{\partial x^2}+\frac{\partial^2 T}{\partial y^2}+\frac{\partial^2 T}{\partial z^2}\right)-\rho c\frac{\delta T}{\delta t}=Q' \tag{2.20}$$

where k is the thermal conductivity, c is the specific heat capacity, ρ is the material density, and Q' is the unit volume heat. In order to facilitate the mesh processing of the simulation object by the finite element method, the model is discretized based on the above model, and the basic finite element formula for temperature calculation can be expressed as:

$$Q = K_T T + C\dot{T} \tag{2.21}$$

where Q is the heat flux vector, K_T is the heat conduction matrix, T is the node temperature matrix, C is the specific heat capacity matrix, and $\dot{T}$ is the node temperature change rate vector. Considering the thermoelastic plastic deformation in the milling process [33], the basic finite element formulation of the deformation field is as follows:

$$\Delta R = K\Delta u \tag{2.22}$$

According to Eqs. 2.24 and 2.25, the basic equation of milling temperature field can be obtained after the simulation object is meshed by finite element method:

$$\begin{bmatrix} F \\ Q \end{bmatrix} = \begin{bmatrix} K & 0 \\ 0 & K_T \end{bmatrix} \begin{bmatrix} u \\ T \end{bmatrix} + \begin{bmatrix} 0 & 0 \\ 0 & C \end{bmatrix} \begin{bmatrix} \dot{u} \\ \dot{T} \end{bmatrix} \tag{2.23}$$

where F is the force vector, u is the node displacement vector, and $\dot{u}$ is the node velocity vector.

Based on the above thermodynamic analysis, the FEM is a numerical calculation method for solving partial differential equations [34]. The FEM is an approximate function assumed in each element to represent the unknown field function to be solved in the whole solution domain. By this method, the approximate functions in the element can be expressed by the values and interpolation functions of unknown field functions or their derivatives sampled at each node of the element. In this way, the numerical value of the unknown field function or its derivative at each node will be changed into a new unknown quantity (i.e., degree of freedom) in the finite element analysis under certain conditions. Thus, the original continuous infinite degree of freedom problem is simplified to a discrete finite degree of freedom problem. A typical finite element analysis includes pre-processing process, simulation process, and post-processing process at the same time, the verification process of simulation results can be completed by combining experiments as illustrated in Fig. 2.14.

According to the above simulation process flow chart, the micro-milling experiment is carried out with Ti-6Al-4V titanium alloy as the milling workpiece material [35] and the simulation results are shown in Fig. 2.15. Based on the simulation result, most of the heat generated during the micro-cutting process diffuses and dissipates with the chip, as shown in Fig. 2.15a, b. Heat diffuses from workpiece to the tool through the tool-chip contact area. Figure 2.15a shows that only a small area around the cutting edge is in contact with the chip. Therefore, heat flow from chip to tool is quite limited leading to the fact that the tool has a lower temperature compared to the chip. Figure 2.15c shows the temperature distribution along the cutting edge. It

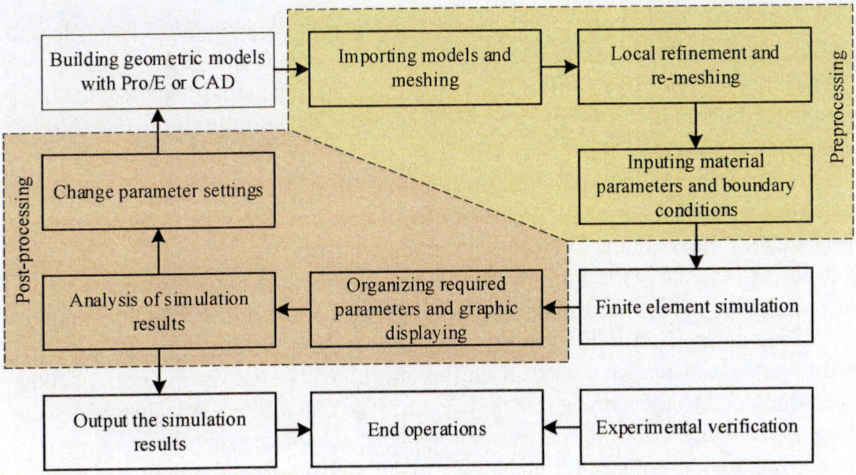

Fig. 2.14 A flow chart of FEM simulation process

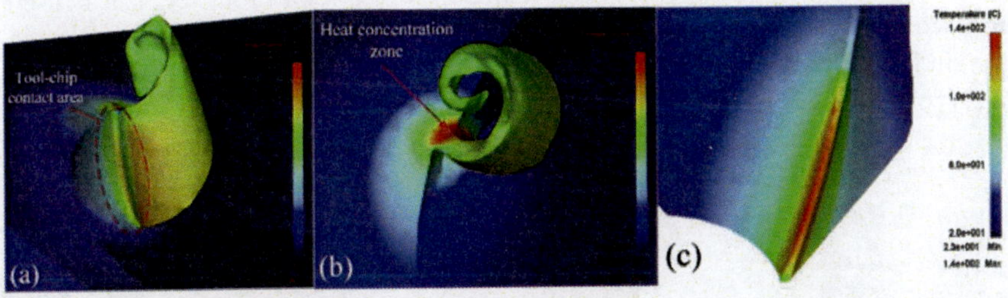

Fig. 2.15 Temperature distribution (**a**). **b** in the chip and **c** along the cutting edge. The workpiece material is Ti-6Al-4V titanium alloy and the cutters are two-flute flat uncoated WC/Co micro-end mills [35]

is interesting that the tool tip is cooler than other sections of cutting edge. This may be a result of heat diffusing and dissipating into the mass of workpiece around the tooltip.

2.5 Milling Process Modeling and Control

2.5.1 Types of Milling Cutters

Milling has been the most studied and applied machining process due to its capability in machining complex geometrical components with high precision. Milling cutters come in a variety of shapes and sizes, but the most common are the cutting edge distributed on the outer edge or end face of a cylinder, cone, or special revolving body, or the multi-tooth cutter with inserted cutter teeth. Each tooth is equivalent to a turning tool, and its cutting characteristics are identical to those of turning.

There are many types of milling cutters, which can be generally divided into three types according to their uses: plane milling cutter, groove milling cutter, and forming surface milling cutter.

The cylindrical milling cutter is mostly used for processing planes on the horizontal milling machine, mainly made of high-speed steel. The spiral cutting edge is distributed on the cylinder surface without an auxiliary cutting edge. Spiral cutter teeth cut into and out of the workpiece gradually, so the cutting process is more stable, generally suitable for machining narrow plane with a width less than the length of the milling cutter. End milling cutter, also known as face milling cutter, is mostly used to process planes on a vertical milling machine. The axis of the milling cutter is perpendicular to the surface to be processed. The end mill is equivalent to a small diameter cylindrical milling cutter with a shank, which is generally composed of three to four teeth. It is used for machining plane, step, groove, and mutually perpendicular plane. It is fixed in the spindle of the machine tool with a taper handle or straight handle. The end mill is equivalent to a small diameter cylindrical milling cutter with a shank, which is generally composed of three to four teeth. It is used for machining plane, step, groove, and mutually perpendicular plane. It is fixed in the spindle of the machine tool with a taper handle or straight handle, as shown in Fig. 2.16b.

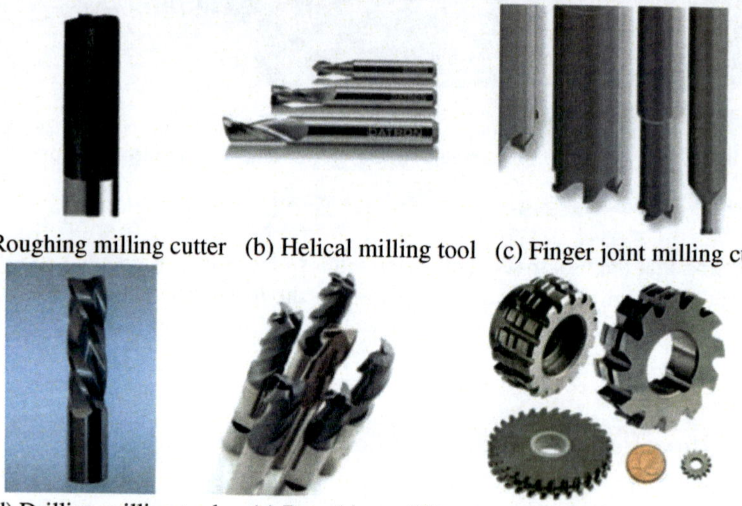

(a) Roughing milling cutter (b) Helical milling tool (c) Finger joint milling cutter

(d) Drilling milling tool (e) Roughing milling cutter (f) Slot milling cutter

Fig. 2.16 Common milling cutters [36]

2.5.2 *Milling Types*

(1) **End milling**

When machining a plane with an end milling cutter, it can be divided into three milling methods according to the relative position (or cutting relationship) between the milling cutter and the workpiece processing surface [37, 38], as shown in Fig. 2.17.

(1) Symmetrical milling. The milling cutter axis is located at the center of symmetry of the milling arc length, indicating that symmetric milling is used when the cutting thickness is the same when cutting in and out. The average cutting thickness of this milling method is quite large. When milling hardened steel with a low feed per tooth, use symmetrical milling to make the teeth cut into the workpiece beyond the chilled layer.

(2) Asymmetric up-milling. Asymmetric up-cut milling has a cutting thickness that is less than the cutting thickness when cutting out. When milling carbon steel and general alloy steel, this milling method can reduce the impact during cutting, allowing the service life of the carbide end mill to be more than doubled.

(3) Asymmetric down milling. The cutting thickness when cutting in is greater than the cutting thickness when cutting out is asymmetric down milling. The practice has proved that when asymmetric down milling is used to process stainless steel and heat-resistant alloys, it can reduce the spalling wear of cemented carbide and increase the cutting speed by 40% to 60%.

(2) **Pheriferal milling**

According to the changing law of cutting layer parameters during milling, pheriferal milling has two forms: up milling and down milling as shown in Fig. 2.18.

(1) Up-milling. When milling, the direction of cutting speed of the milling cutter cuts the workpiece in opposite to the feed direction of the workpiece. This type of milling is called up-milling. When up-milling, the cutting thickness of the teeth gradually increases from zero. When the cutter tooth starts to cut, due to the influence of the blunt radius of the cutting edge, the cutter tooth slips on the working surface, produces squeeze and friction, which causes a

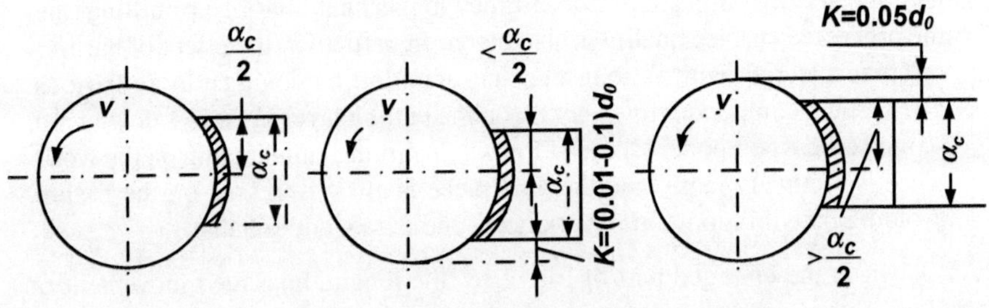

Fig. 2.17 Main procedures of end milling used in workshops and research

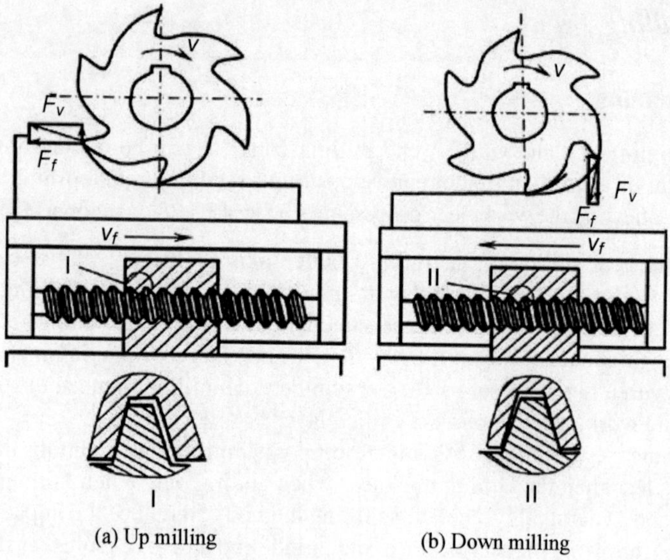

Fig. 2.18 Main procedures of pheriferal milling

seriously chilled layer on the surface. When it slides to a certain extent (i.e., the cutting thickness is equal to or greater than the blunt radius of the cutting edge), the teeth can cut into the workpiece. When the next cutter tooth cuts in, it squeezes and slides on the chilled layer, which makes the cutter tooth easy to wear and at the same time increases the surface roughness of the workpiece. In addition, during up-milling, when the contact angle is greater than a certain value, the vertical milling component will increase, which will easily loosen the clamping of the workpiece and cause vibration.

(2) Down milling. During milling, the cutting speed direction of the milling cutter when it cuts into the workpiece is the same as the feed direction of the workpiece. This milling method is called down milling. In down milling, the cutting thickness of the cutter teeth is the largest when cutting in, and then it decreases gradually to avoid squeezing, slipping, and wearing phenomenon. Moreover, the cutting distance of the teeth is relatively short. The milling cutter wears less, the service life can be 2 to 3 times longer than that of up-milling, and the processed surface quality is also better. In particular, the effect of hard-to-machine materials with a strong tendency of milling hardening is more obvious. The vertical component force acting on the cutting layer is always downward, so the vertical component force of the whole milling cutter acting on the workpiece is relatively large, and the workpiece is always pressed on the fixture, avoiding the vibration of the workpiece, and is safe and reliable.

As shown in the enlarged part of Fig. 2.18, the longitudinal feed movement of the milling table is generally realized by the lead screw and the nut. The nut is fixed, and the worktable is moved by the rotation of the lead screw. When up-milling, the longitudinal milling component force is opposite to the longitudinal force that drives the worktable to move, so that the transmission between the screw and the nut is always tight, the worktable will not move, and the milling process is relatively stable. During down milling, the direction of the longitudinal component of the milling force is always the same as the direction of the longitudinal force that drives the worktable. If there is a gap between the lead screw and the nut drive pair when the longitudinal milling component force is greater than the friction between the worktable and the guide rail, the worktable will drive the lead screw to move, causing the worktable to vibrate, and the worktable will not feed. Even, the phenomenon of hitting the knife will appear in severe cases. Therefore, if down milling is used, it is necessary to require the milling machine worktable to have a device for eliminating the gap in the feed screw nut pair, or to take other effective measures. Therefore, it is advisable to adopt up-cut milling on milling machines without a screw nut clearance elimination device.

2.5.3 Milling Parameters and Cutting Layer Parameters

(1) **Machining parameters**

(1) **Milling speed**. It refers to the linear velocity (in m/s) at the cutting position of the milling cutter tooth, which can be expressed as

$$v = \frac{\pi d_0 n}{1000} \tag{2.24}$$

where d_0 is the diameter (in mm) of the milling cutter, and n is the milling cutter speed (r/s).

(2) **Feed rate**. There are three feed rates in milling. The feed rate per revolution f refers to the distance that the workpiece moves along the feed direction relative to the milling cutter when the milling cutter rotates one revolution, and its unit is mm/r. The feed rate per tooth a_f refers to the distance that the workpiece moves along the feed direction relative to the milling cutter when the milling cutter rotates one tooth, and its unit is mm/z. Feed rate v_f, which refers to the distance of the workpiece relative to the milling cutter along the feed direction in unit time, that is, the feed speed of the milling machine table. Because the feed rate per second is too small, the feed rate per minute is often marked on the machine tool nameplate, and its unit is mm/s or mm/min. The relationship among the three feed rates is

$$v_f = nf = nza_f \tag{2.25}$$

where z is the number of teeth of the milling cutter.

The feed rate of each tooth is selected according to the strength of the cutter teeth, the thickness of the cutting layer, and the chip content. It is used to calculate the milling force and check the strength of the cutter teeth. The feed rate per revolution is closely related to the surface roughness of the machined surface. The feed rate per revolution is selected for finish and semi-finish in milling. Because the main motion and feed motion of the milling machine is driven by two motors respectively, which are different from the horizontal lathe, there is no internal connection between them. Whether the feed rate a_f per tooth or f per revolution is selected, the feed rate v_f must be calculated and called out according to the value mentioned in the catalog of the machine. Generally, only feed rate v_f is marked on the catalog of the milling machine.

(3) **Depth of cut a_p**: the cutting edge through the cutting edge base point and measured in the direction perpendicular to the working plane, as shown in Fig. 2.19. It is the cutting layer dimension measured parallel to the axis of the milling cutter.

(4) **Side feed a_e**: the cutting amount measured in the feed movement direction parallel to the working plane and perpendicular to the cutting edge base point, as shown in Fig. 2.20. It is the cutting layer dimension measured perpendicular to the axis direction and feed direction of the milling cutter.

(2) Milling layer parameters

Similar to that of turning, the two adjacent teeth of the milling cutter respectively form a layer of metal between the two transition surfaces on the workpiece, which is called the cutting layer.

(1) **The cutting layer thickness h_D**. It is the vertical distance between the transition surfaces formed by the main cutting edges of two adjacent cutter teeth on the milling cutter, as shown in Fig. 2.21.

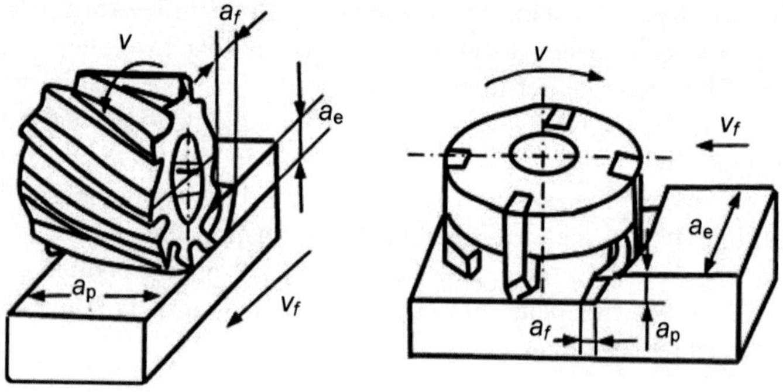

(a) Peripheral milling (b) end milling.

Fig. 2.19 Machining parameters

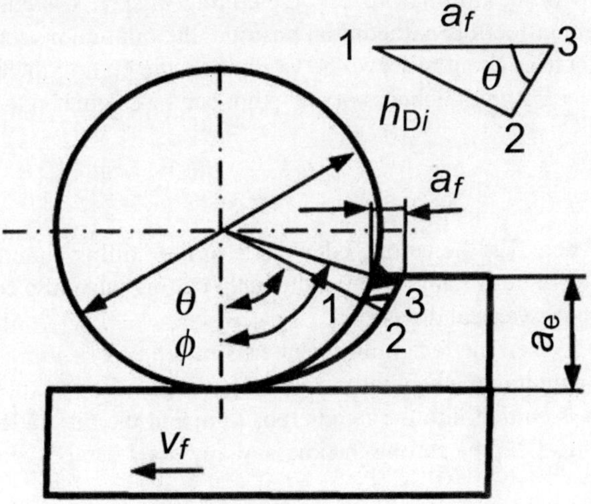

Fig. 2.20 Machining parameters of cylindrical milling

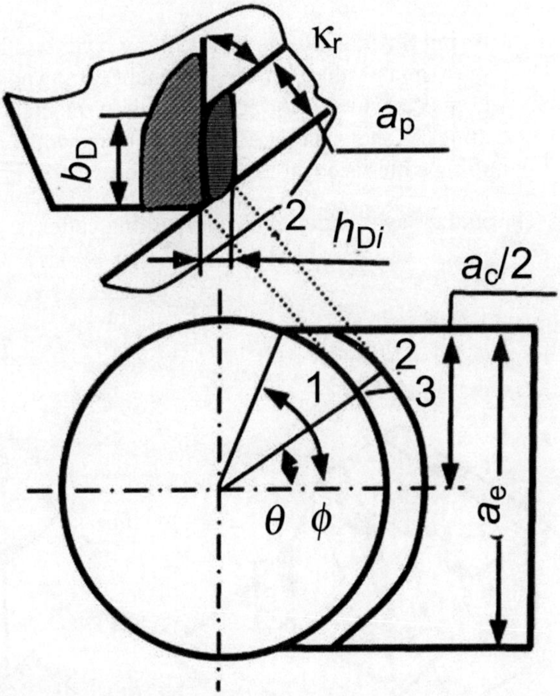

Fig. 2.21 Milling layer parameters of the end mill

In the milling process, the cutting thickness is constantly changing. When the cutter tooth is in the initial position, the minimum value of $h_D = 0$; when the cutter tooth is about to leave the workpiece, the cutting thickness is the maximum value. The cutting thickness of any point can be calculated as

$$h_{Di} = f_z \sin\theta \tag{2.26}$$

where θ is the cutting edge angle of the milling cutter, which is the angle between the radius of any point on the main cutting edge of a cutter tooth and the centerline in the vertical direction.

When the face milling cutter is milling, the cutting thickness of the cutter tooth is minimum when it is just about to cut into the workpiece, and it is maximum when it is cutting into the middle position, and then it gradually decreases. As shown in Fig. 2.22, the cutting thickness of any point can be calculated as

$$h_{Di} = f_z \cos\theta \sin\kappa_r \tag{2.27}$$

(2) **Cutting layer width b_D.** The cutting layer width refers to the contact length (approximate value) between the main cutting edge and the cutting layer of the workpiece, which is the actual length of the main cutting edge participating in cutting. As shown in Fig. 2.21, the cutting width of the straight tooth cylindrical milling cutter is equal to the back feed a_p, that is $a_p = b_D$.

The cutting layer width of the end milling cutter is

$$b_D = \frac{a_p}{\sin\kappa_r} \tag{2.28}$$

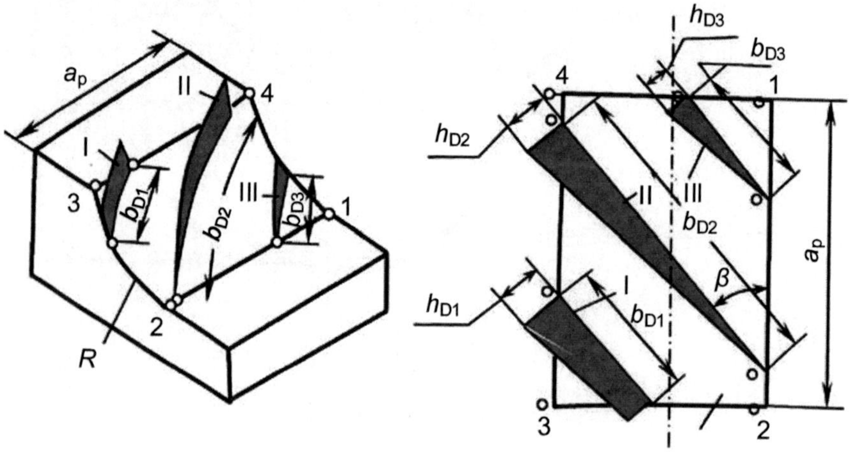

Fig. 2.22 Cutting width of the helical cylindrical milling cutter

The cutting layer width of the helical cylindrical milling cutter varies with the position of cutter teeth. At the beginning of each cutter tooth, theoretically, only a point on the main cutting edge contacts with the workpiece. With the rotation of the tool, the cutter tooth position changes, the cutting width gradually increases. After reaching a certain value, the cutting width gradually decreases with the rotation of the tool as shown in Fig. 2.22. The cutting width of each tooth of the helical cylindrical milling cutter changes slowly, so the total cutting width of the milling cutter also changes gradually. However, the cutting width of each tooth of the straight tooth cylindrical milling cutter varies between a_p and 0, resulting in a large change in the total cutting width of the milling cutter. This is the main reason for the smooth cutting process of spiral gear cylindrical milling cutter.

(3) Cutting area A_D. The instantaneous cutting layer cross-sectional area of a single tooth of the milling cutter is $A_D = h_D b_D$. Because the thickness h_D and width b_D of the cutting layer change at any time during milling, the cross-sectional area A_D of the cutting layer of a single tooth changes with time. This also means that the total cutting area also varies with time. For the convenience of calculation, the average cutting area A_{Dav} can be expressed as:

$$A_{Dav} = \frac{Q}{v} = \frac{a_e a_p v_f}{\pi d_0 n} = \frac{a_p a_e a_f z}{\pi d_0} \tag{2.29}$$

In the milling process, the periodic change of the total cutting layer cross-sectional area $A_{D\Sigma}$ will inevitably lead to the periodic change of milling force, which can easily produce vibration. Therefore, the whole process system should have good rigidity during milling. This is also the reason why cylindrical milling cutters and end mills are generally made into spiral teeth.

2.5.4 Milling Forces

In the milling process, each working tooth of the milling cutter is affected by the cutting force, and the milling resultant force is the sum of the cutting forces on each tooth. The resultant force can be broken down into three mutually orthogonal components as illustrated in Fig. 2.23.

According to the influence on the cutting tool and milling power, it can be divided into three categories.

(1) Main cutting force F_c: the component force in the circumferential direction of the milling cutter, also known as tangential force or circumferential force. The most consumed power is the main cutting force.

(2) Back force F_p: the component force in the direction of milling cutter radius, also known as radial force. Generally, it does not consume power, but it will make the cutter bar bend and deform.

Fig. 2.23 Milling component model

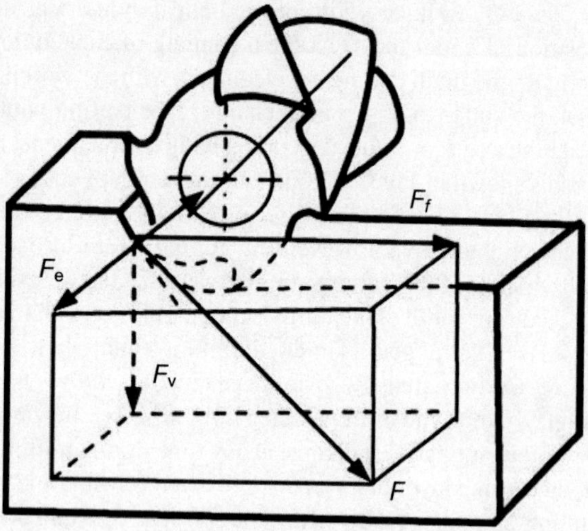

(3) Axial force F_a: it acts on the axis direction of the milling cutter. Its size is related to the helix angle $\beta(\lambda_s)$ for the spiral gear cylindrical milling cutter and is related to the main deflection angle κ_r and the edge inclination angle λ_s for the end milling cutter.

According to the influence on the machining process and machine tool, it is decomposed according to the moving direction of the worktable.

a. Vertical component force F_v: it acts on the vertical direction of the working table of the milling machine. Since the rigidity of the milling machine in this direction is the lowest, when F_v is large, it is easy to cause vibration. In end milling and end milling, the axial force of milling cutter $F_a = F_v$.
b. Transverse component force F_e: it is flat on the milling machine table and perpendicular to the longitudinal movement direction of the table (or parallel to the transverse movement direction of the table). The axial component of cylindrical milling cutter $F_a = F_e$.
c. awesome force F_f: it is parallel to the longitudinal motion direction of the worktable, also called the feed force, which consumes the power of the feed motor.

The main cutting force can be obtained according to $M = F_c d_0/2$, and the other forces can be obtained according to their ratio to the main cutting force F_c. It can also be calculated by the empirical formula of the relevant manual, which is related to the material of the workpiece, the material of the cutter, the form of the milling cutter, and the milling amount.

The cutting power can be expressed as:

$$P_c = \frac{F_c v \times 10^{-3}}{60} \quad (2.30)$$

2.5.5 The Milling System Dynamics

During milling, the direction of milling force and cutting thickness changes with the tool rotation. Because of the periodic intermittent cutting, the dynamic characteristics of the machining system cause the workpiece tool deformation, which has a dynamic impact on the milling force, resulting in the static cutting force model has certain limitations. If the extreme cutting conditions lead to severe deformation (vibration) of the tool or workpiece in the system, the static cutting force model is difficult to accurately predict the actual cutting situation. Therefore, it is necessary to model and simulate the dynamic cutting force in the milling process.

As shown in Fig. 2.24, the machine tool system can be simplified as two degrees of freedom vibration system in two mutually perpendicular directions which are x and y. The milling dynamic equation is expressed in the form of differential equation as:

$$\begin{bmatrix} m_x & 0 \\ 0 & m_y \end{bmatrix}\begin{bmatrix} \ddot{x} \\ \ddot{y} \end{bmatrix} + \begin{bmatrix} c_x & 0 \\ 0 & c_y \end{bmatrix}\begin{bmatrix} \dot{x} \\ \dot{y} \end{bmatrix} + \begin{bmatrix} k_x & 0 \\ 0 & k_y \end{bmatrix}\begin{bmatrix} x \\ y \end{bmatrix} = \sum_{j=1}^{N}\begin{bmatrix} F_{xj} \\ F_{yj} \end{bmatrix} = \begin{bmatrix} F_x(t) \\ F_y(t) \end{bmatrix} \tag{2.31}$$

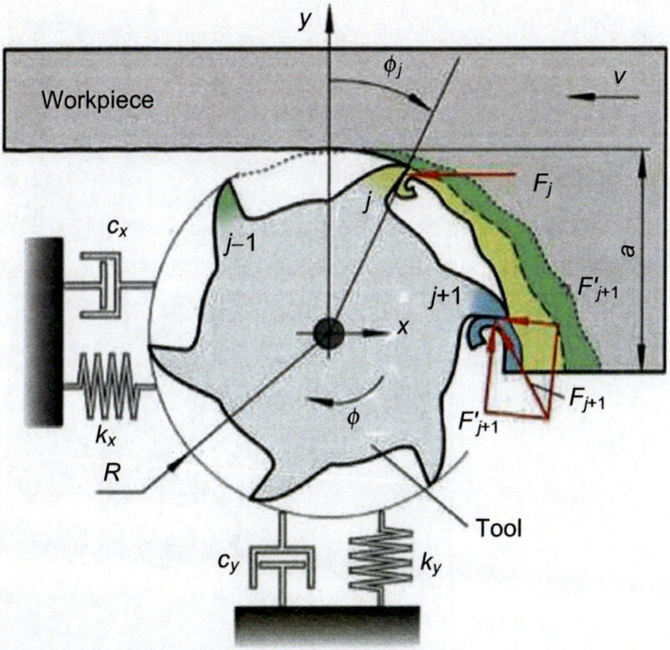

Fig. 2.24 A dynamic model of the milling system [26]

where m_x and m_y are masses of the machine tool system in the x and y directions respectively, c_x and c_y are the dampings of the machine tool system in the x and y directions respectively, k_x and k_y are the stiffness of the machine tool system in the x and y directions respectively, and F_{xj} and F_{yj} are the cutting force components acting on cutter tooth j in x and y directions respectively.

During milling, the cutting force excites the machining system in the feed x and normal y directions and causes the corresponding dynamic displacements x and y respectively. After a coordinate transformation, $v_j = -x\sin\phi_j + y\cos\phi_j$, it acts on the cutting thickness direction of the cutter tooth, where $\phi_j = (j-1)\phi_p + \phi$, and it is the instantaneous radial contact angle of the cutter tooth j. for a milling cutter with evenly distributed cutter teeth, $\phi_p = 2\pi/N$, and ϕ are measured clockwise from the normal y-direction. If the spindle rotates at the angular speed Ω (rad/s), the radial contact angle changes to $\phi_j = \Omega t$ at any time. The final cutting thickness is composed of two parts, one is the static cutting thickness part $f_t\sin\phi_j$ of the tool as a rigid body motion, and the other part is the dynamic cutting thickness change caused by the vibration of the current cutter tooth and the previous cutter tooth. The dynamic cutting thickness change is shown in Fig. 2.25. Since the cutting thickness is measured radially, the total cutting thickness can be expressed as:

$$h(\phi_j) = \left[f_t \sin\phi_j + \left(v_{j,0} - v_j\right)\right]g(\phi_j) \tag{2.32}$$

where f_t is the feed rate per tooth, $v_{j,0}$ and v_j are the dynamic displacements of the tool in the previous tooth cycle and the current tooth respectively, and $g(\phi_j)$ is the unit

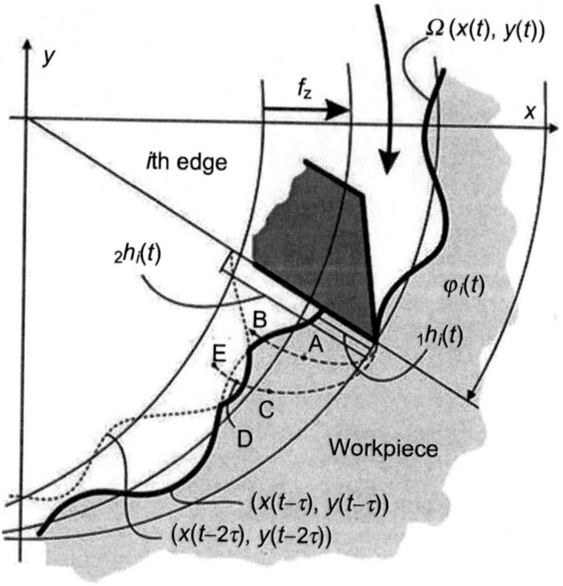

Fig. 2.25 Varying cutting thicknesses in dynamic machining

step function used to determine whether the cutter teeth are in cutting state, which can be expressed as:

$$g(\phi_j) = \begin{cases} 1 \ (\phi_{\text{st}} \le \phi_j \le \phi_{\text{ex}}) \\ 0 \ (\phi_j < \phi_{\text{st}} \text{ or } \phi_j > \phi_{\text{ex}}) \end{cases} \tag{2.33}$$

where ϕ_{st} and ϕ_{ex} are the cutting in angle and cutting out angle of the tool respectively.

Although the static part of the cutting thickness changes with the rotation of the milling cutter, the term can be omitted in the flutter stability analysis because it has nothing to do with the regeneration effect. The dynamic cutting thickness can be expressed as:

$$h_j(\phi) = \Delta s_x \sin\phi_j + \Delta s_y \cos\phi_j \tag{2.34}$$

where the tool regeneration displacement $\Delta s_x = x - x_0$, and $\Delta s_y = y - y_0$. (x, y) and (x_0, y_0) represents the dynamic displacement of the current cutter tooth and the previous cutter tooth cycle respectively. The tangential and radial dynamic cutting forces acting on the cutter tooth J can be expressed as:

$$\begin{cases} F_{\text{t}j} = K_{\text{tc}} a_{\text{p}} h(\phi_j) \\ F_{\text{r}j} = K_{\text{r}} F_{\text{t}j} \end{cases} \tag{2.35}$$

where K_{r} is the ratio of radial force coefficient K_{rc} to cutting force coefficient K_{tc}. By decomposing the cutting force in x and y directions, the cutting force can be given as:

$$\begin{bmatrix} F_{xj} \\ F_{yj} \end{bmatrix} = \begin{bmatrix} -\cos\phi_j & -\sin\phi_j \\ \sin\phi_j & -\cos\phi_j \end{bmatrix} \begin{bmatrix} F_{\text{t}j} \\ F_{\text{r}j} \end{bmatrix} \tag{2.36}$$

By adding the cutting forces acting on the cutter teeth, the total cutting forces acting on the cutter teeth are obtained as:

$$\begin{bmatrix} F_x \\ F_y \end{bmatrix} = \sum_{j=0}^{N-1} \begin{bmatrix} F_{xj}(\phi_j) \\ F_{yj}(\phi_j) \end{bmatrix} \tag{2.37}$$

Substituting Eqs. 2.39, 2.40, and 2.41 into 2.42, the specific relationship between total cutting force and processing parameters can be expressed as

$$\begin{bmatrix} F_x \\ F_y \end{bmatrix} = \frac{1}{2} a_{\text{p}} K_{\text{tc}} A(t) \Delta S(t) \tag{2.38}$$

where the dynamic displacement of cutter teeth $\boldsymbol{\Delta S}(t) = [\Delta s_x, \Delta s_y]^{\mathrm{T}}$. The direction coefficient $\boldsymbol{A}(t)$ with time dependence is

$$A(t) = \begin{bmatrix} \alpha_{xx} & \alpha_{xy} \\ \alpha_{yx} & \alpha_{yy} \end{bmatrix} = \sum_{j=0}^{N=1} g(j)\left\{\begin{bmatrix} -1 & -K_{\mathrm{r}} \\ K_{\mathrm{r}} & 1 \end{bmatrix} \sin 2\phi_j + \begin{bmatrix} K_{\mathrm{r}} & -1 \\ -1 & -K_{\mathrm{r}} \end{bmatrix} \cos 2\phi_j + \begin{bmatrix} -K_{\mathrm{r}} & -1 \\ 1 & -K_{\mathrm{r}} \end{bmatrix}\right\} \tag{2.39}$$

With the rotation of the milling cutter, the direction coefficient changes with time, which is the most fundamental difference between milling and other machining methods with constant cutting force direction. However, for the milling force, the angular frequency of $\boldsymbol{A}(t)$ is $\omega = N\Omega$ and the period is a periodic function of $T = 2\pi/\omega$. Therefore, it can be expanded into Fourier series as:

$$\begin{cases} A(t) = \sum\limits_{r=-\infty}^{\infty} A_r \mathrm{e}^{\mathrm{i}r\omega t} \\ A_r = \dfrac{1}{T}\displaystyle\int_0^T A(t)\mathrm{e}^{-\mathrm{i}r\omega t}\mathrm{d}t \end{cases} \tag{2.40}$$

The Fourier series of periodic function is usually used to solve the periodic system. Altintas and Budak [39] have proved that the high-order harmonics of the periodic function do not affect the prediction accuracy. Therefore, it can be expanded by Fourier series and the DC component can be retained as:

$$A(0) = \frac{1}{T}\int\limits_0^T A(t)\mathrm{d}t \tag{2.41}$$

Because $\boldsymbol{A}(0)$ is only valid when the tool cutting-in angle ϕ_{st} and the cutting-out angle ϕ_{ex}, i.e., $g(\phi_j) = 1$. The Eq. 2.46 can be expressed as:

$$A(0) = \frac{1}{\phi_{\mathrm{p}}}\int_{\phi_{\mathrm{st}}}^{\phi_{\mathrm{ex}}} A(\phi)\mathrm{d}\phi = \frac{N}{2\pi}\overline{A} \tag{2.42}$$

where the average direction coefficient $\overline{\mathrm{A}}$ can be expressed as:

$$\overline{A} = \begin{bmatrix} \overline{\alpha_{xx}} & \overline{\alpha_{xy}} \\ \overline{\alpha_{yx}} & \overline{\alpha_{yy}} \end{bmatrix}$$

$$= \frac{1}{2}\left\{\begin{bmatrix} K_r & -1 \\ -1 & -K_r \end{bmatrix} \sin 2\phi + \begin{bmatrix} 1 & K_r \\ K_r & -1 \end{bmatrix} \cos 2\phi - 2\phi \begin{bmatrix} K_r & 1 \\ -1 & K_r \end{bmatrix}\right\}_{\phi_{st}}^{\phi_{ex}} \quad (2.43)$$

Therefore, the milling dynamic model in Eq. 2.36 can be simplified as:

$$M\ddot{S}(t) + C\dot{S}(t) + KS(t) = \frac{N}{4\pi} a_p K_{tc} \overline{A} \Delta S(t) \quad (2.44)$$

where the dynamic displacement matrix $\boldsymbol{S}(t) = [x, y]^T$. $\boldsymbol{M}$, $\boldsymbol{C}$, and $\boldsymbol{K}$ are the mass, viscous damping, and stiffness matrices of the system, which can be expressed as:

$$M = \begin{bmatrix} m_x & 0 \\ 0 & m_y \end{bmatrix};\ C = \begin{bmatrix} 2m_x\zeta_x\omega_{nx} & 0 \\ 0 & 2m_y\zeta_y\omega_{ny} \end{bmatrix};\ K = \begin{bmatrix} m_x\omega_{nx}^2 & 0 \\ 0 & m_y\omega_{ny}^2 \end{bmatrix} \quad (2.45)$$

In Eq. 2.50, the modal parameters ω_{nx} and ω_{ny} are the undamped natural frequencies of the system in x and y directions respectively, and ζ_x and ζ_y are the damping ratios of the system in the x and y directions respectively. The natural frequency, damping ratio, and mass of the system can be calculated by analyzing the admittance curve of the frequency response function obtained from the impact test, such as amplitude-frequency/phase-frequency curve method and real frequency/virtual frequency curve method.

Based on the established milling dynamic models, the chatter stability of the system can be analyzed by zero-order analytical (ZOA) [40], semi discrete time domain (SDM) [41], multi-frequency method (MFM) [42], time domain numerical method (TNM) [43] and Full discretization method (FDM) [44]. The performance of each algorithm is compared as follows.

(1) ZOA uses the analytical method to solve the flutter stability. The simulation speed is fast and the most widely used method to obtain the flutter stability lobe diagram. However, the high-order harmonics of the periodic function are not considered, and the milling cutter may be separated from the cutting area due to excessive vibration, which leads to low prediction accuracy. Especially in the radial cutting depth, it can't predict the additional stable region and double period bifurcation.
(2) When the radial cutting depth is less than 10% of the cutter diameter, the multi-modal excitation and serious intermittent cutting will cause high-order harmonics. SDM and MFM take into account the influence of higher harmonics in the process of solving, so the prediction accuracy is high, but the calculation efficiency is relatively low.
(3) TNM establishes a relatively real cutting kinematics and dynamics model, considering the cutting tool geometry and cutter tooth eccentricity, as well as the non-linear factors such as the cutter tooth vibration away from the cutting area and the cutting force coefficient of the processing material, the lobe profile

obtained by TNM, has the highest accuracy. However, it needs to solve differential equations in time-domain simulation, which requires a large amount of calculation and time consumption.

(4) Based on SDM and FDM discretizes the state term and delay term in the system dynamics equation by linear interpolation method, and obtains the periodic direction coefficient by interpolating the boundary value of the time, which improves the calculation efficiency greatly while ensuring the accuracy.

2.6 High-Speed Machining

2.6.1 Introduction to High-Speed Machining

The theory of high-speed machining (HSM) was proposed by C. J. Salomon in 1931. The core idea is that there is a certain range of cutting speeds for any materials where machining is not possible due to excessively high temperatures (in US literature this is called "Death Valley") [2, 45]. In the range of conventional cutting speeds, that is, before reaching the critical cutting speed, the cutting temperature and tool wear increase with the increase of cutting speed. However, if the cutting speed continues to increase, the cutting temperature and tool wear rate will decrease. There are temperature variations at different cutting speeds in HSM for nonferrous and ferrous materials, as shown in Fig. 2.26. Therefore, it shortens the cutting time extremely, improves the production efficiency, and obtains better surface quality and accuracy in HSM.

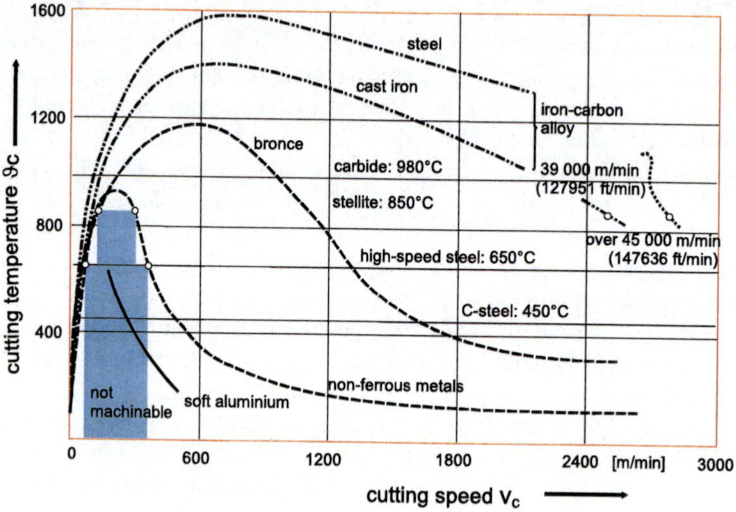

Fig. 2.26 Machining temperatures at high cutting speeds [47]

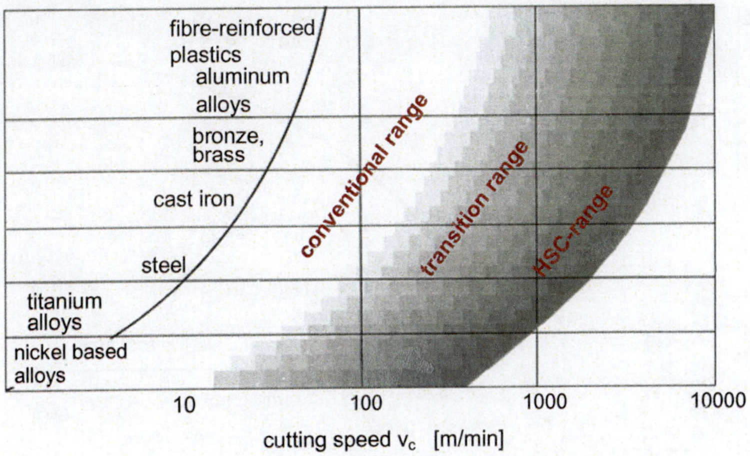

Fig. 2.27 Cutting speed area depends on material [47]

Presently, there is not a clear unified definition of HSM. Generally, the HSM is regarded as a manufacturing process in a higher cutting speed, feed rate, and smaller cutting amount, using tools with super-hard wear- and heat-resistant materials, in which higher processing accuracy and surface quality are obtained. The HSM cutting speed range is not clearly defined. It is generally considered that the cutting speed of HSM is 5–10 times higher than the conventional cutting speed, as shown in Fig. 2.27. The range of feed speeds is generally 2–25 m/min, and sometimes can be as high as 60–80 m/min. Almost every metal material has a critical cutting speed, which is different from each other. The HSM cutting speed range is affected by processing materials and methods [46]. For example, the HSM cutting speed range of cast irons is 900–3000 m/min, while the high-speed cutting speed range of aluminum and titanium alloy materials is 1000–7000 m/min and 100–1000 m/min, respectively. The HSM cutting speed range of turning is generally 700–7000 m/min, while the HSM cutting speed range of milling and grinding are 600–6000 m/min and 5000–10,000 m/min, respectively [2].

In HSM, the cutting time is shortened, the processing efficiency is improved, higher processing accuracy and surface quality are obtained, and the cutting processing requirements of some special materials are satisfied. Schematically, the overall influence of high cutting speeds on the performance of machining is shown in Fig. 2.28 [48]. In recent years, with the breakthroughs of key technologies, such as high-power and high-speed spindle units, high-performance servo control systems, and super-hard wear- and heat-resistant tool materials, HSM is developed rapidly, which is an important development direction of advanced manufacturing technology. It is widely used in equipment manufacturing in the fields of aerospace, mold, optical instruments, precision machinery industries, and many others [2, 5, 47].

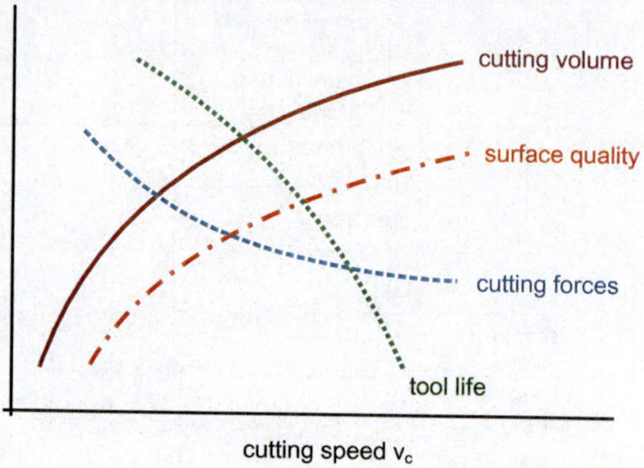

Fig. 2.28 Machining characteristics for HSM technology [48]

2.6.2 Advantages of High-Speed Machining

The HSM technology has become an advanced manufacturing technology integrating high efficiency, high quality, and low consumption. Compared with conventional cutting, HSM has many advantages some are listed below [2]:

(1) Higher productivity. Due to the high cutting speed and feed rate in HSM, compared with conventional cutting, the material removal rate per unit time is greatly improved, which greatly reduces the cutting time.

(2) Smaller cutting force. Compared with conventional cutting, the outflow speed of chips is greatly increased, and the cutting deformation is reduced. Therefore, the cutting force in HSM is 30–90% lower than that in conventional cutting. Presently, the minimum wall thickness of thin-walled parts on airplanes processed by HSM can reach 3–5 μm.

(3) Smaller thermal deformation. Since more than 90% of cutting heat is taken away by the high-speed outflowing chips, and the amount of heat accumulated on the workpiece is minimal. The temperature of the workpiece increases minimally. Therefore, HSM is suitable for machining parts that are deformed easily on heating, have lower processing melting points, and require higher precision.

(4) Higher processing accuracy. The workpiece is basically in a "vibration-free" stable state in HSM. The cutting force and heat are less affected, and the defects such as built-up edges and surface residual stress are effectively controlled. Hence, it is easier to obtain parts with higher machining accuracy in HSM.

(5) Higher machinability of difficult-to-process materials. Due to small cutting force and cutting deformation in HSM, tools are not prone to wear and can be used to process difficult-to-machine materials, such as aluminum, magnesium, nickel, and titanium alloys in the aerospace industry.

2.6.3 Modeling of the Three-Dimensional Instantaneous Milling Force

The tool run-out affects the cutting force greatly and results in pre-mature tool life in HSM. To investigate this relationship, an improved instantaneous milling force per tooth is proposed, with the inclusion of the tool run-out effect. The un-deformed chip thickness considering tool run-out is defined and modeled, according to the geometrical relationships and axial milling ranges per tooth. Meanwhile, instead of studying the conventional average flank wear, tool wear per tooth is studied for a more sensitive correlation with force. A general milling force model is developed, and without loss of generality, it focuses on a 3-flute ball nose end milling cutter with constant helical lead for the convenience of discussion. This is one of the most widely applied types of cutters in HSM. The milling force can be modeled according to the cross-sectional area of instantaneous milling [1, 49].

The main geometric parameters of the cutter edge are shown in Fig. 2.29. The hemispherical center of the cutter is Q, and the Cartesian coordinate system O-xyz is

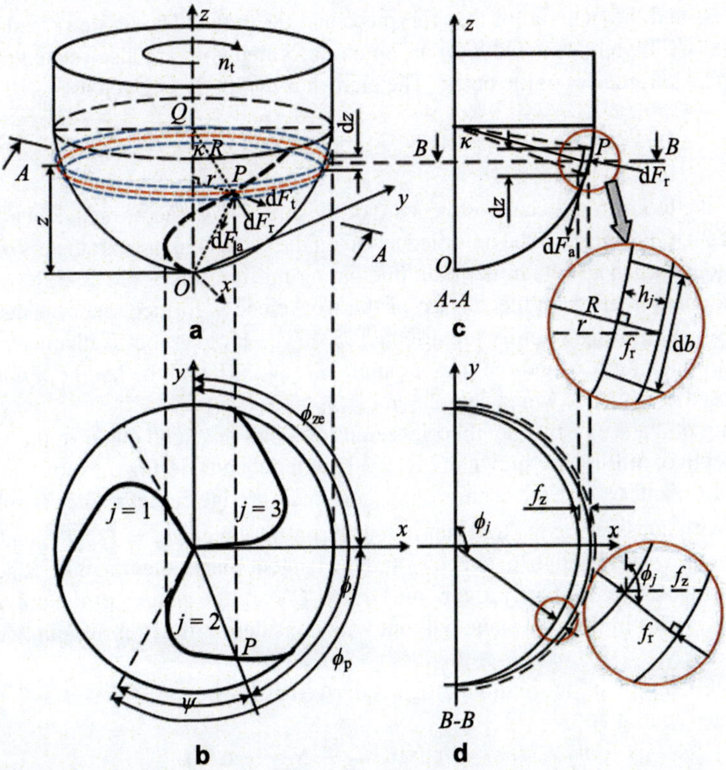

Fig. 2.29 Schematic of a 3-flute ball nose milling cutter. **a** Flutes in front view **b** and top view. **c** Elemental chip in front view **d** and top view

established with a point O, the hemisphere vertex, as its origin. The point P is on the cutter edge of the j-th ($j = 1, 2, \ldots, N_t$) flute, where N_t is the total flute number of the cutter. The flute radius is R, with $\overrightarrow{PQ} = R$ as shown in Fig. 2.29a. The serial numbers of flutes are set in the counter-clockwise direction. The z-coordinate of point P is z ($z > 0$). The spindle speed n_t is set in the clockwise direction.

The neighborhood at the point P in the z-axis direction is chosen as the milling element, whose length is dz. The dynamic Cartesian coordinate system P-tra is established with the point P as its origin, and the t-, r- and a-axes are paralleled to the directions of instantaneous tangential, radial, and axial milling forces, as shown in Fig. 2.29a, c. According to the basic formulas of the instant rigidity forces [50, 51], the elemental instantaneous milling forces $\mathrm{d}\boldsymbol{F}_{\mathrm{M}j}(t, z) = (\mathrm{d}F_{tj}, \mathrm{d}F_{rj}, \mathrm{d}F_{aj})^{\mathrm{T}}$ in the P-tra coordinate system of the point P are given as:

$$\mathrm{d}F_{\mathrm{M}j}(t,z) = Kh_{j(\mathrm{ro})}(t,z)\mathrm{d}b(z) \quad (j = 1, 2, \ldots, N_t) \tag{2.46}$$

where the elements in the matrix $\boldsymbol{K} = (K_{\mathrm{tc}}, K_{\mathrm{rc}}, K_{\mathrm{ac}})^{\mathrm{T}}$ are milling force coefficients in t-, r- and a-axes directions, which can be determined by the identification method based on experimental data. The coefficients in the matrix $\boldsymbol{K}$ are related to milling parameters and the material of the workpiece and the cutter. The un-deformed chip thickness (UCT) $h_{j(\mathrm{ro})}$ and width db of the un-deformed chip are also related to the geometrical parameters of the cutter. The elemental width db is given as:

$$\mathrm{d}b(z) = \mathrm{d}z \cdot \csc\kappa(z) = \mathrm{d}z \cdot \csc[\arccos(1 - z/r)] \ (z < r) \tag{2.47}$$

Since the tool run-out changes the state of the cutter-workpiece engagement, the theoretical UCT h_j needs to be modified based on the shape and the parameters of the tool run-out, which results in periodic fluctuating milling forces [52]. At the same time, the position error on the surface of the workpiece is formed, and the surface roughness and milling stability are affected [53]. The tool run-out is characterized by the run-out length r_0 and the run-out angle α_0 [54, 55], and the height of cutters is expressed as a_{h}, as shown in Fig. 2.30a.

Milling processes of two neighboring flutes of a ball nose end cutter in the same z-axial depth of milling are shown in Fig. 2.30b without considering the tool run-out. The point Q_j and Q'_j are the centers of the current flute edge circles in the xOy plane without/with considering the tool run-out. The run-out length $r_0 = \overline{Q_j Q'_j}$, and the run-out angle α_0 is the angle between r_0 and the tangent line of the first flute edge, as shown in Fig. 2.30c. Similarly, the point Q_i and Q'_i are the centers of the previous flute edge circles in the xOy plane without/with considering the tool run-out, where $i = 1, 2, \ldots, N_t$.

The UCT $h_{j(\mathrm{ro})}$ with tool runout effect is derived as [56]:

$$h_{j(\mathrm{ro})}(t,z) = \max\left(0, h_j + \Delta h_j - \Delta h_i\right) \tag{2.48}$$

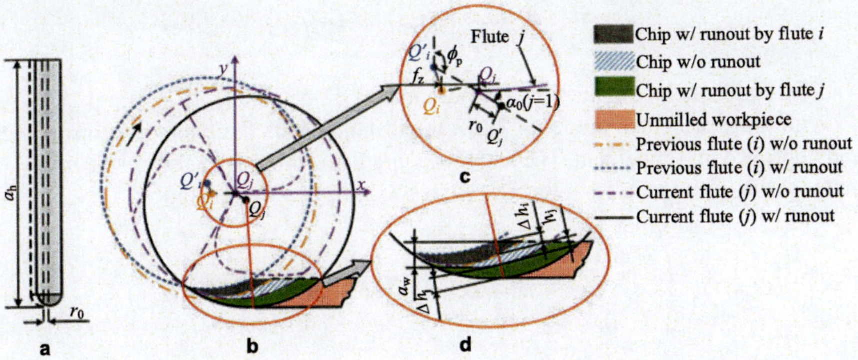

Fig. 2.30 Effect of the tool run-out, where **a** Type of run-out. **b** Modeling of UCT. **c** Run-out parameters, and **d** Extra UCT

where the relation of serial numbers i and j of the current and previous flutes is given, according to the series of flute numbers in Fig. 2.29b. h_j is the theoretical UCT without the tool runout effect at milling time t. Δh_j and Δh_i are the extra UCTs that are caused by the tool runout of the current and previous flutes respectively. These intermediate variables are given as:

$$\begin{cases} i = j - 1 + N \cdot \lfloor (N+1-j)/N \rfloor (j = 1, 2, \ldots, N) \\ h_j(t,z) = f_r \sin(\kappa(z)) = f_z \sin(\phi_j(t,z)) \sin(\kappa(z)) \\ \Delta h_k(t,z) = r_0 \sin(\kappa(z)) \cos[\alpha_0 + \psi(z) + (k-1)\phi_p] (k = i, j) \end{cases} . \tag{2.49}$$

where the integral symbol $\lfloor x \rfloor$ represents the maximum integer, which is not more than the real variable x. The r-axial angle ϕ_j, the a-axial angle κ, the pitch angle ϕ_p, and the lag angle ψ are milling process parameters, as shown in Fig. 2.29, these parameters are given as:

$$\begin{cases} \phi_j(t,z) = \phi_s + 2\pi n_t t/60 - (j-1)\phi_p - \psi(z) \\ \kappa(z) = \begin{cases} \arccos(1 - z/R) \ (z < R) \\ \pi/2 \ (z \geq R) \end{cases} \\ \psi(z) = [z \tan(\beta(z))]/[R \sin(\kappa(z))] \end{cases} \tag{2.50}$$

where ϕ_p is the pitch angle and $\phi_p = 2\pi/N_t$, and ϕ_{ze} is the r-axial angle while the z-coordinate, z, of the first flute ($j = 1$) is equal to 0. The initial angle phase ϕ_s is equal to ϕ_{ze} at the initial sampling time ($t = 0$), and it's related to the sampling selection time merely, given as $\phi_s \in [0, \phi_p)$. n_t is the spindle speed of the cutter. β is the helix angle of the cutter. According to the geometric model in Fig. 2.29, the helix angle β is then expressed as:

$$\beta(z) = \arctan\left[(r(z)\tan\beta_c)/R\right] \tag{2.51}$$

where β_c is the helix angle of the cylindrical part of the cutter.

With the mechanical model of UCT, the instantaneous force model is promoted and improved for refinement. The theoretical predicted milling force $\boldsymbol{F}$ on the integral cutter at the milling time t in the x-, y- and z-axial is expressed as [56]:

$$F(t) = \sum_{j=1}^{N_t} \begin{pmatrix} -\cos\phi_j & -\sin\kappa\sin\phi_j & -\cos\kappa\sin\phi_j \\ \sin\phi_j & -\sin\kappa\cos\phi_j & -\cos\kappa\cos\phi_j \\ 0 & \cos\kappa & -\sin\kappa \end{pmatrix} \mathrm{d}F_{Mj}(t, z) \tag{2.52}$$

where the z-axial upper and lower boundaries of the CWE, z_{ju} and z_{jd}, are determined by the r-axial angle ϕ_j, as shown in Fig. 2.29b, which are correspond to the entry and exit moments of the cutter-workpiece engagement respectively.

Three-flute ball nose end milling cutters are chosen in the experiments to validate the developed approach. The material of the workpiece is Inconel 718, the Ni-based super alloy. The work platform is arranged on the high-speed machining center MIKRON HSM600U. The cutter diameter is 6 mm. The material of the workpiece is Inconel 718, the Ni-based super alloy. The size of the workpiece is 112.5 mm × 40 mm. The sampling frequency is 50 kHz. The feed rate is 1.555 m/min. The spindle speed is 10400 rpm. The force and vibration signals are measured by the Kistler 9119AA2 Quartz 3-component dynamometer and Kistler 8636C accelerometers. Each tool cutter completes 315 cuts with the same workpiece, with identical condition and same cutter.

The stable period in 0.02 s before the end of milling in the 110th step is selected as the statistical sampling period. Milling forces F_x and F_y in each step with time t are shown in Fig. 2.31. The prediction amplitudes and means of milling forces are in good agreement with the experimental data in the x-axial direction. The amplitude and frequency variations increase with the milling time, which is affected by the tool

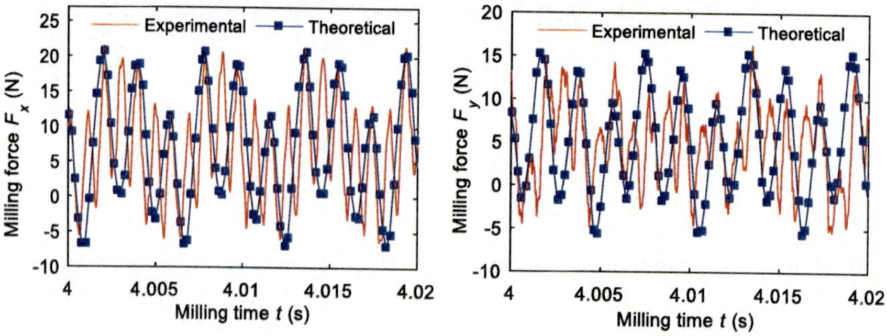

Fig. 2.31 Theoretical and experimental instantaneous milling forces in the feed (x-axial) and normal (y-axial) directions

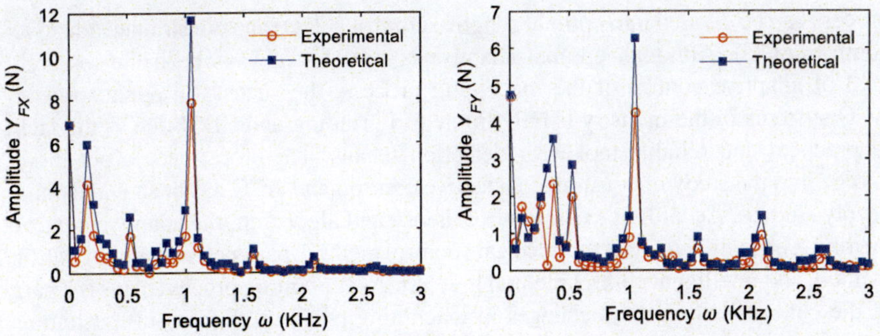

Fig. 2.32 Spectrums of theoretical and experimental instantaneous milling forces in the feed (*x*-axial) and normal (*y*-axial) directions

wear. There are three main peaks in one period on the theoretical milling force curves, which correspond to the time when flutes of the cutter are cutting the workpiece respectively. Theoretical milling forces are related to the tool run-out parameters.

Amplitude frequency characteristic curves of milling forces obtained from both the experiments and the theoretical model are shown in Fig. 2.32. The larger amplitudes are concentrated in the low-frequency region mainly in the initial milling stage, and the theoretical model can accurately describe the responses of milling force in the high amplitude and low-frequency region. Since the CWE hasn't reached the stable state in the initial milling stage, and the cutter is in the period of running-in wear, some high frequency, and low amplitude interfering signals are generated in this stage, which is more obvious in the *y*-axial direction as shown in right side figure of Fig. 2.32. With the milling progressing, the cutter is in the adhesive wear stage and the wear is slowly increased after the initial stage before it becomes severe. As shown in the single-sided amplitude spectrum, the maximum amplitude appears in the middle frequency region, and the interfering signals decreased significantly. The milling force spectrums of the theoretical model agree with the experimental data highly. As the noise interference is not included in the theoretical model, the response amplitudes of instantaneous milling forces in the model are larger than those of the experimental data. To further improve the machining accuracy and surface quality of the workpiece, an on-line monitoring and control system for the high-speed machining need be developed.

2.7 Control of Machining Process

In order to improve production efficiency, machining accuracy, equipment utilization, reduce costs, and safety production of machine tools, the adaptive control technology of the machining process has been paid attention by many researchers since it was put forward in the early 1960s. Generally, the adaptive control system of the machining process can be divided into optimal adaptive control (ACO) and constrained adaptive control (ACC). Although optimal adaptive control has always been the research goal of adaptive control of the machining process, the successful application of ACO systems in the industry is only limited to grinding and EDM due to the lack of practical and reliable tool wear detection means. Therefore, people put more energy into the study of constrained adaptive control, and ACC has been widely and deeply studied. According to the control theory and algorithm, the adaptive control methods can mainly include the fixed gain control method, parameter adaptive control method, and intelligent control method [57, 58]. As the cutting process itself is a part of the control loop, normal changes in machining (such as changes in machining process parameters) will reduce the performance of the controller, and even lead to the instability of the control system [59].

In order to obtain stability and good control performance in a wide range of machining conditions, Koren and Masory [60] developed a variable gain turning adaptive control system. Compared with the early feedback adaptive control system, the system adds a parameter estimation module and a controller adaptation module in the structure. The parameter estimation module provides the estimation of the machining process parameters, the controller adaptation module provides the estimation of the machining process parameters, and the controller adaptation module uses the estimated Parameters to adjust the control gain to maintain a constant open-loop gain ideal value. Lauderbaugh and Ulsoy [61] used the second-order machining process model (including servo loop) to control the instantaneous cutting force, and obtained good transient performance, and made the system suitable for multi-axis milling.

Batel and Shin [62] introduced zero phase error tracking control (ZPETC) into the model reference adaptive control (MRAC) system, and designed an extended model reference adaptive control system which can control non-minimum phase objects; Rober and Shin [63] implemented this method on high-speed milling machine tools to control the cutting force in rough machining.

Altintas and Aslan [64] proposed an adaptive process control method combining virtual machining systems and online monitoring systems to solve the problem of the limited application of machining process control in industrial production due to the lack of practical cutting force sensors. The cutting force is predicted in real-time by a built-in current information of the CNC system, and the controller gain is adjusted according to the cutting force deviation, thus there is no need to install expensive and impractical sensors on the machine tool. In the approach by integrating virtual machining system with the on-line monitoring and control system, the forces are extracted from the feed and spindle drive motor current commands by compensating the distortion caused by the structural dynamic chain between the cutting and motor locations. The use of CNC inherent force-sensing eliminates the need for costly and impractical sensors mount on the machine. Current online monitoring systems are blind to changes in the cutter–workpiece engagement and process, which prevents them to be robust against collision forces and false tool failure alarms. The proposed system brings such critical information from the virtual machining system and enables the on-line monitoring and control algorithms to detect the tool failure, control the process more robustly. The proposed method has wide applications which include the detection of chatter at specific tool path locations, collecting machining load and idle data and matching them to NC program for improved process planning, and auto-calibrating the process simulation system from the on-line force and torque measurements [64]. The measured peak cutting forces predicted from motor current, commanded feed speed, spindle torque received from CNC, and tool breakage detection results are in Fig. 2.33. It can be seen that the peak cutting forces remain constant without any overshoot, the false tool breakage detection does not occur at transient changes in the cutter–workpiece engagement.

The performance of the control system is greatly improved by using the parameter adaptive control method for adaptive control of the machining process. However, to design an adaptive controller based on the traditional adaptive control theory, it

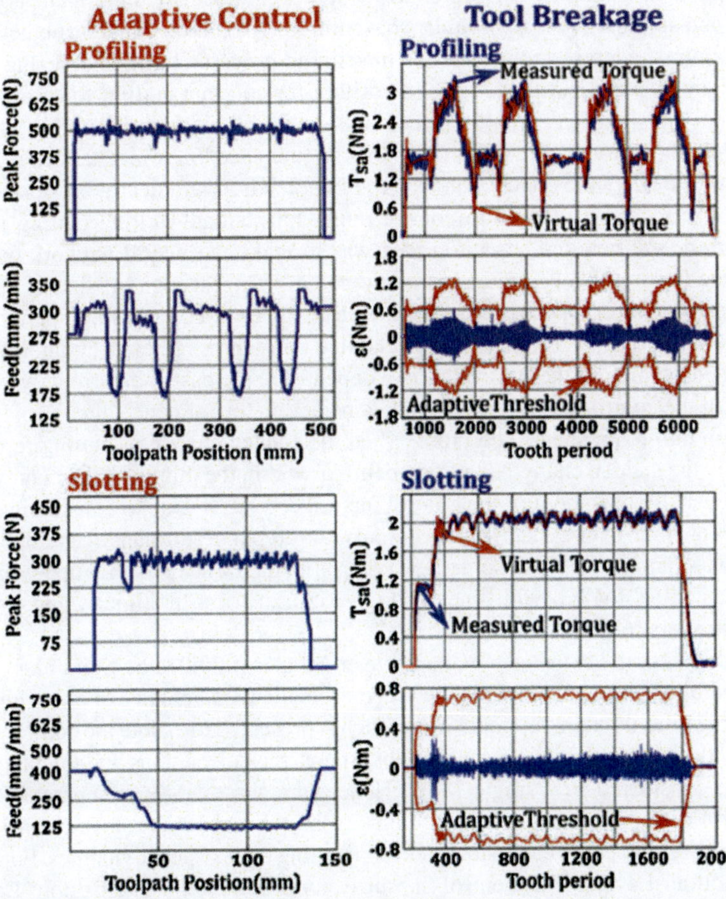

Fig. 2.33 Profiling and slotting operations with virtual feedback: adaptive control and tool breakage results [64]

is necessary to analyze and model the controlled object including the servo loop and machining process dynamics. In the process of cutting, due to the change of cutting parameters (cutting speed, feed rate, and cutting depth), the interference of chip removal process, the influence of dynamic characteristics of machine tool and the drive of machine tool, etc. Due to the interaction of dynamic factors, the machining process is highly time-varying, non-linear, and uncertain, which makes it very difficult to establish the mathematical model of the machining process accurately. Therefore, the adaptive control based on the mathematical model of the object is generally difficult to obtain satisfactory control effect.

Presently, although the theory and technology of adaptive control are developing rapidly in the application of linear time-varying systems, the research progress of adaptive control for nonlinear systems is still very slow. Moreover, many adaptive control rules have a large number of calculations, which makes them difficult to be applied in practice. Therefore, with the development of intelligent control research and application, many scholars have introduced intelligent control methods into constrained adaptive control of the machining process. Tarng and Huang [65] proposed an adaptive learning control algorithm for constant cutting force milling. The system consists of two parts: a feedforward neural network is used to obtain the inverse model of the controlled object, and a fuzzy feedback mechanism is used to adaptively modify the connection weight of the feedforward neural network. Luo et al. [66] proposed a method to control multi-dimensional milling cutting force by using a neural network. Its basic idea was to make the neural network output appropriate feed rate (keeping the cutting force constant) through off-line or online learning and then decompose it into the feed rate along each axis by using a parameter interpolation algorithm. In the existing adaptive constant force control algorithms, the parameter tuning of the controller only depends on the past and current dynamic behavior of the controlled system, without considering the influence of control input and system output prospect. Therefore, when the sudden change of cutting force is induced by the sudden change of cutting depth or width, the output of the controlled system is usually overshoot or the control input is too large. Due to considering the influences of control inputs and system outputs, the output performances and stability of systems are improved with some new control methods, such as the model-based predictive control (MPC) with linearization, log transform, nonlinear, and robust force controllers [67].

Extended with the existing MPC, a 2-Layer-MPC is provided by Stemmler et al. [68]. Regarding the existing MPC, which controls the feed velocity of a machining center in the time-domain, the control approach is presented to enable the control task in the position-domain. In two or more subsystems, the correlations and interactions cannot be considered in a single MPC. Hence, the control task is separated into two control systems to decouple the control task. A first MPC (Top- Layer MPC) determines a reference for each subsystem concerning their system dynamics. Further MPCs (Bottom-Layer MPC) control the subsystems separately. The Bottom-Layer MPC controls the machine with respect to a feed velocity reference $r(t)$ in the time-domain. An additional MPC can be used to serve as a reference generator. Thereby,

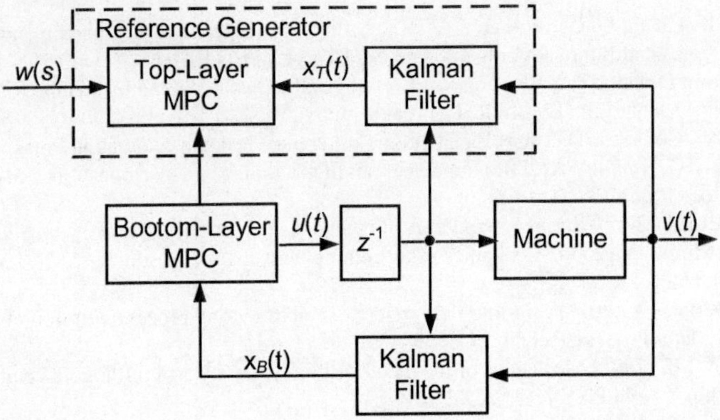

Fig. 2.34 Principle of 2-Layer-MPC [68]

the reference generator transforms the feed velocity reference $w(p)$ from position-domain p to time-domain t. The corresponding functional diagram is depicted in Fig. 2.34. Regarding the estimations of the Kalman Filter, the Top-Layer MPC should maximize the position p to maximize the feed velocity. The maximum allowed feed velocity, which depends on the position, must be taken into account. The Top-Layer MPC uses more process information than the Bottom-Layer MPC. Hence, the tasks of the Bottom-Layer and the Top-Layer can be combined in one single MPC. The application control approach of 2-Layer-MPC is performed as following steps:

(1) modeling of the machine;
(2) identification/parameterization of the model;
(3) development of a Kalman filter;
(4) defining a cost function and the constraints for the control task;
(5) numerical and empirical validation of the closed-loop.

References

1. Altintas Y (2012) Manufacturing automation: metal cutting mechanics, machine tool vibrations, and CNC design, 2nd edn. Cambridge University Press, Cambridge
2. Grzesik W (2017) Advanced machining processes of metallic materials: theory, modelling and applications, 2nd edn. Elsevier
3. Assessment of machining models: progress report. Mach Sci Technol 4(3)
4. Arrazola PJ, Özel T, Umbrello D, Davies M, Jawahir IS (2013) Recent advances in modelling of metal machining processes. CIRP Ann 62(2):695–718
5. Byrne G, Dornfeld D, Denkena B (2003) Advancing cutting technology. Ann CIRP 52(2):483–507
6. Landers RG, Barton K, Devasia S, Kurfess T et al (2020) A review of manufacturing process control. J Manuf Sci Eng 142(11):110814
7. Childs T, Maekawa K, Obikawa T, Yamane Y (2000) Metal machining theory and applications. Wiley

8. Arinez JF, Chang Q, Gao RX et al (2020) Artificial intelligence in advanced manufacturing: current status and future outlook. J Manuf Sci Eng 142(11):110804
9. Sagapuram D, Udupa A, Viswanathan K et al (2020) On the cutting of metals: a mechanics viewpoint. J Manuf Sci Eng 142(11):110808
10. Shaw MC, Cookson JO (2005) Metal cutting principles, 2nd edn. Clarendon Press.
11. Boothroyd G, Knight WA (2006) Fundamentals of machining and machine tools, 3rd edn. CRC Press, Boca Raton
12. Grote A (eds) (2008) Springer handbook of mechanical engineering
13. Ren H, Altintas Y (2000) Mechanics of machining with chamfered tools. ASME J Manuf Sci 122:650–659
14. Jin X, Altintas Y (2011) Slip-line field model of micro-cutting process with round tool edge effect. J Mater Process Technol 211:339–355
15. Merchant ME (1945) Mechanics of the metal cutting process, part I: orthogonal cutting and a type 2 chip. J Appl Phys 16(5):267–275
16. Lee H, Shaffer BW (1951) The theory of plasticity applied to a problem of machining. Trans ASME J Appl Mech 73:405–413
17. Loewen EG, Shaw MC (1954) On the analysis of cutting-tool temperatures. Trans ASME 76:217
18. Palmer WB, Oxley PLB (1959) Mechanics of orthogonal machining. Proc Inst Mech Eng 1–196(173):623–654
19. Wang X, Jawahir IS (2007) Recent advances in plasticity applications in metal machining: slip-line models for machining with rounded cutting edge restricted contact grooved tools. Int J Mach Mach Mater 2:347–360
20. Atlati S, Haddag B, Nouari M, Moufki A (2015) Effect of the local friction and contact nature on the built-up edge formation process in machining ductile metals. Tribol Int 90:217–227
21. Yaguchi H, Onodera N (1988) The effect of tellurium on the machinability of AISI 12L14+ Te steel. Trans Iron Steel Inst Jpn 28(12):1051–1059
22. Fang N, Dewhurst P (2005) Slip-line modeling of built-up edge formation in machining. Int J Mech Sci 47(7):1079–1098
23. Wegener K (2014) Cutting edge influence on machining titanium alloy. In: Laperrière L, Reinhart G (eds) The International Academy for Production Engineering. CIRP Encyclopedia of Production Engineering. Springer
24. Adem KA, Fales R, El-Gizawy AS (2015) Identification of cutting force coefficients for the linear and nonlinear force models in end milling process using average forces and optimization technique methods. Int J Adv Manuf Technol 79(9–12):1671–1687
25. Bassett E, Köhler J, Denkena B (2012) On the honed cutting edge and its side effects during orthogonal turning operations of AISI1045 with coated WC-Co inserts. CIRP J Manuf Sci Technol 5(2):108–126
26. Grzesik W (2014) Machining of spheroidal ductile iron. In: Laperrière L, Reinhart G (eds) The international Academy for Production Engineering, CIRP Encyclopedia of Production Engineering. Springer
27. Ren H, Altintas Y (2000) Mechanics of machining with chamfered tools. J Manuf Sci Eng 122(4):650–659
28. Pawade RS, Joshi SS, Brahmankar PK (2008) Effect of machining parameters and cutting edge geometry on surface integrity of high-speed turned Inconel 718. Int J Mach Tools Manuf 48(1):15–28
29. Abukhshim NA, Mativenga PT, Sheikh MA (2006) Heat generation and temperature prediction in metal cutting: a review and implications for high speed machining. Int J Mach Tools Manuf 46(7–8):782–800
30. Hao G, Liu Z (2020) The heat partition into cutting tool at tool-chip contact interface during cutting process: a review. Int J Adv Manuf Technol
31. Boothroyd G (1988) Fundamentals of metal machining and machine tools, vol 28. CRC Press
32. Zhang Q, Zhang S, Li J (2017) Three-dimensional finite element simulation of cutting forces and cutting temperature in hard milling of AISI H13 steel. Procedia Manuf 10:37–47

33. Fang G, Zeng P (2005) Three-dimensional thermo–elastic–plastic coupled FEM simulations for metal oblique cutting processes. J Mater Process Technol 168(1):42–48
34. Richard Y (2007) Investigation of dry machining with embedded heat pipe cooling by finite element analysis and experiments. Int J Adv Manuf Technol 31(9–10):905–914
35. Thepsonthi T, Özel T (2015) 3-D finite element process simulation of micro-end milling Ti-6Al-4V titanium alloy: experimental validations on chip flow and tool wear. J Mater Process Technol 221:128–145
36. Haynes G (2016) Milling machines & milling operations, 2nd edn: the fundamentals of conventional and CNC milling. Cyber Press
37. Gygax PE (1980) Cutting dynamics and process-structure interactions applied to milling. Wear 62(1):161–184
38. Alauddin MMBMHM, Mazid MA, El Baradi MA, Hashmi MSJ (1998) Cutting forces in the end milling of Inconel 718. J Mater Process Technol 77(1–3):153–159
39. Altintas Y, Budak E (1995) Analytical prediction of stability lobes in milling. Ann CIRP 44(1):357–362
40. Budak E, Altintas Y (1998) Analytical prediction of chatter stability in milling—part I: general formulation Part II: application to common milling systems. Trans ASME J Dyn Syst Measure Control 120:22–36
41. Insperger T, Stephen G (2004) Updated Semi-discretization method for periodic delay—differential equations with discrete delay. Int J Numer Meth Eng 61:117–141
42. Merdol D, Altintas Y (2004) Multi frequency solution of chatter stability for low immersion milling. Trans ASME J Manuf Sci Eng 126(3):459–466
43. Li ZQ, Liu Q (2008) Solution and analysis of chatter stability for end milling in the time-domain. Chin J Aeronaut 21(4):169–178
44. Ding Y, Zhu LM, Zhang XJ, Ding H (2010) Second-order full-discretization method for milling stability prediction. Int J Mach Tools Manuf 50:502–509
45. Toenshoff HK, Denkena B (2013) Basics of cutting and abrasive processes. Springer
46. Schulz H, Moriwaki T (1992) High-speed machining. CIRP Ann 41(2):637–642
47. Schulz H (2003) High-speed machining. In: Manufacturing technologies for machines of the future, pp 197–214
48. Neugebauer R, Bouzakis KD, Denkena B et al (2011) Velocity effects in metal forming and machining processes. CIRP Ann 60(2):627–650
49. Wan M, Zhang WH, Qin GH, Tan G (2007) Efficient calibration of instantaneous cutting force coefficients and runout parameters for general end mills. Int J Mach Tools Manuf 47(11):1767–1776
50. Zhu Z, Yan R, Peng F, Duan X et al (2016) Parametric chip thickness model based cutting forces estimation considering cutter runout of five-axis general end milling. Int J Mach Tools Manuf 101(2):35–51
51. Sun C, Altintas Y (2016) Chatter free tool orientations in 5-axis ball-end milling. Int J Mach Tools Manuf 106(7):89–97
52. Yang Y, Zhang WH, Wan M (2011) Effect of cutter runout on process geometry and forces in peripheral milling of curved surfaces with variable curvature. Int J Mach Tools Manuf 51(5):420–427
53. Schmitz TL, Couey J, Marsh E, Mauntler N, Hughes D (2007) Runout effects in milling: surface finish, surface location error, and stability. Int J Mach Tools Manuf 47(5):841–851
54. Li KX, Zhu KP, Mei T (2016) A generic instantaneous undeformed chip thickness model for the cutting force modeling in micromilling. Int J Mach Tools Manuf 105(6):23–31
55. Wang SB, Geng L, Zhang YF, Liu K, Ng TE (2015) Cutting force prediction for five-axis ball-end milling considering cutter vibrations and run-out. Int J Mech Sci 96–97(6):206–215
56. Zhu K, Zhang Y (2017) Modeling of the instantaneous milling force per tooth with tool run-out effect in high speed ball-end milling. Int J Mach Tools Manuf 118:37–48
57. Liang SY, Hecker RL, Landers RG (2004) Machining process monitoring and control: the state-of-the-art. Trans ASME J Manuf Sci Eng 126–122:297–310

58. Matsubara A, Ibaraki S (2009) Monitoring and control of cutting forces in machining processes: a review. Int J Automation Technol 3(4):445–456
59. Lauderbaugh LK, Ulsoy AG (1988) Dynamic modeling for control of the milling process. ASME J Eng Ind 110(4):367–375
60. Koren Y, Masory O (1981) Adaptive control with process estimation. Ann CIRP 30(1):373–376
61. Lauderbaugh LK, Ulsoy AG (1989) Model reference adaptive force control in milling. ASME J Eng Ind 111(1):13–21
62. Barthel JW, Shin YC (1993) Adaptive control of non-minimum phase processes with application to the end milling processes. In: Proceedings of the American control conference, pp 2449–2454
63. Rober SJ, Shin YC (1996) Control of cutting force for end milling processes using an extended model reference adaptive control scheme. ASME J Manuf Sci Eng 118(3):339–347
64. Altintas Y, Aslan D (2017) Integration of virtual and on-line machining process control and Monitoring. CIRP Ann Manuf Technol 66:349–352
65. Tarng YS, Hwang ST (1995) Adaptive learning control of milling operations. Mechatronics 5(8):937–948
66. Luo T, Lu W, Krishnamurthy K, McMillin B (1998) A neural network approach for force and contour error control in multi-dimensional end milling operations. Int J Mach Tools Manuf 38:1343–1359
67. Landers RG, Ulsoy AG, Ma Y-H (2004) A comparison of model-based machining force control approaches. Int J Mach Tools Manuf 44(7–8):733–748
68. Stemmler S, Abel D, Schwenzer M, Adams O, Klocke F (2017) Model predictive control for force control in milling. IFAC-Papers OnLine 50(1):15871–15876

Chapter 3
Tool Wear and Modeling

During the cutting process, the tool will gradually become dull. When tool wear reaches a certain point, the cutting force increases, the cutting temperature rises, and even vibration occurs. At the same time, the workpiece dimensional may exceed the tolerance range, and the quality of the machined surface degrades significantly. At this point, the tool must be replaced. The tool may be damaged suddenly during the cutting process, resulting in tool breakage. Tool wear, breakage, and useful life (also known as usability or durability) are all related to machining efficiency, quality, and cost, making the tool failure one of the most important cutting problems.

Cutting tool failure can generally be classified into two types: wear and breakage. The former occurs gradually and continuously, whereas the latter occurs suddenly, as in chipping and spalling. Tool wear is primarily determined by the mechanical and physical properties, as well as the cutting conditions of the tool and workpiece materials. The tool wear differs significantly from the common wear of mechanical components in the following way: the chip bottom in contact with the rake face is a new surface with high activity and no oxidation.

3.1 Types of Tool Wear

During the cutting process, the rake face and the chip, the flank surface and the workpiece frequently squeeze and rub against each other, resulting in high temperatures at the tool tip and cutting edges. As a result, wear occurs on the tool's rake and flank face. The rake face wear is shaped like a crescent depression, and the flank wear is shaped like a wear band. Usually, the rake and flank faces wear at the same time and affect each other as illustrated in Fig. 3.1.

K. Zhu, *Smart Machining Systems*, Springer Series in Advanced Manufacturing,
https://doi.org/10.1007/978-3-030-87878-8_3

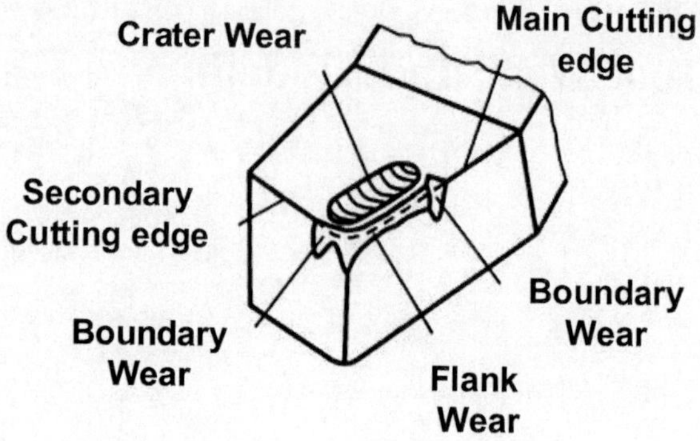

Fig. 3.1 Tool wear pattern

3.1.1 Crater Wear

When machining plastic metal, tool wear frequently occurs on the rake face due to insufficient heat resistance and wear resistance, high cutting speed, and large cutting thickness. Since the chip bottom surface and tool rake face are new surfaces with high chemical activity during the cutting process, more than 80% of the contact area is difficult for air and cutting fluid to enter under the action of high temperature and high pressure on the contact surface. Because the chip slides along the rake face and gradually grinds a crescent-shaped pit on the rake face (as shown in Fig. 3.1), this type of wear is commonly referred to as crater wear.

At first, the crescent crater's leading-edge is still a short distance away from the main cutting edge. The crescent crater gradually expands forwards and backward as the cutting process progresses, and its depth increases. The highest cutting temperature is found in the position with the greatest depth. Its width is determined by the chip's width and does not change significantly during the wear process. When a crescent crater develops to a narrow edge between its leading and cutting edges, the cutting edge's strength decreases, which easily leads to chipping. The maximum depth of the crescent crater KT, as shown in Fig. 3.2, represents the wear value of the rake face.

3.1.2 Flank Wear

When cutting, the freshly machined surface of the workpiece makes contact with the tool flank and rubs against it, causing flank wear. Although the flank has a back angle, the cutting edge is not as sharp as it was before, and there is a certain degree of the blunt circle. The contact pressure between the flank and the workpiece surface is extremely high, and elastic and plastic deformation occurs. As a result, the flank and the workpiece have a small area of contact, and wear occurs on this contact surface, forming a small edge with a zero back angle, and can be seen in Fig. 3.2. This type

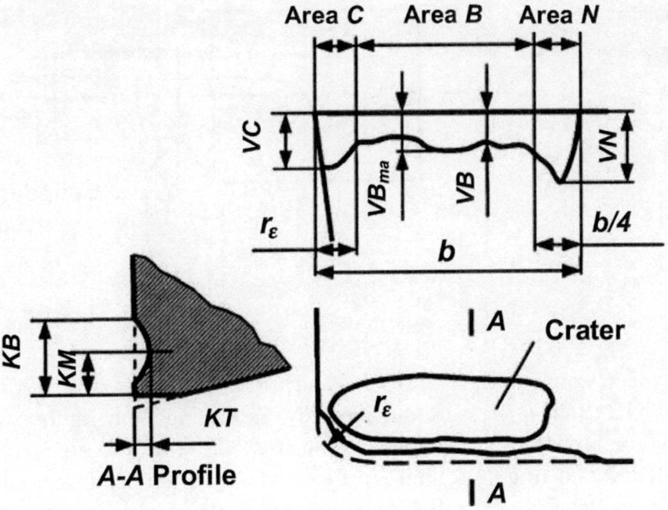

Fig. 3.2 Tool wear measurement

of wear is most common when plastic metals and cast iron are cut at a slower speed and with a smaller cutting thickness (less than 0.1 mm).

Flank wear as shown in Fig. 3.2 is often uneven. The maximum value of tooltip (Zone *C*) has low strength, poor heat dissipation condition, and severe wear which is *VC*. The main cutting edge that gets close to the rear knife surface (Zone *N*) of the workpiece outer skin, which is ground into a serious deep groove, is expressed as *VN*. On the middle part of the worn belt on the rear cutter surface (Zone *B*), the wear is relatively uniform. The average wear width is expressed as *VB*, while the maximum wear width is VB_{max}.

It is easy to wear the rake and flank tool surfaces at the same time when cutting plastic metals with medium cutting speed and medium cutting thickness (0.1–0.5 mm).

3.1.3 Boundary Wear

Deep grooves are typically ground on the flank surface when cutting steel, with the main cutting edge close to the surface to be machined and the secondary cutting edge close to the tip, as shown in Fig. 3.3. These are the two locations where the main and minor cutting edges of the workpiece come into contact with the surface to be machined or the machined surface of the workpiece. There are two primary causes of boundary wear.

(1) During cutting, the compressive and shear stresses are very high on the rake and flank face near the main cutting edge, but the stress on the cutting edge at the

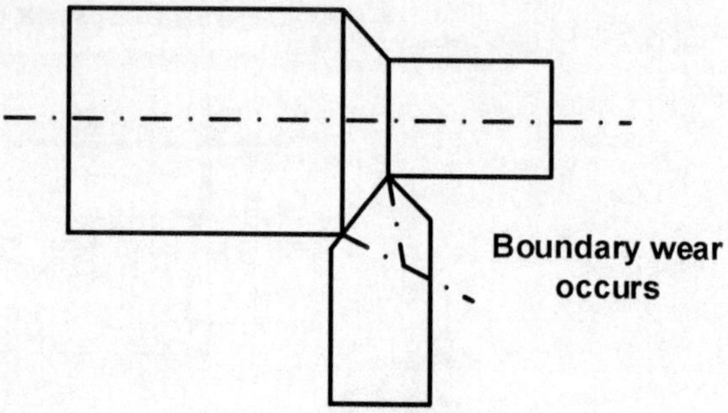

Fig. 3.3 Boundary wears

work piece's outer surface drops suddenly, forming a high-stress gradient and causing large shear stress. Simultaneously, the cutting temperature on the rake face is the highest, and the contact point with the work piece's outer surface is cooled by air or cutting fluid, resulting in a high-temperature gradient and high shear stress. As a result, boundary wear occurs on the main cutting edge's flank.

(2) Work hardening reduces the cutting thickness of the auxiliary cutting edge near the tooltip to zero, causing the cutting edge to slide and causing boundary wear on the flank. Boundary wear is also common when machining rough-skinned castings and forgings.

3.1.4 Tool Wear Criteria

The cutting force, cutting temperature, and machining quality will all be affected as the tool wears. As a result, a maximum allowable wear amount, which is the tool blunt standard, must be specified based on the machining conditions. In general, wear on the flank of general tools has a greater impact on machining accuracy and cutting force than wear on the rake face. Furthermore, the wear on the flank is simple to assess. As a result, in tool management and scientific research, the blunting standard is typically formulated based on the wear amount of the flank, as illustrated in Fig. 3.4.

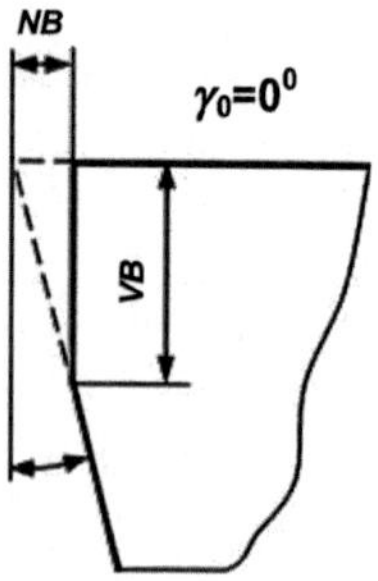

Fig. 3.4 Radial wear of the tool

Table 3.1 Wear standard of carbide turning tools

Processing conditions	Standard for flank wear *VB*/mm
Finish turning	0.1–0.3
Rough turning of alloy steel	0.4–0.5
Rough turning of carbon steel	0.6–0.8
Rough turning of iron castings	0.8–1.2
Rough turning of iron castings low-speed Rough turning of steel and large iron castings	– 1.5

The precision machining tools used in automatic production produce the wear criteria following the accuracy requirements of the workpiece. In this case, the tool wear size along the radial direction of the workpiece is frequently used as the criterion to measure the tool blunt, as illustrated in Fig. 3.4.

The development of wear standards should take into account not only the rational use of cutting tools but also machining accuracy. At the same time, it should consider the impact of numerous factors such as workpiece material machinability, and tool manufacturing difficulty. As a result, the standard of tool wear varies depending on the processing conditions. For example, in finish machining, the bluntness standard can be set lower; in rough machining, the blunting standard can be set higher; and when the rigidity of the processing system is low, whether machining within the blunting standard will produce vibration should be considered.

The recommended values of wear standards for carbide turning tools are shown in Table 3.1 based on investigation data from manufacturing practice. The international standard organization has specified a tool wear standard in the durability test of cylindrical turning tools, which are only used for cutting experiments. Aside from finishing, the standard of tool wear in actual production is usually higher.

3.2 The Formation of Tool Wear

The wear mechanism and wear strength are different under different workpiece materials, tool materials, and cutting conditions, as shown in Fig. 3.5, the proportion of various types of wear at different cutting speeds (cutting temperatures) When cemented carbide tools are used to process steel. It can be seen from the figure that the cutting temperature has a decisive influence on the tool wear for a certain tool material and workpiece material. When the cutting temperature is low, the mechanical wear caused by the hard point in the workpiece material is dominant. When the cutting temperature is high, the thermal and chemical wear which is greatly affected by the cutting temperature is dominant.

Fig. 3.5 The influence of cutting speed on tool wear strength. 1—Mechanical wear, 2—adhesive wear, 3—diffusion wear, 4—thermochemical wear

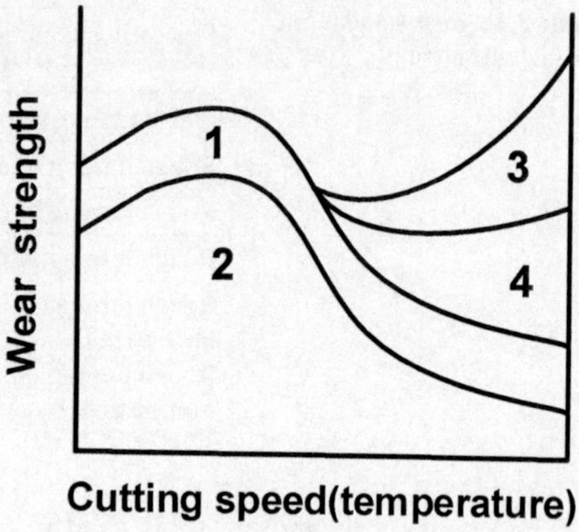

3.2.1 Mechanical Wear

Mechanical wear, also known as hard point wear, is primarily caused by impurities in the workpiece material, hard points in the matrix structure (such as carbides, nitrides, oxides, et.), and debris, among other things, which form a groove on the tool surface. This type of wear is more noticeable on tool steel (including high-speed steel) cutters. This type of wear is relatively minor due to the high hardness of cemented carbide tools.

Hard point wear occurs at various cutting speeds, but it is the primary cause of low-speed tool wear (such as a broach, die, tap, etc.). This is due to the lower cutting temperature at low cutting speed, and other types of wear are minimal. Hard point wear is generally thought to be directly proportional to the relative sliding distance between tool and workpiece or cutting distance.

3.2.2 Adhesive Wear

Adhesion is a cold-welding phenomenon that occurs when the tool and workpiece material comes into contact with the distance between atoms on the actual contact area of the friction surface under sufficient pressure and high temperature. Because of the relative movement, the adhesive point of two friction surfaces will tear and be taken away by the other side. If the adhesive fracture occurs on the tool's side, it will result in tool loss, also known as adhesive wear.

In general, adhesive point fracture occurs on the lower hardness side, i.e. the workpiece material. However, because tool materials frequently have flaws like uneven structure, internal stress, micro-cracks, voids, and local soft spots, the tool surface is usually broken and taken away by the workpiece material, resulting in wear. Adhesive wear occurs in high-speed steel, cemented carbide, ceramic tools, cubic boron nitride, and diamond tools. Adhesive wear is minimal due to the high shear and tensile strength of high-speed steel. Furthermore, the grain size of cemented carbide has a significant impact on bond wear. The slower the wear rate, the smaller the grain size.

The degree of adhesive wear is determined by the cutting temperature, the affinity of the tool and the workpiece material, the hardness ratio of the tool and the workpiece material, the shape and microstructure of the tool surface, and the stiffness of the processing system. For example, the higher the affinity between the tool and the workpiece material, the lower the hardness ratio, and the more severe the adhesive wear.

3.2.3 Diffusion Wear

When the cutting temperature is high, the tool surface is always in contact with the fresh surface, resulting in a high level of chemical activity. When the chemical element concentrations of the tool and the workpiece material differ significantly, they will diffuse into each other in a solid-state, causing changes in the chemical composition of the tool and the workpiece material on both sides of the friction surface. This eventually degrades the tool material's performance and leads to the tool failure. This type of wear is known as diffusion wear. For example, when the cutting temperature reaches 800 °C or higher, the C, W, Co, and other elements in the cemented carbide diffuse into the chips and are taken away, whereas the Fe element in the chips diffuses to the cemented carbide surface to form a new low brittleness hardness composite carbide. The C in the WC in the cemented carbide diffuses out to cause poor carbon and reduce the hardness. Simultaneously, Co diffusion reduces its content, lowering the bonding strength of WC, TiC, and other carbides with the matrix and thus increasing tool wear. However, the diffusion capacity of TiC is not as good as that of WC. At high temperatures, a TiC protective layer will be formed on the surface, which will hinder the progress of diffusion. Therefore, WC-TiC-Co type cemented carbide should be used when high-speed cutting steel parts.

Diffusion wear is the primary cause of cemented carbide tool wear during medium and high-speed cutting. It frequently occurs concurrently with bond wear. The deepest part of the crescent crater on the rake face of the cemented carbide tool has the highest cutting temperature and also the most severe spread. Furthermore, adhesion is more likely to occur here, so the wear rate is the fastest, and the crescent crater is formed by the combined action of diffusion wear and adhesive wear.

The cutting temperature, the chemical composition of the workpiece, and tool materials all influence the rate of diffusion wear. When the cutting temperature of a particular tool material rises, the diffusion speed increases gradually at first, then rapidly. Different elements have different diffusion rates. For example, because the diffusion rate of Ti is lower than that of Co, C, W, and other elements when cutting steel parts with cemented carbide tools, the anti-diffusion ability of WC-TiC-Co cemented carbide is greater than that of WC–Co. The use of TiC and TiN coated tools can significantly improve the chemical stability of the tool surface while also reducing diffusion wear. Furthermore, the diffusion speed is proportional to the flow rate of the chip bottom layer on the tool surface. When the flow speed is high, the concentration difference of the diffusion elements on the friction surface can be kept constant, resulting in a faster diffusion wear speed.

3.2.4 Chemical Wear

At a certain temperature, the tool and certain components of the surrounding medium (such as oxygen in the air, extreme pressure additives in the cutting fluid, sulfur, chlorine, and so on) play a chemical role in forming a layer of lower hardness compounds on the tool's surface. Chips accelerate tool wear, or the tool material is corroded by a specific medium, resulting in tool wear. These are referred to as chemical wear. For example: when turning molybdenum alloys with high-speed steel tools ($a_p = 1$ mm, $f = 0.1$ mm/r, $v_c = 10$–50 m/min), the better the chemical activity of the cutting fluid and surrounding media, the faster the tool wear. Another example: Carbide WC-TiC-Co tool machining stainless steel (including 18% Cr, 9% Ti), when the cutting speed VC = 120–180 m/min, the vulcanized and chlorinated cutting oil is used, the tool usability is lower than that of dry cutting due to the corrosion effect of sulfur and chlorine.

3.2.5 Thermoelectric Wear

Under the influence of high temperatures in the cutting area, the tool and workpiece material form a thermocouple to generate thermoelectric EMF, which forms the thermoelectric flow through the tool-workpiece and tool chip, promoting chemical element diffusion and accelerating tool wear. This type of diffusion wear caused by the thermoelectric force is referred to as thermoelectric wear. The test results show that by insulating the tool-workpiece circuit without changing the stiffness of the tool or the workpiece, tool life can be significantly increased. For example, when the W18Cr4V bit is used to drill stainless steel, if the drill bit is insulated from the drill sleeve, the usability of the drill bit can be increased by 2–6 times.

3.3 Tool Usability and Its Relationship with Cutting Parameters

3.3.1 *Tool Life*

Tool durability is a broad index that describes the cutting performance of various tool materials. The higher the durability under the same cutting conditions, the better the wear resistance of the tool material. Tool durability is an important indicator when comparing the machinability of different workpiece materials. The better the machinability of the workpiece material, the greater the tool durability.

(1) **Definition of tool life**

The pure cutting time (excluding non-cutting time such as tool-setting, measurement, fast forward, and return) from the start of using the sharpened tool until the wear reaches the blunt standard is called tool durability, and it is represented by T. In some cases, the tool cutting stroke lm before reaching the blunt standard can also be used to determine tool durability. The product of the cutting speed vc and the durability T is equal to l_{m}. The number of finished workpieces or passes during finishing can also be used to determine tool durability.

(2) **Relationship between tool life and cutting parameters**

The total cutting time of a new tool from the time it is put into use until it is scrapped, including multiple regrinds, is referred to as tool life, and it is equal to the product of tool durability and the number of regrinds.

3.3.2 *Tool Life Equation*

There is a close relationship between the cutting amount and tool durability, which directly affects the production efficiency and processing cost of machining. Therefore, the impact of the three elements of cutting amount on tool durability is discussed.

(1) **The relationship between cutting speed and tool life**

The tool wear experiment demonstrates that when the workpiece material, tool material, and tool geometric parameters are determined, the cutting speed is the most important factor influencing tool durability. Increase the cutting speed, and the tool's durability will decrease; the relationship can be obtained in the experiment using the single-factor method (other cutting conditions remain unchanged, only the cutting speed is changed). As shown in Fig. 3.6, several curves of tool wear VB versus time corresponding to different cutting speeds are obtained through experiments. According to the required blunt standard, the tool durability corresponding to different cutting speeds can be calculated. Then in the double logarithmic coordinate

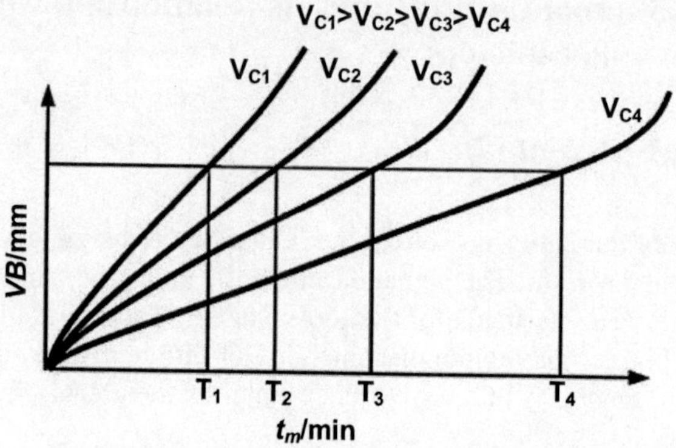

Fig. 3.6 Tool wear curves at different cutting speeds

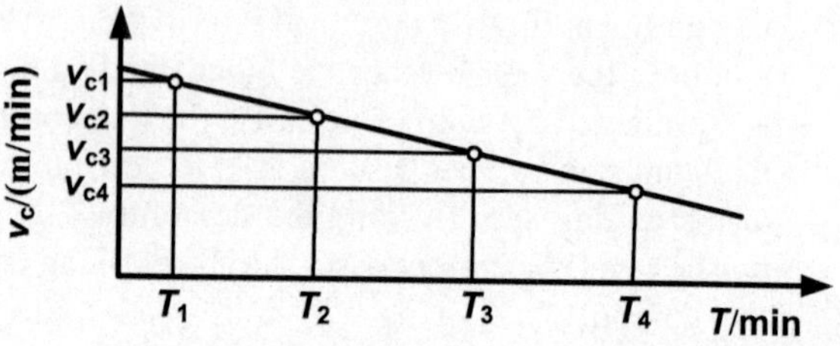

Fig. 3.7 Vc − T curve in the double logarithmic coordinate system

system, mark the points determined by the cutting speed and the corresponding tool durability, as shown in Fig. 3.7. Within a certain range, it can be found that these points are basically on a straight line. The linear equation is

$$\lg v_c = -m \lg T + \lg C_0 \tag{3.1}$$

where v_c is the cutting speed (m/min); T is the tool life (min); m is the slope of the straight line, indicating the influence of v_c on T, independent of the tool material; C_0 is the intercept of the line on the ordinate axis, which is related to the workpiece material and cutting conditions.

Both the exponent m and the coefficient C_0 in the Eq. 3.1 can be measured on the double logarithmic coordinate graph. Equation 3.2 can be expressed in exponential form as

$$v_c T^m = C_0 \tag{3.2}$$

Equation 3.2 is a tool durability formula, also known as Taylor's formula, that is used to determine cutting speed. Figure 3.8 depicts the durability curves of three different tool materials when processing the same workpiece material (nickel–chromium-molybdenum alloy steel). The lower the heat resistance, the lower the m value of the tool, and the smoother the straight line, the greater the effect of cutting

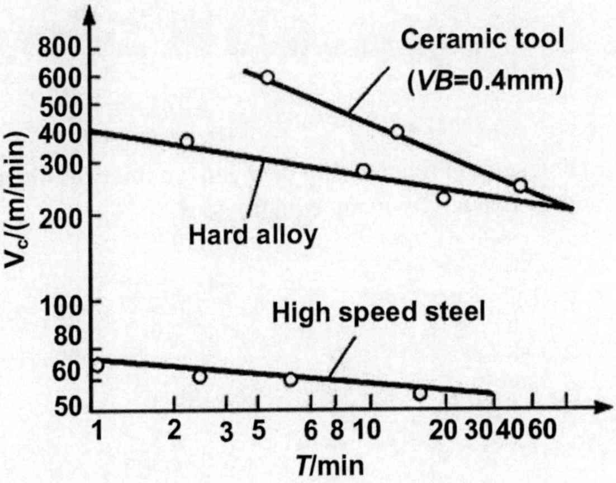

Fig. 3.8 Durability comparison of different tool materials in machining NiCrMo alloy steel

speed on tool durability. For example, high-speed steel tools have poor heat resistance, generally m = 0.1–0.125; Cemented carbide and ceramic tools have better heat resistance, and the slope of the straight line is relatively large, cemented carbide tools m = 0.2–0.3, ceramic tools $m \geq 4$.

It should be noted that the Eq. 3.2 is based on the normal wear of the tool, which is not applicable for brittle tool materials or tool breakage under interrupted cutting.

When the experiment is performed in a wide cutting speed range, due to the influence of built-up edge, the $v_c - T$ relationship is no longer a monotonic function but forms a hump-shaped curve as shown in Fig. 3.9, which corresponds to the rising part of the curve. The formula no longer applies.

(2) **The relationship between the federate and the tool flank wear**

Refer to the experimental procedure for obtaining v_c - T, fix other factors that affect cutting and only change f or a_p to obtain the relational expressions (3.3) and (3.4) of $f - T$ and $a_p - T$ respectively.

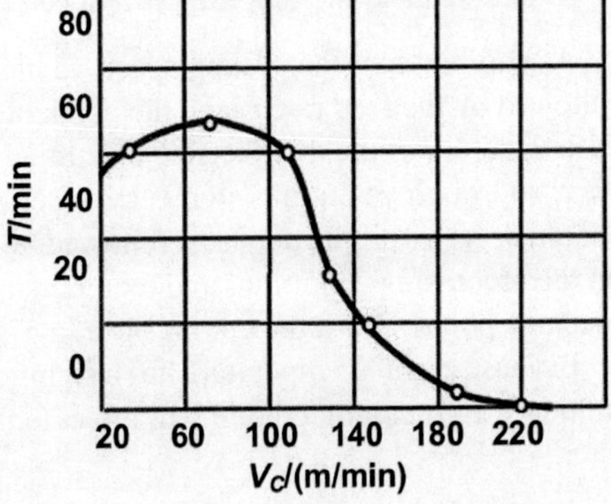

Fig. 3.9 $v_c - T$ hump curve

$$fT^{m_1} = C_1 \tag{3.3}$$

$$a_p T^{m_2} = C_2 \tag{3.4}$$

The general relationship between the three cutting parameters and tool durability can be obtained by using equations:

$$T = \frac{C_T}{v_c^{\frac{1}{m}} f^{\frac{1}{m_1}} a_p^{\frac{1}{m_2}}} \tag{3.5}$$

Let $x = 1/m$, $y = 1/m_1$, $z = 1/m_2$, then

$$T = \frac{C_T}{v_c^x f^y a_p^z} \tag{3.6}$$

where C_T is the tool life coefficient, which is related to the tool, workpiece material, and cutting conditions; x, y, and z are the exponential variables, which respectively represent the influence degree of v_c, f, and a_p on the tool life, generally $x > y > z$.

When using an YT15 carbide turning tool to turn carbon steel with $\sigma_b = 0.637$GPa ($f > 0.7$ mm/r), the following relationship exists between cutting parameters and tool life.

$$T = \frac{C_T}{v_c^5 f^{2.25} a_p^{0.75}} \tag{3.7}$$

As can be drawn from Eq. 3.7:

(1) When all other conditions remain constant, the cutting speed vc is doubled and the tool life t is reduced by 3%.
(2) When all other conditions are held constant, the feed rate f is doubled and the tool life t is reduced to 21% of its original value.
(3) When all other conditions remain constant, the depth of cut capacity ap is doubled and the tool life t is reduced to 59% of its original value.

As can be seen, the cutting speed *vc* has the greatest impact on tool durability, followed by the feed rate f, and the depth of cut a_p has the least, which is consistent with the order of the three factors on cutting temperature. As a result, to reduce tool wear and improve cutting efficiency, when selecting the cutting amount, the order should be large milling depth a_p, followed by the largest possible feed amount f based on the processing conditions and requirements. Finally, when the tool durability and machine power allow, the cutting speed *vc* is selected.

Because cutting temperature has a significant impact on tool wear, all factors that influence cutting temperature will affect tool durability.

3.3.3 Tool Breakage

The tool will not wear out naturally during the cutting process but will be damaged suddenly and fail in a short period. This is referred to as breakage. Breakage is another common type of tool damage, which occurs most often when using brittle tool materials for intermittent cutting or processing high-hardness materials. According to statistics, 50–60% of the damage to cemented carbide knives is broken, while ceramic knives have a higher proportion.

According to nature, tool breakage can be classified as plastic damage or brittle damage, and according to time, it can be classified as early damage or late damage. Early breakage occurs at the start of a cut or after a short period of cutting. It is primarily due to tool manufacturing flaws and stress caused by impact loads that exceed the tool material's strength. The number of impacts on the tool during cutting is generally less than or close to 10^3 times, and the rake and flank surfaces are not significantly worn ($VB = 0.1$ mm). The later breakage is that the tool material is damaged due to mechanical fatigue and thermal fatigue caused by mechanical impact and thermal shock after a certain period of processing.

3.4 Modeling of Tool Wear

Tool life prediction is critical to the actual machining and production process. Accurate tool wear prediction not only improves machining efficiency and benefits, but also helps to avoid losses caused by excessive tool wear. Tool wear during cutting is typically the result of multiple wear mechanisms. The foundation of tool wear

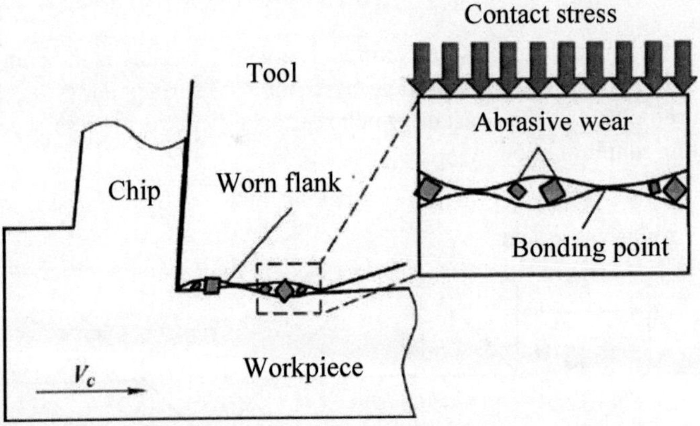

Fig. 3.10 Ketch diagram of flank wear

prediction is the development of tool wear rate models based on various wear mechanisms, as illustrated in Fig. 3.10. The following are some examples of common wear rate models:

3.4.1 Abrasive Wear Rate Model

The schematic diagram of the abrasive wear model is shown in Fig. 3.11, which expresses the contact between the tool rake face and the hard particles in the chips during the cutting process, that is, under the action of pressure, the hard particles in the chips will be embedded into the tool surface, causing the tool surface to wear. Because the chips will slide on the tool surface at a relative speed v within the sliding distance ΔL, the volume of the hard particles sliding on the tool surface is the amount of abrasive wear on the tool at the moment.

Within the sliding length ΔL, the wear volume dv of the tool is as follows:

$$dv = 2r^2 \Delta L \cot\theta \tag{3.8}$$

When the contact load between the tool and the chips is W and the tool surface hardness is H, it can be obtained:

$$W = H\Delta A = H\pi r^2 \tag{3.9}$$

Then Eq. 3.9 is brought into Eq. 3.8 to obtain:

$$\frac{dw}{dt} = dv/dt = \frac{2W\cot\theta}{\pi H} v_c \tag{3.10}$$

where W is the contact load between chip and tool surface, ΔA is the contact area between the tool and single hard particle, H is the tool surface hardness, v_c is the relative slip velocity, G is the constant which is equal to 5 in the process of cemented carbide cutting titanium alloy.

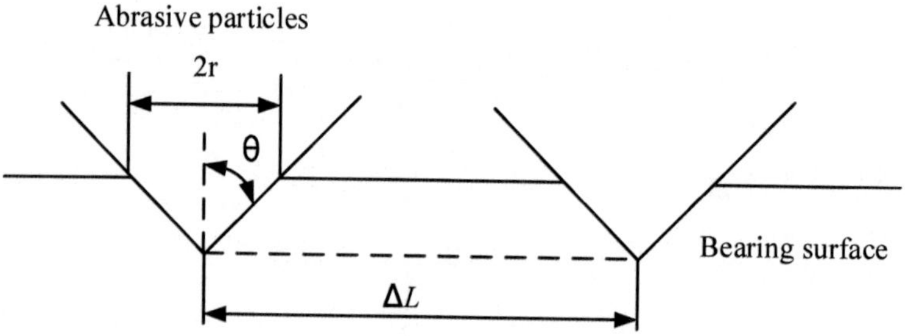

Fig. 3.11 Schematic diagram of abrasive wear principle

3.4.2 Adhesive Wear Rate Model

Adhesive wear is a very common type of wear. Adhesion and fracture of the joint between the sliding surfaces cause the typical adhesive wear mechanism. After metal contact, wear occurs at the local micro-welded joints, and the joints are torn as a result of the gradual relative movement between the contact interfaces. The adhesive wear rate model formula's specific solution is as follows: The tool adhesive wear volume dv within the slip length L is as follows:

$$dv = \frac{\sigma_n}{h}(c/b)ZDL \tag{3.11}$$

where σ_n is the normal stress of the interface between the tool and the chip, c and b are the height of the hard particles on the chip, and the average distance between two adjacent hard particles. Due to the size effect, c and b in the above equation can be approximately regarded as a constant, Z is the probability of bonding wear of hard particles on the contact interface (Holm probability), and$h$ is the hardness of rough particles.

Since the hardness h of the rough particles is related to the overall performance of the tool-chip contact surface pair, it mainly depends on the strain, strain rate, and temperature on the contact surface, and because the strain and strain rate is also affected by temperature, h can be expressed by the temperature on the tool-chip contact surface, as shown in Eq. 3.12; Since the occurrence of bonding wear is a process highly dependent on temperature activation, the Holm probability Z can be expressed by the following Eq. 3.13.

$$h = A_1 \exp(A_2/T) \tag{3.12}$$

$$Z = B_1 \exp(-E/\lambda T) \tag{3.13}$$

where T is the contact surface temperature, E is the activation energy, λ is the Boltzmann constant, A_1, A_2 and B_1 are the constants.

Substituting Eqs. 3.12 and 3.13 into Eq. 3.11, it results in:

$$\frac{dv}{\sigma_n dL} = C_1 \exp[-(\Delta E + \lambda A_2)/\lambda T)] \tag{3.14}$$

In the above formula, C_1 is a constant, $2(\Delta E + \lambda A_2)$ mainly depends on the diffusion layer structure and element concentration on the contact surface, but when the types of cutting conditions are limited, it can be regarded as a constant that is denoted as C_2; Put into the Eq. 3.14, the bond wear rate formula of the tool can be obtained, as shown below.

$$\frac{dw}{dt} = dv/dt = C_1 \sigma_n v_c \exp(-C_2/T) \tag{3.15}$$

where v_c is the relative slip velocity, C_1 and C_2 are the wear characteristic constants.C_1 and C_2 can be measured by wear test. In cemented carbide tools, $C_1 = 4 \times 10^{-4}$ and $C_2 = 7000$ respectively. Because the model is simple in form, and the parameters in the model are easily obtained through experiments. Therefore, this model is widely used in tool wear simulation to predict tool wear.

3.4.3 Diffusion Wear Rate Model

The contact area between the rake face, the chip, the flank face, and the machined workpiece is the concentrated area of the tool diffusion and wear phenomenon, as shown in Fig. 3.12.

In Fig. 3.12, while considering the diffusion of the tool C_0 element to the workpiece, the loss of C_0 element due to the continuous movement of the chip and the workpiece relative to the tool is also considered. According to the analysis of Fick's diffusion law and its solution, the model of the diffusion wear rate of the rake face of the tool is as follows:

$$\frac{dw_s}{dt} = \frac{J_{y=0}}{\rho} = \frac{2C_0}{\rho}\left[\frac{v_c D_0}{\pi x}\right]^{1/2} \exp(-Q/2R(273+T)) \tag{3.16}$$

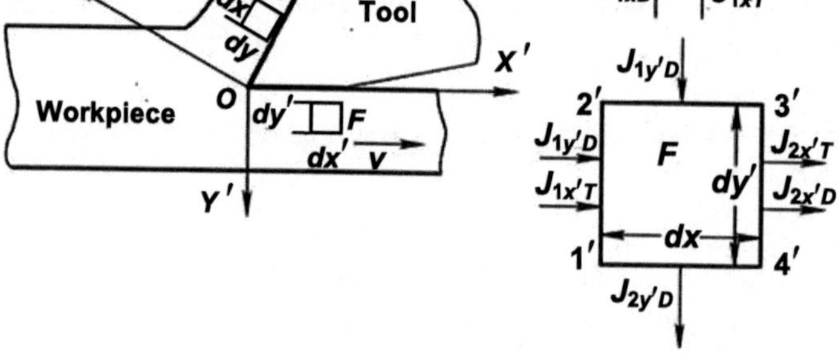

Fig. 3.12 Schematic diagram of diffusion and movement unit in the cutting process

Similarly, the model of tool flank diffusion wear rate is obtained as follows:

$$\frac{dw_s^{'}}{dt} = \frac{J_{y'=0}}{\rho} = \frac{2C_0}{\rho}\left[\frac{vD_0}{\pi x^{'}}\right]^{1/2} \exp(-Q/2R(273+T)) \tag{3.17}$$

where C_0 is the concentration of diffusion material, ρ is the tool density, D_0 is the equation coefficient, Q is the activation energy, R is the gas constant, t is in centigrade, VC is the chip flow velocity, v is the relative sliding velocity between the flank and the machined surface, x is the distance between any point in the tool chip contact area and the cutting edge, and $x^{'}$ is the distance between any point in the tool-work contact area and the cutting edge.

3.4.4 Comprehensive Wear Rate Model

The tool wear process is complex and the wear mechanism interacts during the cutting process due to the coupling effect of cutting force and cutting heat. When the temperature is low, abrasive wear and adhesive wear are the main wear mechanisms of the tool rake and flank; when the temperature rises to a certain point, diffusion wear and adhesive wear occur simultaneously, making it easy to form the crescent groove on the front cutter surface and wear belt on the back face. As a result, it can provide scientific and reasonable technical assistance in monitoring tool wear state, predicting tool life, optimizing tool structure parameters, and process parameters.

According to the findings of numerous scholars, the abrasive wear and bond wear mechanisms will play a significant role in the wear of cemented carbide tools. Hard Element diffusion occurs on the bonding interface between the high-quality alloy tool and the titanium alloy Ti6Al4V when the cutting temperature exceeds 600 °C. At the moment, diffusion wear and bonding wear are the most important. Because abrasive wear has little effect on tool wear, it will not be taken into account in the wear model. Based on the wear mechanism and taking into account the effect of thermal coupling on tool wear, a tool wear model with a cutting temperature of 600 °C as the temperature threshold is constructed, as shown in Eq. 3.18.

$$\begin{cases} \frac{dw}{dt} = \frac{dw_m}{dt} + \frac{dw_n}{dt}\ T < 600\,^\circ\mathrm{C} \\ \frac{dw}{dt} = \frac{dw_n}{dt} + \frac{dw_s}{dt}\ T \geq 600\,^\circ\mathrm{C} \end{cases} \tag{3.18}$$

where dw/dt is the tool volume wear rate, dw_m/dt is the abrasive wear rate, dw_n/dt is the bonding wear rate, dw_s/dt is the diffusion wear rate, T is the tool cutting temperature.

Jawahir [1] summarized various tool life models and concluded that the tool life prediction model was divided into two parts: empirical prediction model and theoretical calculation model. Tables 3.2 show the main models and application

Table 3.2 Tool wear and tool life model

Author(s)	Tool wear model	Scope of application
Molinari [2]	$T = \left[\frac{K_{\gamma(\lim)}\rho^0_{tool}\sqrt{\pi}}{2\sum_{i=1}^{n}(C^0_{i1}-C^0_{i2})\sqrt{D_{i1}(\Theta_{\max})}}\right]^2$ where C^0_{i1}, C^0_{i1} are the concentration D_{i1} is the diffusion coefficient ρ^0_{tool} is the initial density of the tool material	The model describes the diffusion model of rake face based on thermal–mechanical coupling, but it can not directly calculate the wear rate of each point
Usui [3]	$\frac{dw}{dt} = C_1\sigma_n V_C \exp(\frac{-C_2}{T})$ where C_1 and C_2 are the material constant, dw/dt is the volume wear rate	The model only considers adhesive wear
Takeyama [4]	$\frac{dw}{dt} = G(v, f) + D\exp(\frac{-C_2}{T})$ where D and G are the material constant	The model includes abrasive wear and bond wear
Childs [5]	$\frac{dw}{dt} = \frac{A}{H}\sigma_n V_C$ where A is the material constant, H is the tool material hardness	The model only considers the abrasive wear mechanism
Sun [6]	$\frac{dw}{dt} = \frac{2C_0}{\rho}(\frac{V_C D_0}{\pi x})^{1/2}e^{-Q/2TR_Q}$ where C_0 is the Concentration of diffusing substance, ρ is the tool density, D_0 is the equation coefficient or partial factor, V is the relative slip speed, Q is the activation energy, R is the gas constant	The model establishes a temperature-based diffusion wear model
Arsecularatne [7]	Life model based on flank temperature $T = AT_f^B$ where T_f is the flank temperature, A, B is the constant	The model is mainly a tool life model based on thermal wear. Only consider the temperature benefit

scope.

3.4.5 Intelligent Tool Wear Model

The amount of tool wear can be expressed by the dynamic signal during the milling process in the metal milling process, and the wear state of the tool can be monitored by collecting, processing, and identifying the dynamic signal during the milling process. Tool wear monitoring is primarily accomplished through the pattern classification method, with the feature quantities extracted from various processing conditions and processing conditions serving as one aspect and the tool wear status serving as the other, and these two aspects being combined using mathematical modeling and other methods. Display their nonlinear mapping relationship. In recent years, this kind of mathematical modeling usually uses artificial intelligence. The widely studied and applied intelligent tool wear modeling methods include: multiple linear regression models [8], artificial neural networks (ANN) [9], support vector machines (SVM) [10], Hidden Markov Model (HMM) [11], deep learning, and many other methods [12–20]. These topics will be discussed in details in Chap. 5.

3.5 Tool Wear Modeling in High-Speed Milling

Tool wear is an important factor in high-speed milling because it directly affects machining precision and part quality. It is critical to look for a simple way to model and predict tool states. Over the last decade, there has been a great deal of research into wear mechanisms and evolution laws. The main tool wear phenomena are abrasion, adhesion, and attrition [21]. The temperature has been found to have a large influence on tool wear, and the main tool wear mechanism for tungsten-carbide tools is diffusion [22]. Aside from wear mechanics, modeling tool wear is critical for improving process monitoring, control, and system automation. Over the course of machining, typical tool wear evolution consists of three stages: the initial wear stage, the steady wear stage, and the accelerated wear stage [23]. Researchers have developed a variety of approaches for modeling tool wear, including empirical and physical wear modeling approaches [24]. Tool wear conditions can be induced using these methods based on practical experience and technological measurements.

3.5.1 Tool Flank Wear Conditions

The tool flank wear arises due to both adhesive and abrasive wear mechanisms from the intense rubbing action of the clearance face of the cutting tool and the newly formed surface of the workpiece. Its rate of increase at the beginning of the tool life is rapid, settling down to a steady-state then accelerating rapidly again at the end of tool life [25]. A typical tool flank wear curve VB (μm) with different wear conditions is shown in Fig. 3.13. The wear conditions over different machining times are obtained from milling experiments. It shows that the tool is worn soon and the wear is noticeable in the initial wear stage after the machining starts.

To analyze the various tendencies of the wear curve in Fig. 3.13, the wear rate VB' (μm/min) and acceleration rate VB'' (μm/min^2) [26] are calculated as:

$$\begin{cases} VB'(t) = \Delta VB \big/ \Delta t = [VB(t+\Delta t) - VB(t)] \big/ \Delta t \\ VB''(t) = \Delta VB' \big/ \Delta t = \left[VB'(t+\Delta t) - VB'(t)\right] \big/ \Delta t \end{cases} \tag{3.19}$$

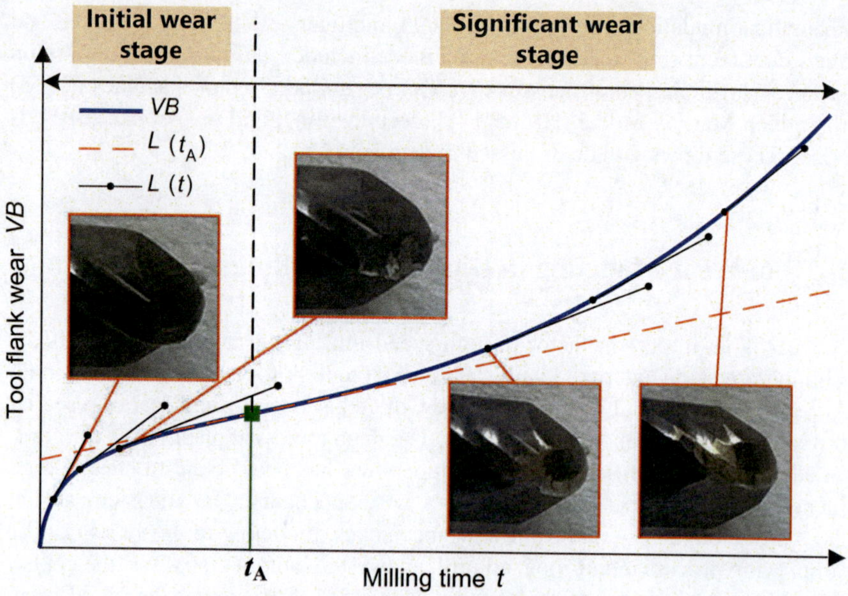

Fig. 3.13 Tool flank wear curve with different wear conditions

where Δt (min) is an interval sampling milling time of the wear curve. Particularly, it is the time when one pass of the workpiece is milled, while the machining length of one pass is bs (mm) in milling experiments.

3.5.2 *Modeling of Tool Flank Wear*

According to the variation of the acceleration rate VB'', the wear curve can be primarily divided into two periods by the critical time t_A, which are regarded as the initial and significant wear stages respectively, as shown in Fig. 3.13. It can be found that there are completely different wear acceleration rates in the two milling stages, before and after the critical time t_A. An overall tool flank wear model $w(t)$ is proposed to fit the wear curve, of which two main requirements need to be satisfied.

Based on these two requirements, two transition functions w_E and w_L are proposed, which are represent principal polynomial fitting curves of the tool wear in the initial wear zone and significant wear stage respectively. To satisfy **Requirement 1** (Continuous differentiability), the wear function $w(t)$ can't be composed of two segmentation functions directly. Hence, the stable point at milling time t_A needs to pay attention. Without loss of generality, the wear model $w(t)$ can be composed by the transition functions, which is given as:

$$w(t) = w_E(t) + w_L(t) \tag{3.20}$$

where the transition functions $w_E(t)$ and $w_L(t)$ represent principal polynomial fitting curves of the tool wear, in the initial wear zone and significant wear stage respectively which is illustrated in Fig. 3.13

For refinement of the transition functions w_E and w_L, combining the experimental wear data, we attempted the polynomial, exponential, logarithmic, multiplicative forms, and mix combinations to fuse the wear curve in the initial and significant wear stages, separately. To satisfy **requirement 2** (Good agreements), there is little effect of the transition functions $w_E(t)$ or $w_L(t)$ in each stage. It means that the overall wear model error with practical tool wear data should be met with the precision request. As shown in Ref. [27], the current functions have good determination coefficients and simpler forms, which are as follows:

$$\begin{cases} w'_E(t) = a_1/(t + b_1) + c_1 \\ w''_L(t) = a_2 t + b_2 \end{cases} \tag{3.21}$$

where a_1 (μm), a_2 (μm/min^3), b_1 (min), b_2 (μm/min^2) and c_1 (μm/min) are the basic transition coefficients, which are determined with experimental fitting.

Rechecking the requirements of the wear model $w(t)$, **Requirement 1** is completed. The remaining requirement is needed to be substantiated with the experimental wear data. The differential and integral of w'_E and w''_L over the milling time t are deducted individually as:

$$\begin{cases} w''_E = (w'_E)' = -a_1/(t + b_1)^2 \\ w'_L = \int w''_L dt = a_2 t^2 + b_2 t \end{cases} \tag{3.22}$$

Given the results of the two different stages, removing low order variables that have little effect on the tool wear in the significant wear stage, the basic prediction model of the tool flank wear is expressed as:

$$w(t) = w_E(t) + w_L(t) = A\ln(Bt + 1) + Ct^3 \tag{3.23}$$

where A (μm), B (min^{-1}), and C (μm/min^3) are the fitting coefficients, which are all positive. There is $w(0) = 0$ apparently.

It is well known that tool wear curves are multiform with different machining conditions, tool and workpiece materials, making studying tool wear conditions difficult. According to the tool wear types in machining processes, the tool flank wear curve with machining time can be divided into three regions by critical times, namely, the initial stage (I), the steady-state region (II), and the accelerated wear zone (III), as shown in Fig. 3.14. The three zones' main forms are sequentially running-in wear, adhesive wear, and three-body abrasive wear, which are consistent with different wear rates. In this paper, a general tool wear model is tried to established

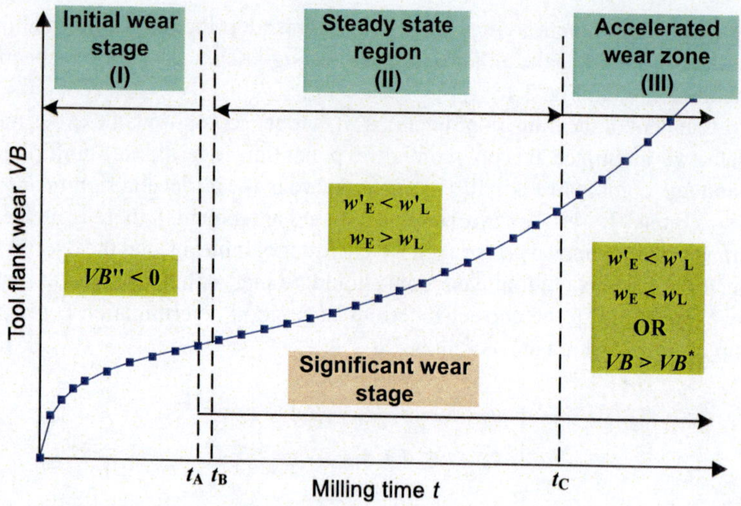

Fig. 3.14 Tool flank wear curve in different stages

as much as possible, which could include all wear stages in an entire machining process.

The wear transition functions $w_E(t)$ and $w_L(t)$ show different wear gain rates by various wear forms. The tool wear status is in the initial rapid wear stage when VB'' < 0. After this stage, when the cubic polynomial function w_L has more influence on tool flank wear than w_E, or when the tool flank wear is more than the critical flank wear limit VB^* (μm), the wear status is into the accelerated wear zone. The effective tool life is over because of the machining precision requirement. It is necessary to determine the critical times t_A, t_B, and t_C, which match with the division conditions of different wear zones.

Furthermore, tool fracture is a significant issue in machining, which is primarily caused by insufficient machining conditions. It directly causes tool failure, and most of the time the milling tool must be replaced with a new one. As a result, the discussion of the damaged tool wear that follows is pointless. The fluctuation errors of tool wear are an unavoidable phenomenon in the milling machining process, which is primarily caused by system chatter, run-out, unevenness platform, inhomogeneous workpiece, and so on. These factors cannot be described specifically and simply, and it is inappropriate to consider these factors primarily in the basic general model.

3.5.3 Generalization of the Tool Wear Model

It is known that the wear rate difference of the models is increasing over the milling time. In the specific wear model $w(t)$ in Eq. 3.23, the wear rate is mainly related to the coefficient C and the exponent of the milling time t. In the specific machining

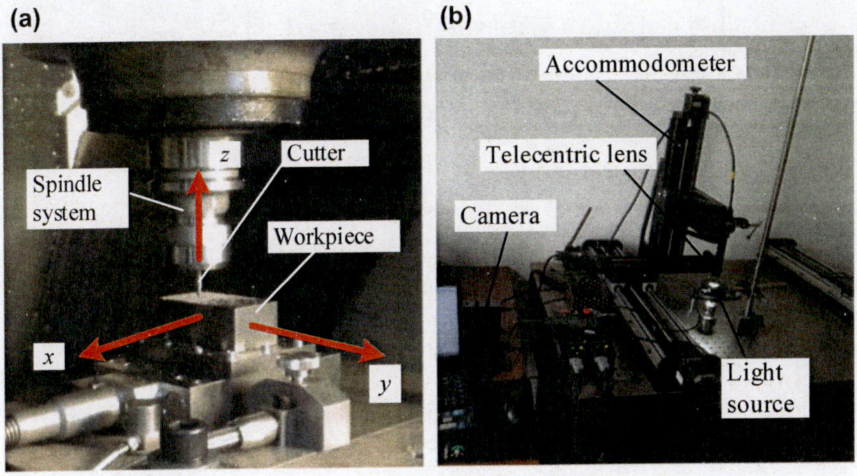

Fig. 3.15 Experimental setup: **a** milling machining platform; **b** tool wear observation and measurement platform

processing, the tool wear is also affected by other factors, such as the system chatter, run-out, unevenness platform, inhomogeneous material, etc. To improve the flexibility of the model and the wear numerical result accuracy with various milling conditions, especially in the significant wear stage II and III, a variable exponent x is introduced into the wear model in Eq. 3.23. Hence, the tool wear model is generalized as:

$$w(t) = A\ln(Bt + 1) + Ct^x \quad (x > 1) \tag{3.24}$$

where the exponent $x > 1$ for that the wear acceleration rate in the significant wear stage is positive, as explicated in Fig. 3.15. To distinguish these two functions, the symbol of tool wear in Eq. 3.23 is modified to w_P.

To investigate a general approach for determining the exponent x of the generic wear model in Eq. 3.24, multiple sets of the tool flank wear data over the machining time need to be obtained, which are measured in varied milling conditions, as shown in Sect. 3.5.4. Based on the experimental data, optimization of the exponent x and analysis of the generic wear model are achieved in the following sections. The optimum solution of the exponent x can be expressed as:

$$x = \arg\max_{x \in \mathbb{R}^+} \left(R^2\right)$$

$$
= \underset{x \in \mathbb{R}+}{\arg\max} \left\{ \frac{\left\{ \sum_{n=1}^{N} [w(n\Delta t) - \overline{w}]\left[VB(n\Delta t) - \overline{VB}\right] \right\}^2}{\sum_{n=1}^{N} [w(n\Delta t) - \overline{w}]^2 \sum_{n=1}^{N} \left[VB(n\Delta t) - \overline{VB}\right]^2} \right\} \quad (n = 1, 2, \cdots, N) \tag{3.25}
$$

where the function $w(\cdot)$ is consistent with the wear model in Eq. 3.24 with the exponent x. The mean wear $\overline{w}$ and $\overline{VB}$ are calculated as $\sum_{n=1}^{N} \frac{w(n\Delta t)}{N}$ and $\sum_{n=1}^{N} \frac{VB(n\Delta t)}{N}$ respectively. The constant N is the number of sampling points in the corresponding tool wear test.

The wear rate w' (w'_E, w'_L) and acceleration rate w'' (w''_E, w''_L), the first and second-order differentiation of the generic wear model in Eq. 3.24 can be calculated as:

$$
\begin{cases} w'(t) = w'_E(t) + w'_L(t) = AB(Bt+1)^{-1} + xCt^{x-1} \\ w''(t) = w''_E(t) + w''_L(t) = -AB^2(Bt+1)^{-2} + x(x-1)Ct^{x-2} \end{cases} \tag{3.26}
$$

Equation 3.26 shows that $w' > 0$ when $\forall t \geq 0$. Based on specific characteristics of wear curves and the division conditions of wear zones shown in Fig. 3.15. The milling time t is denoted as t_A when the change rate of the wear curve reaches the first minimum value. The time t is denoted as t_B when effects on the wear rate of the non- and multi-order functions in the w' expression are equal.

It is known that the flank wear exponentially increases after the tool wear VB reaches a critical limit VB^*. It is recommended that $VB^* = 300$ μm for the average flank wear of all tool teeth [28], as shown with a typical tool wear curve in Fig. 3.14, while the tool must be replaced to avoid catastrophic tool failure. When tool flank wear is less than the critical wear VB^*, the time t is denoted as t_C which effects on the wear rate of logarithmic and multi-order functions w_E and w_L in Eq. 3.24 are equal; otherwise, the critical time t_C is identified when $w(t) = VB^*$, i.e., the tool effective useful life T, which is expressed as:

$$
T = \left\{ T \middle| T^x + \frac{A}{C} \ln(BT + 1) = \frac{VB^*}{C} \right\} \tag{3.27}
$$

where the parameters A, B, and C are obtained from the proposed wear model, as shown in Eq. 3.24. The model is fitted with the specific wear data in the early machining process, and it predicts the wear conditions in the later, where the sample data and prediction result are applied in the same processing environment.

$$\begin{cases} w''(t_A) = 0 \\ w'_E(t_B) - w'_L(t_B) = 0 \\ t_C = \min\left\{T, \arg\min\limits_{t\in[t_B,+\infty)} |w_E(t) - w_L(t)|\right\} \end{cases} \tag{3.28}$$

With the generic wear model in Eq. 3.24, the cutter wear in milling machining can be predicted, in which the wear coefficients *A*, *B*, *C*, and *x* need to be determined.

3.5.4 Analysis of Tool Wear Model

The machining conditions and parameters, the tool milling depth a_p and the spindle speed n_t, etc., are listed in Table 3.3. The 23 experimental tests were conducted with

Table 3.3 Conditions and parameters in milling experiments

Test No	a_p (mm)	f (mm/rev)	n_t (rpm)	Types of milling
1	0.2	0.090	8000	Slotting
2	0.2	0.030	10,000	Slotting
3	0.2	0.090	10,000	Slotting
4	0.15	0.150	10,000	Slotting
5	0.2	0.150	10,000	Slotting
6	0.25	0.150	10,000	Slotting
7	0.2	0.090	20,000	Slotting
8	0.2	0.090	30,000	Slotting
9	0.15	0.150	10,000	Half immersion
10	0.2	0.150	10,000	Half immersion
11	0.25	0.150	10,000	Half immersion
12	0.15	0.030	10,000	Slotting
13	0.25	0.090	8000	Slotting
14	0.25	0.030	10,000	Slotting
15	0.25	0.090	10,000	Slotting
16	0.2	0.150	10,000	Slotting
17	0.25	0.150	10,000	Slotting
18	0.3	0.150	10,000	Slotting
19	0.25	0.090	20,000	Slotting
20	0.25	0.090	30,000	Slotting
21	0.2	0.150	10,000	Half immersion
22	0.25	0.150	10,000	Half immersion
23	0.3	0.150	10,000	Half immersion

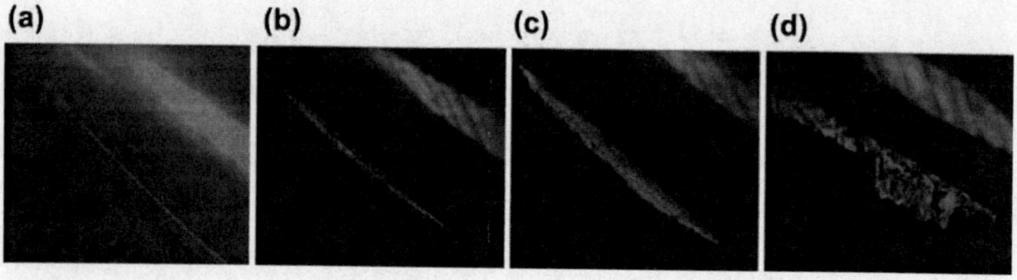

Fig. 3.16 Microscopic images of tool flank edge of test #22 **a** before cutting and after **b** 64, **c** 192, and **d** 315 passes

different spindle speeds, depth of cut, and feed rates for different diameters tools, in which the generalization capability is reflected with varied machining conditions. The material of ball-nose end milling tools is tungsten carbide. The parameters in Table 3.3 are selected according to the recommended data in the manual: UnionTool Tungsten Carbide End Mills UNIMAX Series Vol. 17. The design of the experimental approach in machining experimental tests is followed Taguchi methods. The work platform is arranged on the machining center MIKRON HSM600U, as shown in Fig. 3.15a. Tool flank wear per tooth is measured by the LECIA MZ12.5 stereo-microscope after finished each machining pass, as shown in Fig. 3.15b. The real-time value of tool wear is obtained by the changed ratio of tool geometric dimension in photographs. The tool wear measurement platform is composed of a three-component dynamometer, industrial camera, light source, and a telocentric lens.

For example, the different tool flank wear conditions in experimental processes of test #22 are shown in Fig. 3.16, in which the cutter diameter is 6 mm. Tool flank wear on each flute is different from each other, which is caused by the inevitable tool run-out effect and milling system dynamics in machining. The wear on each flute is fairly consistent in sets #1 and #2. The wear curve variations in Fig. 3.17c indicate that the flank wear on the first flute of set #3 is relatively smaller than others. At the same milling conditions and time, the acceleration zone of set #3 is more fluctuated than set #1 and #2, and a higher wear value in set #3 is achieved than set #1 and #2, which decreases the tool life.

The fitting results of the generic wear model $w(t)$ in Eq. 3.24 with different exponents x are shown in Fig. 3.18. The parameters in the proposed tool wear model are obtained by fitting the theoretical model and the sampling data of tool wear. The determination coefficient R^2 of the fitted theoretical data and experimental raw data of each set with different exponents x is shown in Fig. 3.18d, in which the integers between 2 and 6 are chosen as the optional exponents x. The optimal number x is determined when the determination coefficient R^2 of the wear model and the experimental data achieves the highest value. It can be found that the wear $w(t)$ with different exponents x is closely in the initial stage (I) and the steady-state region (II). The optimum solution of the exponent x is close to the integer 3 or 4, except the fitting result with the experimental data of set 2, of which the determination coefficient R^2 are all more than 98%. It is proved that the overall determination coefficient of the

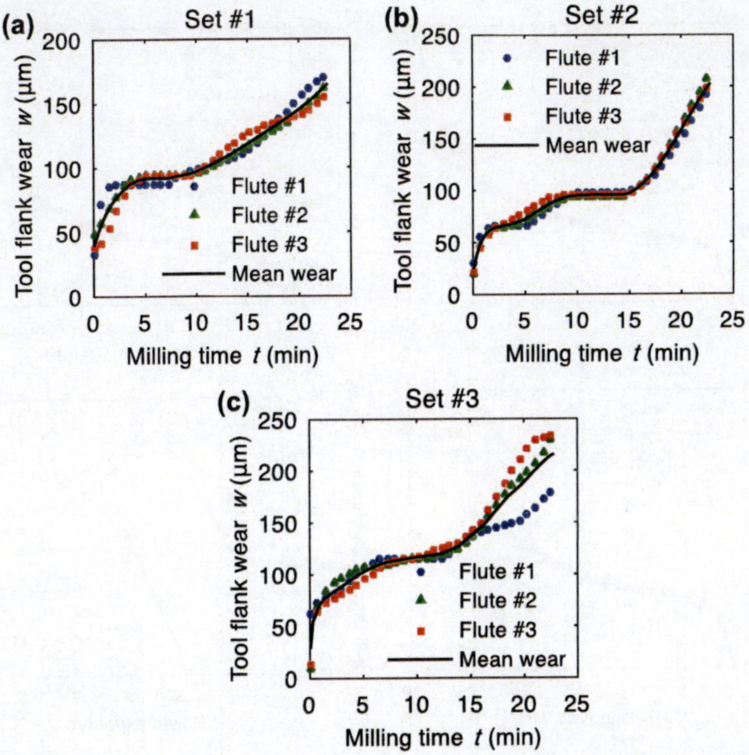

Fig. 3.17 Tool flank wear on each flute of test #22 and mean wear over milling time of the set **a** #1, **b** #2, and **c** #3

generic wear model with a variable exponent x is raised by 2–3.5%. If a higher determination coefficient (>98%) of the model was requested, the generic wear model would be supposed to be adopted.

There is only one set of a positive real solution in Eq. 3.24 when $t \geq 0$. The analytical solutions of critical times t_A, t_B, and t_C with the wear data in experiments are listed in Table 3.4 were based on the milling conditions of test #22.

The wear rate w' and acceleration rate w" over the milling time with different exponents x are shown in Fig. 3.19. The other coefficients A, B, and C of the wear model in Fig. 3.19 is obtained from the set #1 experiment, as listed in Table 3.4. It can be found that the change of the exponent x has no effect on wear rate in the initial wear stage (I, the milling time $t < t_A$). Based on Eq. 3.28, the critical times t_A, t_B, t_C, and the tool life T with the generic wear model in Eq. 3.24 are shown in Fig. 3.20.

According to these analytical results, the characteristics of tool wear conditions are discussed with different exponent x of the generic wear model in Eq. (3.24).

a. When $1 < x < 2$, the critical times are sharply positive (all times are more than 50 min), and the wear rate is decreasing mostly in an entire machining process ($w'' < 0$). This range is not suitable for the wear model generally

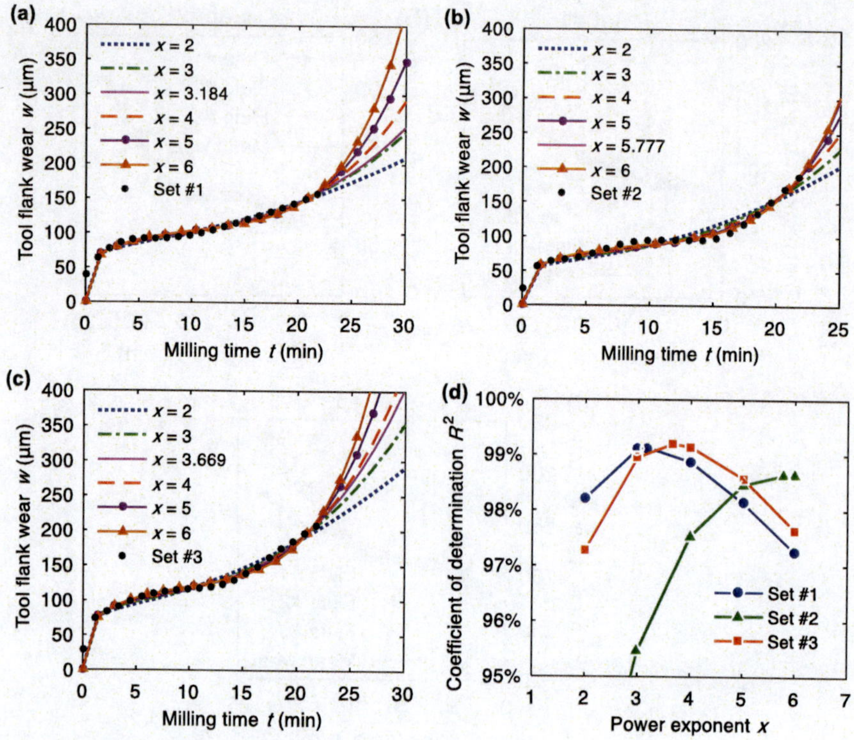

Fig. 3.18 Tool wear model of different power exponents x with the experimental data of the set **a** #1, **b** #2, and **c** #3. **d** The determination coefficient R^2 over different exponents x

Table 3.4 Coefficients, critical times, and the tool life with the generic wear model

Set No	A (μm)	B (min^{-1})	C ($\mu\text{m}/\text{min}^x$)	x	t_A (min)	t_B (min)	t_C (min)	T (min)	R^2 of w(%)
1	13.50	125.5	2.834×10^{-3}	3.184	7.768	9.932	27.63	32.67	99.13
2	15.75	26.50	1.624×10^{-6}	5.777	9.107	11.95	22.32	25.10	98.64
3	17.20	65.10	1.044×10^{-3}	3.669	7.564	9.892	24.31	26.41	99.20

b. When $2 \leq x < 3$, the critical times decrease gently, and the wear rate is increasing (in the II, III wear stages) slowly after sharp decreasing. Therefore, this range applies to the machining conditions with the low wear rate and long tool life relatively.

c. When $x \geq 3$, the initial wear stage is short, and the tool life is decreasing quickly. There is a significant acceleration rate in the II, III wear stages. So, this range applies to machining conditions with a high wear rate and short tool life. It is a

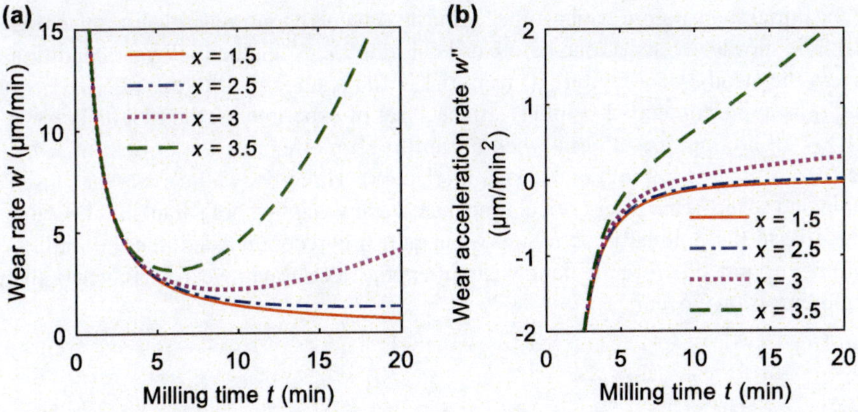

Fig. 3.19 **a** The wear rate w' and **b** acceleration rate w'' over the milling time with different exponents x

Fig. 3.20 Critical times t_A, t_B, t_C, and the tool life T over different exponents x

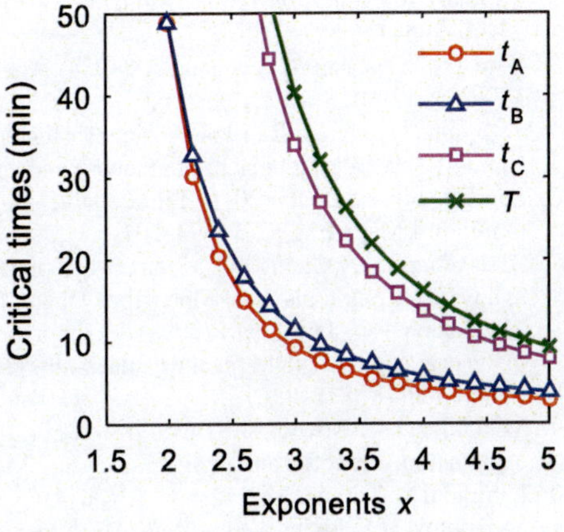

critical condition to a regular wear characteristic curve when the exponent x = 3.

As a result, the specific value of exponent x could be used as an index for tool wear conditions and tool life, allowing appropriate machining conditions to be chosen. According to the proposed model, the exponent x is obtained given certain processing conditions. The wear rate and acceleration rate can be obtained based on the size of x, as shown in Fig. 3.19. To keep the wear rate and acceleration rate as low as possible, the processing conditions must be appropriately adjusted to decrease the value of the exponent x. As a result, the tool life is extended, as shown in the schematic figure of the relationship between exponent x and tool life in Fig. 3.20.

In particular, the critical times of each set of data are close to and consistent with the physical machining processes, as shown in Table 3.4. After the initial rapid wear in the initial stage before critical time t_A, the increment of tool wear remains relatively stable in the steady-state zone before critical time t_C. When the tool wear condition enters the steady-state region (II) from the initial stage, the difference between t_A and t_B is approximately 2–3 min (I). In each set of experiments, the tool flank wear w(t) is all-around 200–250 m when the milling time reaches the critical time t_C, demonstrating the stability of the wear model w(t). Three-body abrasive wear causes rapid and severe wear after critical time t_C. It means that the tool's flank has become sensitive to the gradually increasing wear rate. It is recommended that the milling tool be regrinded before the flank wear enters the accelerated wear zone, where the rapid breakdown occurs.

References

1. Jawahir IS, Ghosh R, Fang XD et al (1995) An investigation of the effects of chip flow on tool-wear in machining with complex grooved tools. Wear 184(2):145–154
2. Molinari A, Nouari M (2002) Modeling of tool wear by diffusion in metal cutting. Wear 252(1):135–149
3. Usui E, Shirakashi T, Kitagawa T (1984) Analytical prediction of cutting tool wear. Wear 100(1–3):129–151
4. Takeyama H, Murata R (1963) Basic investigation of tool wear. J Eng Ind 85(1):33
5. Childs T (2000) Metal machining: theory and applications. Arnold
6. Sun YJ, Sun J, Li JF (2016) Finite elementanalysis on prediction of tool wear in milling titanium. J Mech Eng 52(5):193–201
7. Arsecularatne JA, Zhang LC, Montross C (2006) Wear and tool life of tungsten carbide, PCBN and PCD cutting tools. Int J Mach Tool Manuf 46(5):482–491
8. Marksberry PW, Jawahir IS (2008) A comprehensive tool-wear/tool-life performance model in the evaluation of NDM (near dry machining) for sustainable manufacturing. Int J Mach Tool Manuf 48(7–8):878–886
9. Mikolajczyk T, Nowicki K, Bustillo A et al (2018) Predicting tool life in turning operations using neural networks and image processing. Mech Syst Signal Process 104:503–513
10. Lamraoui M, Thomas M, Badaoui ME (2014) Cyclostationarity approach for monitoring chatter and tool wear in high speed milling. Mech Syst Signal Process 44(1–2):177–198
11. Wang G, Feng X (2013) Tool wear state recognition based on linear chain conditional random field model. Eng Appl Artif Intel 26(4):1421–1427
12. Twardowski P, Wiciak-Pikuła M (2019) Prediction of tool wear using artificial neural networks during turning of hardened steel. Materials 12(19):3091
13. Sick B (2002) On-line and indirect tool wear monitoring in turning with artificial neural networks: a review of more than a decade of research. Mech Syst Signal Process 16(4):487–546
14. Niaki FA, Feng L, Ulutan D et al (2016) A wavelet-based data-driven modelling for tool wear assessment of difficult to machine materials. Int J Mech Manuf Syst 9(2):97–121
15. Lu MC, Wan BS (2013) Study of high-frequency sound signals for tool wear monitoring in micromilling. Int J Adv Manuf Technol 66(9–12):1785–1792
16. Wang J, Yinghao et al (2017) A virtual sensing based augmented particle filter for tool condition prognosis. J Manuf Process 28:472–478
17. Balazinski M, Czogala E, Jemielniak K et al (2002) Tool condition monitoring using artificial intelligence methods. Eng Appl Artif Intel 15(1):73–80
18. Boutros T, Liang M (2011) Detection and diagnosis of bearing and cutting tool faults using hidden Markov models. Mech Syst Signal Process 25(6):2102–2124
19. Lecun Y, Bengio Y, Hinton G (2015) Deep learning. Nature 521(7553):436

20. Hinton GE, Salakhutdinov RR (2006) Reducing the dimensionality of data with neural networks. Sicence 313(5789):504
21. Brinksmeier E, Preuss W, Riemer O, Rentsch R (2017) Cutting forces, tool wear and surface finish in high speed diamond machining. Precis Eng 49(7):293–304
22. Arsecularatne JA, Zhang LC, Montross C (2006) Wear and tool life of tungsten carbide, PCBN and PCD cutting tools. Int J Mach Tools Manuf 46(5):482–491
23. Altintas Y (2012) Manufacturing automation: metal cutting mechanics, machine tool vibrations, and CNC design, 2nd edn. Cambridge University Press, New York, pp 57–60
24. Grzesik W (2015) Advanced machining processes of metallic materials: theory, modelling and applications, 1st edn. Elsevier, Amsterdam, p 213
25. Snr DED (2000) Sensor signals for tool-wear monitoring in metal cutting operations—a review of methods. Int J Mach Tool Manuf 40(8):1073–1098
26. Zhu KP, Wong YS, Hong GS (2009) Multi-category micro-milling tool wear monitoring with continuous hidden Markov models. Mech Syst Signal Process 23(2):547–560
27. Zhu KP, Zhang Y (2019) A generic tool wear model and its application to force modeling and wear monitoring in high speed milling. Mech Syst Signal Process 115:147–161
28. Shao H, Wang HL, Zhao XM (2004) A cutting power model for tool wear monitoring in milling. Int J Mach Tool Manuf 44(14):1503–1509

Chapter 4
Mathematical Foundations of Machining System Monitoring

4.1 Machining System Monitoring

4.1.1 The Content of Machining System Monitoring

To ensure the safety and processing quality of high investment automation processing equipment, machining process monitoring is becoming an urgent problem to be solved in the modern machining system. It has been reported that with the process monitoring system, the machine system can avoid 70% of downtime due to human and machining system failures, which guarantees the system safety, and improve the productivity [1–3].

The main contents of machining system monitoring are:

(1) Machine running status monitoring, including the spindle component operation monitoring, servo drive system monitoring and machine operation safety monitoring, etc.
(2) Condition monitoring of machining process, including vibration monitoring in the cutting process, cutting force monitoring, cutting status monitoring, machining process adaptive monitoring, cutting temperature monitoring, coolant system monitoring, coolant system monitoring and process identification, etc.
(3) Tool condition monitoring, including tool identification, tool adjustment, tool wear, and breakage monitoring, tool compensation, and tool life management, etc.
(4) Quality monitoring of workpiece in the machining process, including automatic identification of the workpiece, dimensional accuracy monitoring, surveillance shape accuracy monitoring, surface roughness monitoring, and workpiece mounting position monitoring, etc.

The working condition of the machining system is a comprehensive reflection of the interaction between the machining process and the machine tool. As shown in Fig. 4.1 [4], this interaction determines the quality of workpiece processing, tool wear,

K. Zhu, *Smart Machining Systems*, Springer Series in Advanced Manufacturing,
https://doi.org/10.1007/978-3-030-87878-8_4

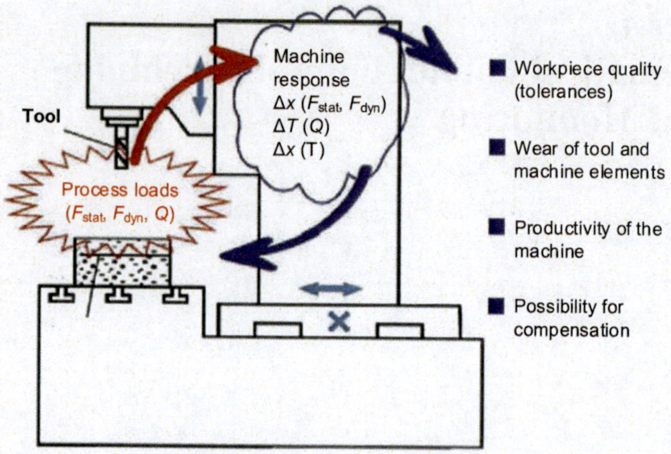

Fig. 4.1 Interaction between machining process and machine tool [4]

wear of other mechanisms or components, output, compensation possibility, etc. Hence, it is necessary to monitor and control the working condition of the machining system.

In the modern machining system, due to the complexity of the cutting process and the machining conditions of the variability, the use of traditional monitoring method based on a single threshold is difficult to meet the requirements, it is necessary to explore more effective intelligent monitoring under the knowledge support.

4.1.2 The System of Machining Process Monitoring

The task of system analysis is to study and describe the dynamic characteristics of the system, identify, control, and optimize the process of the machining system according to the analysis results. The core of the machining system analysis is to establish the system model, which must contain the state information reflecting the dynamic characteristics of the system. Therefore, in a broad sense, any functional relationship or curve that can reflect the dynamic characteristics of the system can be called the system model [5].

To get the dynamic characteristic of the accelerometer, we can establish the dynamic model of the accelerometer according to the measured frequency equation. It is a mathematical model to describe the dynamic performance of the system in the frequency domain. Generally speaking, the theoretical modeling of complex electromechanical equipment is difficult. In contrast, experimental modeling is easy to implement and has higher credibility, especially when only the low-order approximate model of the system is needed. Therefore, experimental modeling is still the main way of complex system dynamic analysis.

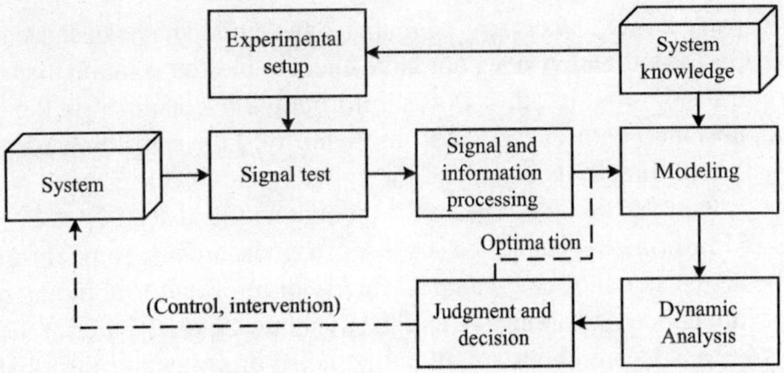

Fig. 4.2 Basic tasks of machining system analysis

The practice of mechanical and electrical engineering is inseparable from experimental modeling. The primary problem of experimental modeling is to obtain the input and output signals reflecting the system state information through the correct experimental design. The state information reflecting the dynamic characteristics of the system is the main object of our research, and the acquisition and utilization of the information is the key to the dynamic analysis of the system. Therefore, the essence of system dynamic analysis is to study the extraction and application methods of system dynamic information. The main task of system dynamic analysis can be summarized as the block diagram shown in Fig. 4.2.

For the structure illustrated in Fig. 4.2, the following basic concepts need to be explained:

(1) System

In this book, "system" refers to a whole with specific functions composed of several interacting and interdependent things. A processing system, for a given input (excitation), will complete the given output function (response). For the actual mechanical and electrical equipment, the system is a relative concept. Taking the numerical control machine tool as an example, when studying its overall performance, "system" includes the processing program carrier, numerical control device, servo drive device, machine tool body, and other auxiliary devices; when only studying the cutting performance, "system" only includes the cutting parameters and tool-related information. In terms of control performance, it can be divided into "numerical control system" and "servo drive system". In a word, the system can be divided into many ways and levels. How to determine the boundary of the system depends on the structure of the system and the purpose of the research.

(2) Signals and Information

From the viewpoint of information theory, any system consists of three elements: material, energy, and information. Information is the soul of the system because it reflects the state and characteristics of the system. The problems we are faced with are information acquisition, transmission, conversion, storage, processing, analysis, and utilization.

However, why do we always say "signal and system"? This is because the information itself does not matter, does not have energy, and the signal is the carrier of information transmission, in other words, information is contained in the signal. A signal is a substance with energy and observability. In system dynamic analysis, what we deal with directly is signal, so the signal is the direct research object, and information is the real research object of system dynamic analysis.

Signals and information cannot be confused. Signals are just some form of information. The actual signal often contains a variety of information components, some of which we are concerned about, we call the dynamic characteristics of the system, and those we do not care about are all called noise or redundant information. For example, in the tool wear monitoring, the cutting force signal contains not only the wear state change information but also other dynamic information, such as system stability information. In system dynamic analysis, we usually spend a lot of energy extracting the information we need from the signal, and this process is called feature extraction.

(3) **Signal Processing**

How to get the system characteristic information of interest through the signal? Therefore, it is necessary to process the signal. The so-called signal processing is to process or transform the signal in order to extract the information of interest from the signal and make use of it. By processing or transforming the signal, the redundant information in the signal can be weakened, the uninteresting noise or interference can be filtered, and the signal can be transformed into a pattern that is easy to analyze and recognize.

In the field of signal processing, "signal processing" sometimes refers to signal processing techniques such as amplification, modulation, demodulation, and filtering. The transformation of signal and statistical processing of the random signal is usually called "signal analysis", in which the processing and transformation of the discrete signal sequence are also called "data processing" and "data analysis".

(4) **Information Processing**

Information processing includes information processing and utilization, and its research object is the system. The concrete content of information processing is to process the system information such as dimension reduction (purification), fusion, and classification, and use the processed information to serve the description and control of the system. All kinds of system pattern recognition methods and system control theory belong to the category of system information processing, and system feature information extraction and processing is the core task of system dynamic analysis.

Dynamic analysis of machining process systems is a new subject-based on system theory, information theory, and cybernetics. It involves a wide range of knowledge, such as signal testing, signal and information processing, system identification, pattern recognition, and computer application. This reflects the characteristics of interdependence, mutual penetration, mutual intersection, and mutual promotion of various disciplines in modern manufacturing engineering research.

4.2 The Content of the Machining Process Monitoring System

The traditional machining process monitoring system generally consists of signal detection, feature extraction, state recognition, decision-making, and control of four parts, as shown in Fig. 4.3.

4.2.1 Signal Detection

In the machining process, many state signals reflect the changes in machining status from different views, monitoring signal selection is often the key to determine the success or failure of the monitoring system [6, 7]. Monitoring signals should have the following characteristics:

(1) Response sensitive and rapid to changes in machining status;
(2) Easy online measurement;
(3) The signal is less disturbed by the environment and has a higher signal-to-noise ratio.

Commonly used machining status detection signals are cutting force, power, acoustic emission (AE), vibration and cutting temperature, etc. The monitoring signal is acquired and pre-processed with the corresponding sensor. Signal pre-processing includes isolation, amplification, filtering, and A/D conversion, etc.

4.2.2 Feature Extraction

Feature extraction uses statistical methods to extract the characteristic parameter from the detection signal, which can most reflect the machining state changes. Feature

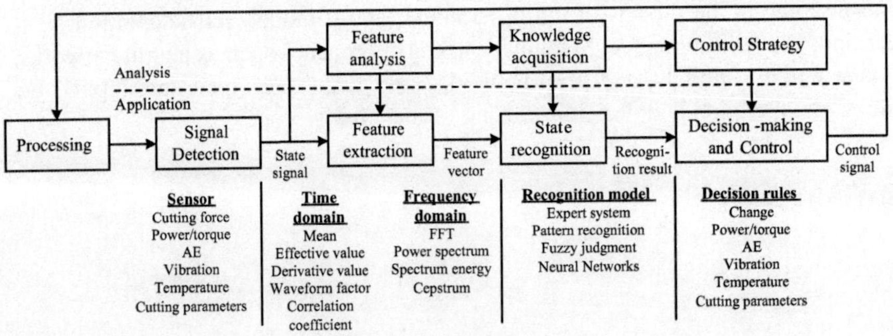

Fig. 4.3 The content of machining process monitoring systems

extraction methods are mainly time domain analysis and frequency domain analysis. The feature extraction effect is to improve the signal-to-noise ratio and to increase the anti-jamming capability of the system. The quality of the extracted feature parameters will directly affect the performance and reliability of the whole monitoring system.

4.2.3 State Recognition

State recognition is classified according to the machining status, which relied on the obtained feature parameters through the identification model, its core is the model used. The function of the model is to achieve the mapping from the feature space to the state space. The model can be established by physical relationships or empirical formulas. According to the characteristics of the model, the model can be divided into the fixed-parameter model, adaptive model, self-learning model, etc. A multi-model system refers to the machining status of the detection signal through several models for analysis, to obtain more monitoring information. Under conditions without increasing the cost of equipment, this system obtains more state information through software processing, so which makes monitoring more accurate and reliable.

4.2.4 Decision-Making and Control

According to the result of state recognition, decision-making, and control under the guidance of the decision-making knowledge base to make decision-making for the machining process of the fault, and perform the corresponding adjustment and control. The main requirements for decision-making and control methods are:

(1) Data processing speed, judgments, and decision-making quickly.
(2) Decision-making accuracy, to determine the error as little as possible.
(3) Adapt to a variety of processing.

To meet these requirements, the research of the machining process monitoring is developing in the aspects of multi-parameter, multi-model, self-adaptation, self-learning, and fault tolerance. Machine learning approaches such as neural networks, hidden markov models, and deep networks have been developed and applied for online identification and decision-making of machining processes.

4.3 The Methods of Machining Process Monitoring

4.3.1 Introduction

The modern machining system is a complex process that is coupled and interacted with various factors. It has the flowing characteristics:

(1) The co-existence of continuity of machining process and discontinuity of data information.
(2) The co-existence of the gradient of state feature and abruption of fault.
(3) The MIMO characteristic of the machining process.
(4) The time-variant characteristic of the machining process.
(5) The non-linear relationship between input parameters and output signals.

As a result, condition monitoring and fault detection are necessary for the normal operation of a modern manufacturing system. Firstly, it is necessary to collect the observation data (which can reflect the operation state of the manufacturing process) in a certain way with some priors of the manufacturing characteristics. Then corresponding mathematical model should be built to judge whether the system is normal and to detect the fault type. Finally, based on the monitoring results, proper solutions need to be proposed.

The analyzing of the condition of the manufacturing system must resort to a referring mathematical model which can well represent the characteristics of the system. Fault detection is to detect that whether the current state of the system is normal and to identify the abnormal degree if the system is abnormal. The change of system state is reflected by the change of inherent dynamical characteristics. In this sense, detection is to identify the real state of the system at every time. From the theoretical view of point, detection is just a kind of system identification.

Utilizing the output and input signals of the manufacturing system, the system identification method builds a mathematical model for the system with certain principles. The system model describes the process characteristic and the relationship between the input and the output. Work condition is then analyzed and predicted based on the built model. The corresponding mathematical background is necessary for the understanding of the whole process of modern manufacturing system monitoring and the deployment of monitoring approaches. Hence, before introducing the monitoring methods for modern manufacturing systems, we give the corresponding mathematical knowledge. To data, there exist many mathematical methods adopted in modern manufacturing system monitoring and fault detection. The main methods involve time-series analyzing, parameters estimation (least square estimation, sequential least square estimation, and maximum likelihood estimation), and neural networks etc.

4.3.2 Stochastic Process Based Methods

The characteristics of modern manufacturing systems determine that the fault in the modern manufacturing system is various and stochastic. Therefore, learning corresponding stochastic process theory before is necessary for the understanding of the mechanic of the fault and the understanding of monitoring and fault detection of the machining process.

(1) **Definition of Stochastic Process**

From the theory of mathematical statistics, it could be known that the stochastic process is a set of ordered random variables. Namely, when there is a stochastic process $\{x(t), t \in T\}$ (T represents time range), $x(t)$ is an ordinary random variable for a fixed time node. The series of random variables constitute a stochastic process.

(2) **Moment Functions of Stochastic Process**

Statistical characteristic is the basic characteristic of the stochastic process, it describes the essence of the stochastic process and reveals inevitability from the contingency. The fundamental statistic characteristic of stochastic process is the distribution function: probability density function for continuous stochastic process and probability function for discrete stochastic process. In engineering practice, the most common used characteristics are two moment functions: first order moment function (namely mean function) and second order moment function (the most important is auto covariance function). All of these functions are deterministic and easily applicable.

(1) First order moment function (mean)

Discrete stochastic process is denoted by $\{x_t\}, t = 0, \pm 1, \pm 2, \ldots$, first order moment function (Mean function) is defined as:

$$\mu_{x,t} = E[x_t] = \int_{-\infty}^{+\infty} x dF_t(x) \tag{4.1}$$

(2) Second order moment function

For discrete stochastic process $\{x_t\}$, $t = 0, \pm 1, \pm 2, \ldots$, auto covariance function is defined as:

$$\begin{aligned}\gamma(t, s) &= E\left[(x_t - \mu_{x,t})(x_s - \mu_{x,s})\right] \\ &= \int_{-\infty}^{\infty}\int_{-\infty}^{\infty} (x - \mu_t)(y - \mu_s) dF_{t,s}(x, y)\end{aligned} \tag{4.2}$$

where $F_{t,s}$ (x, y) is the joint distribution of (x_t, x_s).

In the same way, the autocorrelation function could be written as:

$$\rho(t, s) = \gamma(t, s) \Big/ \sqrt{\gamma(t, t)\gamma(s, s)} \tag{4.3}$$

The variance function could be defined as:

$$\sigma_{x,t}^2 = E\left[(x_t - \mu_{x,t})^2\right] \tag{4.4}$$

The mean square function could be defined as:

$$\psi_{x,t} = E\left[(x_t)^2\right] \tag{4.5}$$

It worth noting that the auto covariance function is the most important function among all of the second-order moment functions for stochastic process $\{x_t\}$. It describes the auto correlation structure of the process and has a symmetrical nonnegative definite characteristic. The other three functions could be seen as the special cases of the auto covariance function.

(3) **Special Stochastic Process**

For the general stochastic process, to descript the characteristics completely, the higher moments functions such as the third and fourth-moment function need to be computed. This is very tedious. In engineering practice, with some realistic assumptions, a special stochastic process is always adopted to simplify the computation.

(1) **Stationary Process**

If all of the moment functions are independent of the starting time, then the process is strictly stationary. When only the first and second-order moment functions are independent of starting time, the process is named as a widely stationary process.

Obviously, the mean function, variance function, and mean square value function of the stationary process are constants that are independent of time t. These three constants are denoted by μ_x, σ_x^2 and ψ_x. The auto correlation function $\gamma_{xx,k}$ and auto correlation coefficient $\rho_{xx,k}$ are only related to the delay step k.

(2) **Normal Process**

The random variables in the stochastic process all obey normal distribution. Namely, all of the moment functions are determined by the first and second functions. The stochastic process is named as normal stochastic process or Gauss stochastic process.

(3) **Ergodic Stochastic Process**

If any sample of a stochastic process could be represented by the delaying form of another sample, then the stochastic process is ergodic. Namely, for any samples xi, t, xj, t, i $\neq$ j, i = 1, 2, …, there exists xi, t $\neq$ xj, t − k.

The ergodic process has an important characteristic: the distribution of a random variable at a certain time node is the same as the distribution of a sample in the time axis. Furthermore, it could be inferred that:

(1) The ergodic process must be stationary, but the converses are not all true.
(2) For the ergodic process, the average of moment functions on set could be replaced by the average of moment functions on the time axis.
(4) **Zero Mean Stochastic Process**

Zero mean means that μx = 0. This assumption could be realized easily by subtracting $\widehat{\mu}_x$ in process {xt} if the mean function $\widehat{\mu}_x$ is not zero.

With the four assumptions above, we give a stationary, normal, ergodic stochastic process with zero means. As a classic example of this kind of stochastic process, the white noise process has been widely used in engineering practice.

(5) **White Noise**

White noise process $\{\varepsilon_t\}$ is a special stochastic process with mean function and auto covariance function as:

$$\begin{aligned} \mu_0 &= E[\varepsilon_t] = 0 \\ \gamma_{\varepsilon,k} &= E\left[\varepsilon_t \varepsilon_{t-k}\right] = \sigma_\varepsilon^2 \delta_k \end{aligned} \tag{4.6}$$

where constant σ_ε^2 is the variance of $\{\varepsilon_t\}$. δ_k is discrete δ function or Kroneker-δ function with definition as:

$$\delta_k = \begin{cases} 1 & k = 0 \\ 0 & k \neq 0 \end{cases} \tag{4.7}$$

This implies that the random variables in the white noise process $\{\varepsilon_t\}$ are not correlated. The white noise process plays an important role in process monitoring.

4.4 Parameter Estimation Methods

In the monitoring and fault detection of modern mechanical manufacturing, a required amount of observation data is collected firstly. Then with the data, the fault mechanism will be researched and the monitoring model will be built. After the modeling, the parameter of the monitoring model needs to be estimated.

Observation data $x_1, x_2, ..., x_n$ are assumed to be produced by the model:

$$x_t = \alpha_1 x_{t-1} + \alpha_2 x_{t-2} + \cdots + \alpha_p x_{t-p} + \varepsilon_t \tag{4.8}$$

where $t = p + 1, p + 2, .., n$ p is a given non-negative parameter and $\alpha_1, \alpha_2, ..., \alpha_p$ are unknown parameters denoted by a vector $\alpha = [\alpha_1, \alpha_2, ..., \alpha_p]^T$. Residual $\{\varepsilon_t\}$ is a sequence with an independent identical distribution. And there are $EX_s\varepsilon_t = 0, E\varepsilon_t = 0, E\varepsilon_t^2 = \sigma^2, E\varepsilon_t^4 < \infty$. The parameter α is stationary.

We can also rewrite Eq. 4.8 with matrix form:

$$Y = X\alpha + \varepsilon \tag{4.9}$$

where

$$Y = \begin{bmatrix} x_{p+1} \\ \dots \\ x_n \end{bmatrix} \qquad X = \begin{bmatrix} x_p & x_{p-1} & \dots & x_1 \\ \dots & \dots & \dots & \dots \\ x_{n-1} & x_{n-2} & \dots & x_{n-p} \end{bmatrix}$$

The following sections will discuss the estimation of parameters a and σ^2.

4.4.1 Least Square Estimation

With multiple groups of data that are linearly sampled from the system, parameters α and σ^2 could be estimated via least square estimation.

The least-square estimation defines a function of α with p variables:

$$\alpha_{l,s} = \sum_{t=p+1}^{n} \{x_t - \alpha_1 x_{l-1} - \alpha_2 x_{l-2} - \dots - \alpha_p x_{l-1}\}^2 \tag{4.10}$$

To minimize the function is to minimize the sum of the square of residual $\{\varepsilon_t\}$. The estimated $\widehat{\alpha}$ corresponding to the minimum value of the function is named as the least square estimate. The minimum value of the function is denoted by $\hat{\alpha}_{l,s}$. The minimum sum of the square is:

$$J = \sum_{i=1}^{n} \varepsilon^2(i) = \varepsilon^T \varepsilon \tag{4.11}$$

Least square means:

$$\left.\frac{\partial J}{\partial a}\right|_{\hat{a}=\hat{a}_{l,s}} = 0$$

And the least square estimate could be obtained as:

$$\hat{\alpha}_{l,s} = \left(X^T X\right)^{-1}\left(X^T Y\right) \tag{4.12}$$

This is the least square estimate with batch computation.

On the other hand, for the discrete stochastic process, the iteration least square method could be utilized to estimate the parameters on-line. The iteration method modifies the constant weights in the criterion function to exponential weights. The criterion function is defined as:

$$J_n = \sum_{t=1}^{n} \lambda^{n-t} \varepsilon_t$$

where the λ is weighting coefficient or forgetting factor. The iteration form of the on-line method is:

$$\begin{aligned} \hat{\alpha}(n+1) &= \hat{\alpha}(n) + K(n+1)[y(n+1) - X^T(n+1)\hat{\alpha}(n)] \\ P(n+1) &= \frac{1}{\lambda}[P(n) - K(n+1)X^T(n+1)P(n)] \\ K(n+1) &= P(n)X(n+1)[1 + X^T(n+1)P(n)X(n+1)]^{-1} \end{aligned} \tag{4.13}$$

And the estimate of σ^2 is:

$$\begin{aligned} \hat{\sigma}^2 &= \frac{1}{n-p}\hat{\alpha}_{l,s} = \frac{1}{n-p}\{Y^TY - Y^TX(X^TX)^{-1}X^TY\} \\ &= \frac{1}{n-p}\sum_{t=p+1}^{n}(x_e - \hat{\alpha}_1 x_{t-1} - \hat{\alpha}_2 x_{t-2} - \ldots - \hat{\alpha}_p x_{t-p})^2 \end{aligned} \tag{4.14}$$

where $\hat{\alpha}_1, \hat{\alpha}_2, \ldots, \hat{\alpha}_p$ represent the p weights of $\hat{\alpha}$.

4.4.2 *Yule-Walker Estimation*

According to the observation sequence $x_1, x_2, \ldots, x_n$, the estimated auto covariance functions $\hat{\gamma}_0, \hat{\gamma}_1, \ldots, \hat{\gamma}_p$ could be computed as:

$$\hat{\gamma}_k = \hat{\gamma}_{-k} = \frac{1}{n}\sum_{t=1}^{n-k} x_t x_{t-k} \quad k = 0, 1, \ldots \tag{4.15}$$

Now, we use another method to obtain the iteration relationship of $\hat{\gamma}_k$. If the sequence is a stationary auto regression sequence, then:

$$\begin{cases} \gamma_1 = a_1\gamma_0 + a_2\gamma_1 + \ldots + a_n\gamma_{n-1} \\ \gamma_1 = a_1\gamma_1 + a_2\gamma_0 + \ldots + a_n\gamma_{n-2} \\ \quad\ldots \\ \gamma_n = a_1\gamma_{n-1} + a_2\gamma_{n-2} + \ldots + a_n\gamma_0 \end{cases} \tag{4.16}$$

This equation is named the Yule-Walker equation. Some notations are defined as:

$$A = \begin{bmatrix} \gamma_0 & \gamma_1 & \cdots & \gamma_{n-1} \\ \gamma_1 & \gamma_0 & \cdots & \gamma_{n-2} \\ \cdots & \cdots & \cdots & \cdots \\ \gamma_{n-1} & \gamma_{n-2} & \cdots & \gamma_0 \end{bmatrix} \quad B_n = \begin{bmatrix} \gamma_1 \\ \gamma_2 \\ \cdots \\ \gamma_n \end{bmatrix} \quad \alpha = \begin{bmatrix} \alpha_1 \\ \alpha_2 \\ \cdots \\ \alpha_n \end{bmatrix}$$

And there is:

$$Aa = B_n \tag{4.17}$$

In Eq. 4.16, $\gamma_0, \gamma_1, ..., \gamma_n$ are replaced by $\hat{\gamma}_0, \hat{\gamma}_1, ..., \hat{\gamma}_n$ and $a_1, a_2, ..., a_n$ are replaced by $\hat{a}_1, \hat{a}_2, ..., \hat{a}_n$. Then Eq. 4.17 can be modified as:

$$\hat{A}_p \hat{a} = \hat{B}_p \tag{4.18}$$

where $\hat{A}_p$ and $\hat{B}_p$ are samples of A and B_n. According to Eq. 4.18, there is:

$$\hat{a} = \hat{A}_p^{-1} \hat{B}_p \tag{4.19}$$

This is the estimate via Yule-Walker method. The parameter σ^2 could be estimated by:

$$\begin{aligned} \hat{\sigma}^2 &= \hat{\gamma}_0 - a_1\hat{\gamma}_1 - a_2\hat{\gamma}_2 - ... - a_p\hat{\gamma}_p \\ &= \hat{\gamma}_0 - a^T \hat{B}_p \\ &= \hat{\gamma}_0 - \hat{A}_p^{-1} \hat{B}_p \hat{B}_p \end{aligned} \tag{4.20}$$

4.4.3 Maximum Likelihood Estimate

If the residual is assumed to obey normal distribution, then the observation sequence is a normal stationary auto regression sequence. Then $[x_1, x_2, ..., x_n]^T$ obey normal distribution $N(0, A_n)$. The notation A_n is the covariance matrix of $[x_1, x_2, ..., x_n]^T$. Then the likelihood function of $[x_1, x_2, ..., x_n]^T$ is:

$$l(a, \sigma^2 | x_1, x_2, ..., x_n) = -\frac{n}{2} \lg 2\pi - \frac{1}{2} \lg |A_n| - \frac{1}{2} x_n^T A_n^{-1} x_n \tag{4.21}$$

where $|A_n|$ is the determinant of A_n and $X_n = [x_1, x_2, ..., x_n]^T$. Hence Eq. 4.21 is the function of a and σ^2. Estimates $\hat{a}$ and $\hat{\sigma}^2$ that maximize likelihood $l(a, \sigma^2 | x_1, x_2, ..., x_n)$ are named maximum likelihood estimates. To compute the maximum likelihood is not easy work in practice. The approximate likelihood is always computed. Compared to the maximum likelihood estimate, the Yule-Walker method and least square method have wider applications in practice.

4.5 Time Series Analysis in Condition Monitoring

Time series analysis methods build different equations for the observation sequence of the system. Then the system is analyzed by the differential equation model. From a systematic perspective, observation data could be seen as the response of certain systems excited by external action. On one hand, we could dynamically analyze the system based on the model. On the other hand, we could also predict the future state and tendency of the system based on the model. The model especially integrated with machining conditions is significant to the monitoring and fault detection of the modern machinery manufacturing system.

Common time series models [8] are necessary for the processing of the data with time series and the following system identification. The two basic time series models, AR and ARMA models are the most widely used. And in the monitoring and fault detection of modern machinery manufacturing systems, the ARMA model is commonly adopted [9–11].

4.5.1 The Auto-Regression Model AR(N)

(1) **An Auto-regression Model with Order *n***

Definition 2.1: The model below is named regression model with order n, denoted by AR(n):

$$x_t = a_0 + a_1 x_{t-1} + a_2 x_{t-2} + \ldots + a_n x_{t-n} + \varepsilon_t \tag{4.22}$$

where $\{\varepsilon_t\}$ is white noise sequence, it reflects the influence of random elements on $\{x_t\}$, and $E\{x_s \varepsilon_t\} = 0$ for $s < t$, a_t is the coefficient. The $\{x_t\}$ is named auto regression sequence with order n. When $a_0 = 0$, the model is a centralized auto regression model.

Polynomial:

$$a(u) = 1 - a_1 u - a_2 u^2 - \ldots - a_n u^n \tag{4.23}$$

Equation 4.23 is the coefficient polynomial of the auto regression coefficient. When $a(u)$ satisfies stationary condition (all the roots of $a(u)$ are outside the unit circle), Eq. 4.22 represents a stationary AR(n) model. Otherwise, the model is nonstationary (or generalized AR(n)).

If a stationary sequence satisfies Eq. 4.22, then according to $Ex_t = c$, it could be obtained:

$$Ex_t = a_0 + a_1 Ex_{t-1} + a_2 Ex_{t-2} + \ldots + a_n Ex_{t-n} + E\varepsilon_t$$

Namely

$$c = a_0 + a_1 c + a_2 c + \ldots + a_n c$$

$$Ex_t = c = a_0 \big/ (1 - a_1 - a_2 - \ldots - a_n)$$

The stationary centralized model is:

$$\begin{aligned} w_t &= x_t - c \\ w_t &= a_1 + a_2 w_{t-1} + \ldots + a_n w_{t-n} + \varepsilon_t \end{aligned} \tag{4.24}$$

The centralized stationary AR (n) is commonly applied when there is no need to consider $Ex_t = c$.

(2) **First-order Auto Regression Model AR (1)**

When the stationary normal observation with zero mean is only related to the preceded value, the model is first order auto regression model AR(1).

$$x_t = a_1 x_{t-1} + \varepsilon_t \tag{4.25}$$

where a_1 is the model parameter of the auto regression model (namely regression parameter). Equation 4.25 reflects the correlated relationship between random variables at different time nodes and it describes the system dynamically. Hence, AR(1) has extension ability. It can predict the future state of the system.

On the other hand, it is necessary to test the AR(1) model after it is fitted via the given stationary sequence $\{x_t\}$. Namely, it is necessary to examine whether the AR(1) model is contrary to the assumption. The most fundamental task is to examine whether the residual ε_t is white noise.

4.5.2 The Auto Regression Moving Average Model ARMA(n, m)

Definition 2.2 The model below is the mixture model of the auto regression model with order n and moving average model with order m (concisely denoted by ARMA(n, m)).

Stationary normal time series $\{x_t\}$ with zero mean is not only dependent on its n previous value $x_{t-1}, x_{t-2}, ..., x_{t-n}$ but also dependent on the m previous disturbance $\varepsilon_{t-1}, \varepsilon_{t-2}, ..., \varepsilon_{t-m}$. Namely:

$$x_t = \alpha_0 + \alpha_1 x_{t-1} + \alpha_2 x_{t-2} + ... + \alpha_n x_{t-n} + \varepsilon_t - \beta_1 \varepsilon_{t-1} - \beta_2 \varepsilon_{t-2} - ... - \beta_m \varepsilon_{t-m} \tag{4.26}$$

where $\{\varepsilon_t\}$ is white noise series and $Ex_s\varepsilon_t = 0$ for all $s < t$. α_i is auto regression parameter and β_i is moving average parameter. Their coefficient polynomials are:

$$\alpha(u) = 1 - \alpha_1 u - \alpha_2 u^2 - ... - \alpha_n u^n$$
$$\beta(u) = 1 - \beta_1 u - \beta_2 u^2 - ... - \beta_m u^m$$

There do not exist common roots between them.

When the roots of $\alpha(u)$ and $\beta(u)$ are all inside unite cycle, Eq. 4.26 represents a reversible ARMA(n,m). The corresponding solution is named as reversible series. Likewise, by operating a delay operation on ARMA(n, m), Eq. 4.26 can be written as:

$$\alpha(B)X_t = \beta(B)\varepsilon_t \tag{4.27}$$

where $\beta(B)\varepsilon_t = \varepsilon_t - \beta_1\varepsilon_{t-1} - \beta_2\varepsilon_{t-2} - .... - \beta_m\varepsilon_{t-m}$.

If there exists a stationary normal series $\{x_t\}$ with zero mean that does not fulfill AR(1) model. Then ARMA(2, 1) model could be utilized. The model could be represented as:

$$x_t = \alpha_0 + \alpha_1 x_{t-1} + \alpha_2 x_{t-2} + \varepsilon_t - \beta_1 \varepsilon_{t-1} \tag{4.28}$$

It implies that:

(1) The x_t is linearly related with x_{t-1} and x_{t-2}. x_t has the auto regression form presented by a second-order differential equation. α_1 and α_2 are the parameters of auto regression.
(2) The x_t is also linearly related with ε_t and ε_{t-1}. The linear relationship is represented by a moving average form with second-order difference.
(3) The $\{\varepsilon_t\}$ is white noise sequence. $Ex_s\varepsilon_t = 0$ for all $s < t$.

The ARMA(2, 1) decomposes x_t into determinant and stochastic parts. The determinant part is the mean value $E[x_t]$ at time t. It is determined by $\alpha_1 x_{t-1} + \alpha_2 x_{t-2} - \beta_1\varepsilon_{t-1}$. The stochastic part is determined by ε_t. Similar to AR(1) model, Eq. 4.28 can be written as:

$$\varepsilon_t = x_t - \alpha_1 x_{t-1} - \alpha_2 x_{t-2} - \beta_1 \varepsilon_{t-1} \tag{4.29}$$

Hence, ARMA(2, 1) model could be seen as a mechanism that transforms related time series into an independent series.

Likewise, it is necessary to conduct practical tests for ARMA(2, 1). The fundamental of the practical test is also to test whether the residual ε_t is white noise.

(1) Examining whether ε_t is uncorrelated with $\varepsilon_{t-2}, \varepsilon_{t-3}, \ldots$

The two-step auto correlation of ε_t is computed as:

$$\rho_{\varepsilon_t^2,2} = \frac{\sum_{t=2}^{N} \varepsilon_t \varepsilon_{t-2}}{\sum_{t=3}^{N} \varepsilon_t^2}$$

If there is $\rho_{\varepsilon_t^2,2} \to 0$, then the ARMA(2, 1) model is fitted to the process.

(2) Examining whether ε_t is correlated with $x_{t-3}, x_{t-4}, \ldots$

The three-step correlation function between x_t and ε_t is:

$$\rho_{\varepsilon_t,x_t,3} = \frac{\sum_{t=3}^{N} \varepsilon_t x_{t-3}}{\sqrt{\sum_{t=4}^{N} \varepsilon_t^2 \sum_{t=4}^{N} x_{t-3}^2}}$$

Likewise, If there is $\rho_{\varepsilon_t,x_t,3} \to 0$, then the ARMA(2, 1) model is fitted to the process.

With the AR(n) and ARMA(m, n) models, the coefficients and model parameters can be further extracted, and applied with discriminant functions for state identification, which are to be discussed in the next chapter.

4.6 The Machining State Description

Machining status feature information sensing includes both state signal detection and state feature extraction. The purpose is to detect a variety of machining status signals through the sensor and extract the characteristic information that can effectively reflect the nature of the machining state from the detected signal data. From the perspective of pattern recognition to describe the machining process of monitoring, the detection of a variety of state signals can be called process state mode, and access to the characteristics of information called feature mode. The machining state feature information sensing process is through the detection of machining status signals, after pre-processing to obtain the state model, and then the state model is processed to obtain the feature pattern. The state and feature patterns are represented by state vectors and feature vectors, as shown in Fig. 4.4.

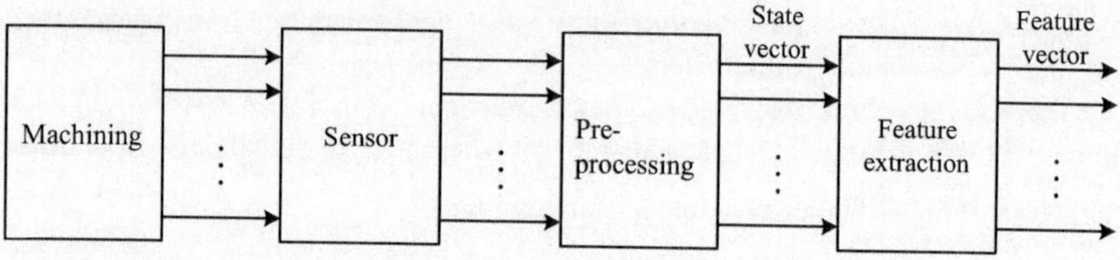

Fig. 4.4 The extraction of machining state feature information

4.6.1 Typical Anomaly State of the Machining Process

Typical anomaly state in the machining process can be divided into two categories: one is a sudden failure, such as tool breakage, sharp tool wear, etc.; the other is the slowness of the fault, such as tool wear and so on. These two types of faults can be described by mathematical models. The machining process model can be expressed as:

$$Y(t) = f(X(t), \theta) + \varepsilon_t \tag{4.30}$$

where $X(t) \in R^n$ and $Y(t) \in R^m$ are the input vector and output vector of the process respectively; θ is the model parameter vector and ε_t is the model noise.

(1) **Abrupt Change Description**

A abrupt change in the machining state can be expressed as:

$$Y(t) = f(Z(t), \theta) + \varepsilon + \gamma\delta_{t,\tau} \tag{4.31}$$

where $\gamma \in R^n$ is the unknown quantity of the fault size; τ is the time at which the fault occurred; $\delta_{t,\tau}$ is the unit step function:

$$\delta_{t,\tau} = \begin{cases} 1 & (i \leq j) \\ 0 & (t \geq \tau) \end{cases} \tag{4.32}$$

The time at which the machining process occurs and the degree of failure is determined based on the values of τ and γ.

Abrupt change (or failure) due to its sudden and drastic, its production will seriously damage the machine processing performance and the quality of the workpiece, requiring such a fault monitoring is fast and accurate. The corresponding fault detection method is the residual detection method, that is, according to the machining of the actual output and the model forecast output changes between the residual to detect the machining process of sudden failure.

(2) **Slow Variation Fault Description**

The slow variation fault of the machining process can be expressed as:

$$\begin{cases} Y(t) = f(X(t), \theta(t)) + \varepsilon_t \\ \theta(t) = \theta_0 + \Delta\theta(t) \end{cases} \quad (4.33)$$

The machining process of slow variation fault is also called parameter deviation type fault. The corresponding fault detection method is the model parameter estimation method, that is, according to the deviation between the estimated value and the normal value, to determine the system failure and size.

4.6.2 *Process Model Based State Feature Extraction*

The traditional machining state feature extraction method mainly adopts the nonparametric method based on the statistical analysis of the data, which obtains the time-domain feature or the frequency domain directly from the detected state signal data. These features belong to the shallow knowledge of the machining process, and only to reflect the changing characteristics of the processing state, and cannot further provide the total state change in which the number of changes caused by the machining process, and what is caused by cutting conditions such deep knowledge. The core idea of state feature extraction based on process model identification is through the effective identification of the machining process, and from the model structure, model parameters, model functions, and other characteristics of the processing process to obtain the deep knowledge, to constitute the feature vector. There are two types of state extraction methods based on the machining process identification: one is based on the process model parameter estimation method; the other is based on the model prediction residual estimation method.

(1) **Feature Extraction Based on the Model Parameter Estimation**

This method applies to the case where the process model is known, and the state feature information is obtained from the identified model parameters by identifying the process model. The input of the model is the cutting parameter and the output is the machining status signal. In the real-time monitoring, the machining state is forecasted by the process, and the model parameters are corrected according to the forecast error and identification algorithm, and the model automatically adapts (traces) the input and output characteristics of the machining process, and the characteristics of the process are also condensed into the model parameters. In addition, the state feature obtained from the model parameters is not affected by changes in the amount of cutting.

The extraction process of the tool wear feature based on the parameter estimation of the adaptive model is:

(1) Establishing the cutting process input and output model, and testing the validity of the model;

(2) Detecting the input and output status signals of the model;
(3) According to the model and the input and output values, the model parameters are estimated by the least-squares method;
(4) Detect changes in model parameters and as tool wear feature. The amount of feature self-learning/recognition system classifies or estimates tool wear.

(2) **Feature Extraction Based on Model Prediction Residual Detection**

When the machining process is too complex and the model structure is unknown, the black box model is used to identify the machining process. In this case, the machining state features are extracted from the model prediction residuals. Based on obtaining the normal machining state input and output observation data, establish the equivalent machining process input and output intelligent model. The model is also called the reference model, which reflects the normal input and output characteristics of the machining process. When the machining process is in the normal state, the residual between the actual process and the model prediction will be small; when the machining process is in an abnormal state, the residual between the actual process output and the model prediction will become larger. Therefore, the feature information of the machining process state can be obtained from the characteristics of the residuals predicted in the machining process model. This is the basic idea of residual detection of machining state feature extraction, the principle is shown in Fig. 4.5.

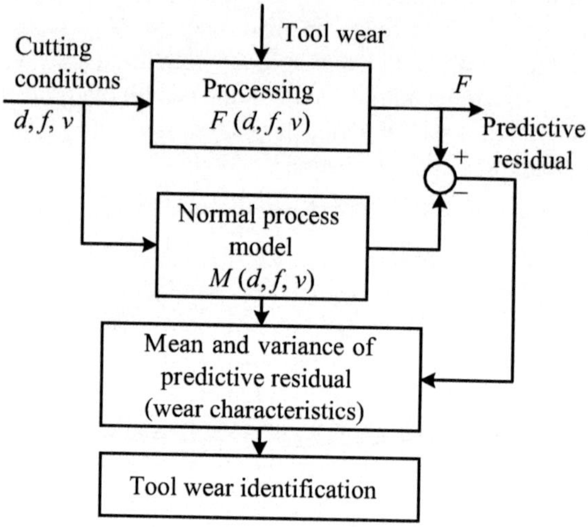

Fig. 4.5 Tool wear feature extraction method based on model prediction residual detection

4.7 Identification of Machining Process

4.7.1 Overview of Process Modeling

The essence of the machining process is the process of the interaction between the machine tool and the workpiece under certain conditions (cutting parameters). The input is cutting parameters such as cutting depth, feed speed and cutting speed, etc. The output is a variety of machining status signals, such as cutting force, torque, acoustic emission (AE), vibration and temperature changes, etc. When the machining process fails, such as tool wear or breakage, will lead to changes in the processing status signal. The traditional monitoring system obtains the fault feature directly from the change of the characteristic parameters (such as time domain or frequency domain parameters) of the detection state signal and realizes the recognition and decision of the machining state fault. However, changes in the machining parameters due to changes in the cutting parameters will result in changes in the machining state signals, and this change is often much greater than the change in machining status signals due to machining failures (such as tool wear). The traditional monitoring system, which only considers the effect of the cutting condition on the processing state signal, can only monitor the machining process under the single working condition (the cutting condition is fixed), so it cannot be used in the modern mechanical advanced manufacturing system which cutting condition is changeable. A reasonable approach should be to detect the input and output of the machining process state while analyzing the relationship between the input and output characteristics to distinguish the process state. This is the basic idea of feature extraction based on process model identification.

The machining process of modern manufacturing system is a multi-factor interaction and coupling of the complex process, its performance in:

(1) Continuity and discrepancy. For the machining process, a cutting process in one pass process is continuous; for the part manufacturing process, the data information, such as part size, surface roughness, etc. are discrete.
(2) Degeneration and mutation. In the fixed processing conditions, the machining process state characteristics have degenerated, such as machine temperature, tool wear, dimensional accuracy changes, etc.; but the tool damage and break, etc. are mutated.
(3) Multi-input and multi-output. As mentioned earlier, the machining process is a multi-input and multi-output process.
(4) Time variability. In the machining process, since the characteristics of tool wear, dimensional accuracy and surface quality, etc. vary with time, the characteristics of machining status have significant non-stationarity.
(5) Nonlinearity. In the machining process, there is a complex nonlinear relationship between the input parameters and the output status signal, and the monitoring system is also typical nonlinear system.

Machining process modeling, also known as system identification, refers to the process that through the detection process of input and output signals obtains the mathematical model of the process under certain criteria. Through the establishment of the mathematical model of the machining process, you can describe the characteristics of the process and the relationship between the system input and output and can achieve the machining status of the forecast and analysis.

To meet the complex, highly nonlinear, and uncertain requirements machining process of modern advanced manufacturing system, the machining process monitoring model must have a high degree of intelligence, that is self-learning, self-organization, and adaptive ability, as well as robustness and fault tolerance in a multi-factor, strong interference environment. Modeling of machining process and identification must be effective in integrating classical system identification methods and advanced artificial intelligence technologies.

4.7.2 *Model of Machining Process and Identification Method*

(1) **Equivalence Model and Criterion Function**

The mathematical model of the machining process is to use mathematical structure to reflect the behavioral characteristics of the actual process. Commonly used mathematical models are algebraic equations, differential equations, difference equations, and state equations. The characteristics of the machining process are linear and nonlinear, dynamic and static, deterministic and random, and the mathematical model that describes the process characteristics must also have these types. There are two modeling methods for the machining process, which are the theoretical modeling method and the model identification method. The theoretical modeling method is based on the analysis of the movement law of the process, using the known laws, theorems and the principle to establish the mathematical model of the process; The model identification method is based on the fact that the dynamic characteristics of the process are necessarily reflected in the input and output data of the process, using the information provided by the input and output data of the process to establish the mathematical model of the process. Often theoretical modeling is called a "white box" approach, identification modeling is called a "black box" approach, combining the two methods as a "gray box" approach.

The machining process is a complex nonlinear process under the influence of multiple factors. Therefore, the process modeling mainly adopts the identification method. The three basic elements of identification are data, model classes, and guidelines. Identification is following certain criteria in a group of models to select a model with the best data fitting, and the model is an equivalent system with the actual process on the input and output characteristics.

The criterion function is a standard used to measure a model close to the actual process, usually expressed as a functional error, referred to as

$$J(\theta) = \sum_{k=1}^{L} f(\varepsilon(k)) \tag{4.34}$$

In the formula, $f(\cdot)$ is a function of $\varepsilon(k)$; $\varepsilon(k)$ is an error function defined on (0, 1), which generally takes the norm of the deviation vector of the deviation between the identified object and its model output. In the case of discrete data, the criterion function is

$$J(\theta) = \sum_{i=1}^{N} |y_i - y_{mi}|^2 = \|\varepsilon\|^2 \tag{4.35}$$

In the formula, y and y_{mi} represent the outputs of the identified objects and models at the same input respectively; y_i and y_{mi} represent their ith observations respectively; N is the number of observations. The criterion function J is also called the output error criterion, as shown in Fig. 4.6.

When the model structure of the identified system is constant, the criterion function is a function of the model parameters. The principle of machining process model identification is shown in Fig. 4.7.

The common identification method is for the linear model. For the nonlinear model, first, the identification process is linearized, then the linear model identification method is used to estimate the parameters, and finally back into the form of the nonlinear model. A static model of machining process can generally be used polynomial function, such as:

$$Y(U) = a_0 + a_1U + a_2U^2 + \cdots + a_nU^n \tag{4.36}$$

Through mathematical transformation, it can be transformed into a linear model.

Fig. 4.6 Output error criteria for process identification

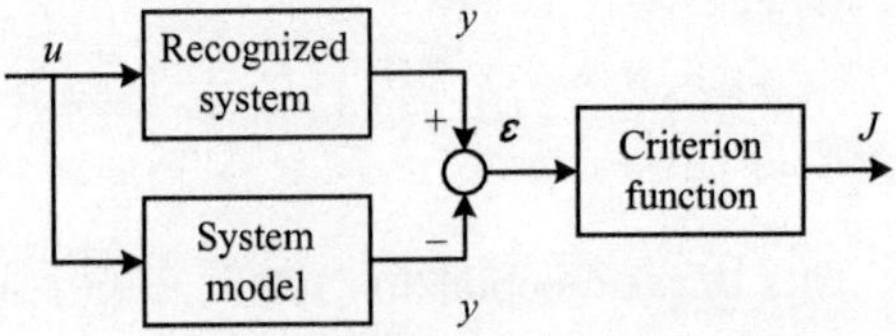

Fig. 4.7 Identification principle of the machining process model

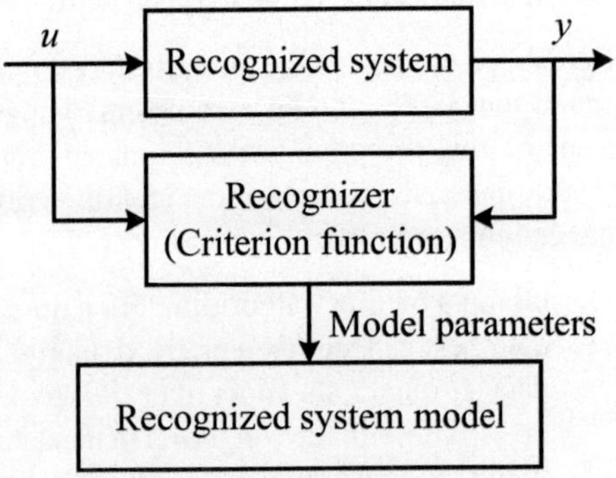

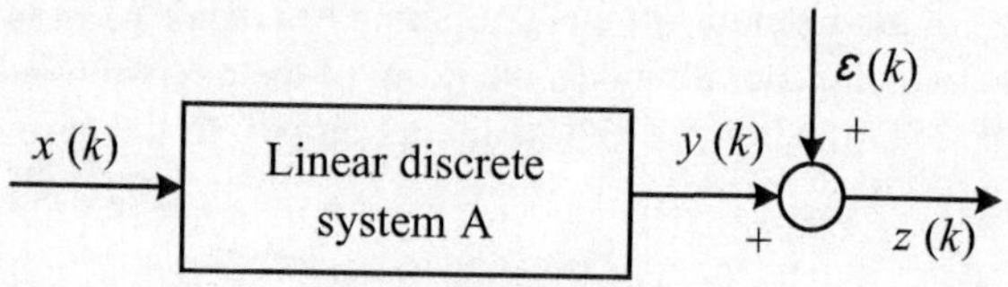

Fig. 4.8 Mathematical expression of linear discrete model

The dynamic model of the machining process is generally expressed as a difference equation:

$$y(t) + a_1 y^{(1)}(t) + \cdots + a_n y^{(n)}(t) = b_0 u(t) + b_1 u^{(1)}(t) + \cdots + b_m u^{(m)}(t) \quad (4.37)$$

(2) **Identify the Model of the System Expression**

A linear discrete model means that one or more variables can be represented as linear combinations of other variables at discrete points in time or space, as shown in Fig. 4.8.

The $x(k)$ and $z(k)$ shown in Fig. 4.8 are the input and output variables of the model respectively, which must be observable at discrete points. $\varepsilon(k)$ is the model noise, and A is the unknown model parameter. Recall that,

$$X(k) = [x_1(k)\, x_2(k)\, \cdots\, x_n(k)]^T$$

$$A = [a_1\, a_2\, \cdots\, a_n]^T$$

Then the output of the linear discrete model can be expressed as:

$$Z(k) = Y(k) + \varepsilon(k) = \sum_{i=1}^{N} a_i x_i(k) + \varepsilon(k) = X^T(k) + \varepsilon(k) \quad (4.38)$$

This linear combination relationship is the basic expression of the machining process identification problem, also known as the least-squares format.

(3) **Estimation of Time-varying Model Parameters**

The model of machining process mostly belongs to the time-varying parameter model, that is, the model parameters change with time. Therefore, it is necessary to ensure that the parameters estimated from the observed information reflect the most recent characteristic of the identified object so that the estimation can track the change of the time-varying parameter. The measures adopted for this purpose are:

(1) Gradual memory algorithm. For a time-varying system, the currently observed data best reflects the current dynamic state of the identified object, the more data is "older", the more likely it is to deviate from the current object characteristics. Through the weighting of the data to highlight the role of the current data. In the process of parameter identification, each time a new data is taken, all the

previous data is multiplied by a weighting factor λ, that is,$0 < \lambda < 1$. After the introduction of the weighting factor, the influence of the historical observation data on the parameter estimation exists, but the influence is constantly weakened with the accumulation of the observed data, and the effect of the parameter estimation is mainly observed in the most recent period. Therefore, the parameter evaluation obtained in this way can reflect the current characteristics of the identified object.

(2) Defined memory algorithm. The main characteristic of the algorithm is that the observed data based on the estimated parameters always maintain the latest fixed group data. The number of groups of observations, such as group N, is used in advance to estimate the number of groups of observations, and when a new set of observations is added each time the parameters are estimated, a set of oldest data is discarded, so that the estimated parameters that the data used is always the latest N group of data.

4.7.3 The Time Series Identification of the Machining State

In the machining actual process, there are many random signals (such as machine vibration, workpiece size accuracy data, etc.), a random relationship between these other signals, and the values of variables often can't be described by any functional relationship. A time-series AR model can be established for this type of process signal with only output and no significant input. That is, the conventional relationship between the observed data of the stochastic process itself is used to describe the regularity of the process. The autoregressive AR(n) model forms:

$$x_t = \sum_{i=1}^{N} a_i x_i + \varepsilon_t, \quad \varepsilon_t \sim NID(0, \sigma^2) \tag{4.39}$$

In the formula, $a_i = (i = 1, 2, \ldots, n)$ is the model parameter, which is independent of x_t; In NID, N means that when t is fixed, ε_t is a random variable of the normal distribution; ID means that when t changes, $\{\varepsilon_t\}$ are independent of each other.

The AR model requires that $\{x_t\}$ be a time series of stationary, normal, and zero mean values. If the data does not meet these conditions, it must be pre-processed to meet the requirements of the AR model. There are many methods for estimating the AR model parameters. The principle of the least-squares estimation method is that the time series $\{x_t\}$ is directly substituted into the above equation to obtain the following linear equations

$$\begin{cases} x_{n+1} = a_1 x_n + a_2 x_{n-1} + \cdots + a_n x_1 + \varepsilon_{n+1} \\ x_{n+2} = a_1 x_{n+1} + a_2 x_n + \cdots + a_n x_2 + \varepsilon_{n+2} \\ \cdots\cdots\cdots\cdots \\ x_N = a_1 x_{N-1} + a_2 x_{N-2} + \cdots + a_n x_{N-n} + \varepsilon_N \end{cases} \tag{4.40}$$

Expressed as a matrix

$$Y = XA + \varepsilon \tag{4.41}$$

where

$$Y = \left[x_{n+1}\, x_{n+2} \cdots x_N\right]^T,\ \mathrm{A} = [a_1\, a_2 \cdots a_N]^T$$

$$\varepsilon = \left[\varepsilon_{n+1}\, \varepsilon_{n+2} \cdots \varepsilon_N\right]^T,\ x = \begin{bmatrix} x_n & x_{n+1} & \cdots & x_1 \\ x_{n+1} & x_n & \cdots & x_2 \\ \cdots & \cdots & \cdots & \cdots \\ x_{N-1} & x_{N-2} & \cdots & x_{N-n} \end{bmatrix}$$

The least-squares estimate of the parameter matrix A is

$$A = \left(X^T X\right)^{-1} X^T Y \tag{4.42}$$

In the actual process monitoring, the AR model parameters can be estimated using the real-time recursive least squares method, so that $\mathrm{Y_N = Y, X_N = X, A_N = A}$, then

$$A_N = \left(X_N^T X_N\right)^{-1} X_N^T Y_N \tag{4.43}$$

The subscripts of the matrices and vectors are denoted by N, which means that they are based on the N observed data $\{x_t\}$, $(t = 1, 2, \ldots, N)$, so that the Eq. 4.39 is the least squares estimator based on N observations. In the matrix X_N, $x_{(i)}$ is the n-dimensional row vector, $\mathrm{k = N - n}$, if z is in the Eq. 4.39

$$P_N = \left(X_N^T X_N\right)^{-1} \tag{4.44}$$

There are

$$A_N = P_N X_N^T Y_N \tag{4.45}$$

Let the existing observation timing be $\{x_t\}$, $(t = 1, 2, \cdots, N)$, estimate the model parameter $\mathrm{A_N}$ according to $\{x_t\}$ based on Eq. 4.41, and now see a new data x_{N+1}, then add it to the original sequence, get $\{x_t\}$, $(t = 1, 2, \ldots, N, N+1)$, and then parameter estimation. According to the matrix theory, it is easy to obtain the least-squares estimation based on N + 1 data:

$$A_{N+1} = A_N + K_{N+1}\left(x_{N+1} - x_{(k+1)}, A_N\right) \tag{4.46}$$

where

$$K_{N+1} = \frac{1}{1 + x_{(k+1)} P_N x_{(k+1)}^T} \tag{4.47}$$

$$P_{N+1} = \left(I - P_N \frac{x_{(k+1)}^T x_{k+1}}{1 + x_{k+1} P_N x_{(k+1)}^T} \right) P_N \tag{4.48}$$

where I is an N-order unit matrix.

The Eq. 4.42 has a profound meaning, which means that the new estimate A_{N+1} of the model parameter is equal to the original estimate A_N to be corrected. The correction term is $K_{N+1}\left(x_{N+1} - x_{(k+1)}\varphi_N\right)$, which is a weighting process of the difference between the new data x_{N+1} and the estimated value $x_{(k+1)}\varphi_N$ of the new data, and the weighting coefficient is K_{N+1}(also referred to as a correction coefficient).

From the point of view of information theory, the observation sequence $\{x_t\}$ used to establish the time series AR model can be regarded as the output of a system, but a_t and σ^2 are the model parameters that are estimated based on $\{x_t\}$ according to a method. Therefore, all the information of the system characteristic and the working state of the system is contained in $\{x_1, x_2, \ldots, x_N\}$ of the size of the N data and the order of the N data, and thus it is included in the M and N parameters. The aggregation of this information is also called data compression. Therefore, the modeling method of the machining process time series AR (n) model is the effective state feature extraction method in the machining process monitoring.

4.7.4 Identification of the Cutting Force

From the principle of metal cutting, we can see that the main cutting force and cutting amount in cutting machining can be expressed in the empirical formula

$$F = C d^{a_1} f^{a_2} v^{a_3} \tag{4.49}$$

where a_1, a_2, a_3 are the model parameter, F is the cutting force, C is constant, d is the cutting depth, f is the feed,v is the cutting speed.

The model parameters C, a_1, a_2 and a_3 reflect the degree of influence of each change of the cutting parameters on cutting force, which reflects the characteristics of the machining process. Therefore, in the case of flexible machining (frequent changes in cutting conditions), the machining state characteristics can be obtained from the cutting force model parameters. Equation 4.45 represents the static nonlinear model of the cutting force. After linearization, the model becomes

$$y = a_0 + a_1 x_1 + a_2 x_2 + a_3 x_3 = X^T A \tag{4.50}$$

where

$$y = \lg F,\ a_0 = \lg C,\ x_1 = \lg d,\ x_2 = \lg f,\ x_3 = \lg v$$

$X^T = \begin{bmatrix} 1 \ x_1 \ x_2 \ x_3 \end{bmatrix}$, $A = \begin{bmatrix} a_0 \ a_1 \ a_2 \ a_3 \end{bmatrix}^T$.
A series of machining conditions were measured under varying cutting conditions. $S_i = \{d_i, f_i, v_i, F_i\}$,$i = 1, 2, \cdots, n$
The resulting model is

$$y(i) = X^T(i)A + \varepsilon(i),\ \ i = 1, 2, \cdots, n \tag{4.51}$$

where ε is the model error. Equation 4.47 represents the form of matrix and vector

$$\underline{Y} = \underline{X}A + \varepsilon \tag{4.52}$$

The square sum of the model residuals is

$$J = \sum_{i=1}^{n} \varepsilon^2(i) = \varepsilon^T \varepsilon \tag{4.53}$$

The least-squares estimate of the parameter can be expressed as

$$\tilde{A} = \left[\underline{X}^T \underline{X}\right]^{-1} \underline{X}^T \underline{Y} \tag{4.54}$$

where $\tilde{A} = \begin{bmatrix} \tilde{a}_0 \ \tilde{a}_1 \ \tilde{a}_2 \ \tilde{a}_3 \end{bmatrix}$. In the monitoring, the parameter estimation is used in real-time recursive least squares.

4.7.5 Neural Network Identification of Machining Process

The artificial neural network (NN) is one of the most studied and applied machine learning methods (which is be studied in the next chapter) for it strong nonlinear feature representation capablilities. The NN method of machining process identification is to establish the neural network model of the machining process to realize the mapping characteristics of the input and output of the machining process. The neural network modeling method is a black box method, and the neural network can approximate any nonlinear mapping with arbitrary precision.

In the actual processing, there is a specific mapping between the input vector such as cutting amount and output vector such as cutting force, temperature and so. However, due to the high complexity and non-linearity of the machining process, it is difficult to establish a mathematical model of mapping using conventional methods. The NN has very good versatility because it has the excellent quality of approximating any complex nonlinear function and is not influenced by the working point

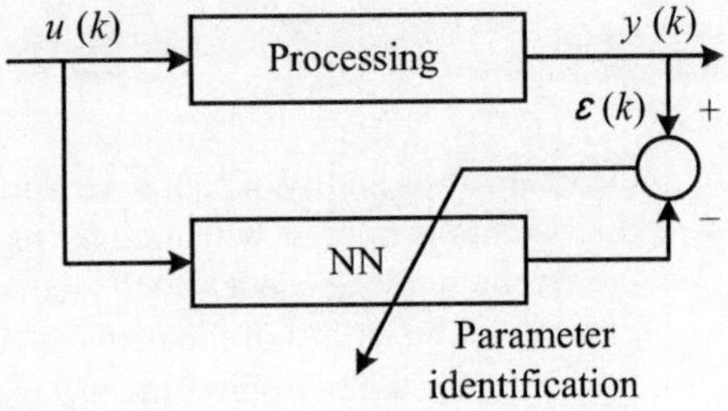

Fig. 4.9 Neural network identification of machining process

range, and the NN parameter estimation algorithm has a strong versatility, almost no prior knowledge of the model, so NN is very suitable for the machining process of modeling.

If the input vector of the machining process is $X(t)$ and the output vector is $Y(t)$, the mapping relationship between the input and output states of the machining process is ϕ:

$$Y(t) = \phi[X(t)] \tag{4.55}$$

The forward neural network is used to approximate this mapping, which is

$$Y(t) = NN[X(t)] \tag{4.56}$$

The method of establishing the machining process model with the forward neural network is based on the input and output of the sample data, and the parameter estimation method is used to determine the connection value of the neural network. As shown in Fig. 4.9.

The steps of identifying the machining process characteristics using the multi-layer forward neural network model are:

(1) Construction machining process model: Determine the machining process input and output vector of the various components, the process characteristics described in the form of Eq. 4.56.
(2) Establishing the corresponding neural network model: The number of model input layers and output layer nodes is determined by the process input and output vector dimensions.
(3) Access to the machining process input and output of a batch of sample data, and the data preprocessing, such as removing noise, normalization, and so on.
(4) Neural network parameters (including connection weights and thresholds) identification. The identification method uses the BP algorithm; the criterion function of the model is to predict the residual of the square sum function. When the model accuracy meets the requirements, the iterative operation of the parameter identification is stopped.

$$J = \sum_{i}^{n} \left(y_i - \hat{y}_i\right)^2 = \varepsilon^T \varepsilon \tag{4.57}$$

(5) Generalization ability of neural network model to test. The purpose of modeling the machining process with multi-layer forward neural network is not only to fit the given sample data but also to predict the unknown data correctly. This is the generalization ability of the network. Under the condition of ensuring certain precision, the use of as few hidden elements as possible can improve the generalization ability of the network. The indicators that measure the generalization capabilities of the neural network model are:

The average of the prediction error ε:

$$MF = \frac{1}{m} \sum_{i=1}^{m} |\varepsilon_i| \tag{4.58}$$

The mean square value of the prediction error ε:

$$MS = \sqrt{\frac{1}{m} \sum_{i=1}^{m} \varepsilon_i^2} \tag{4.59}$$

4.8 The Common Measurement Methods and Characteristics

The sensor is a physical variable-digital measurement conversion device. Its function is to convert the force, displacement, speed, acceleration, temperature, pressure, and other parameters of measured objects into signals that can be detected, transmitted, and processed (such as voltage signals and current signals, etc.) [12]. Therefore, it is also called a transducer or detector, or a transducer in acoustics. The sensor that measures vibration signals is also called a vibration pick-up. The sensor in modern test technology is no longer an independent mechanical measuring device in traditional concepts. It is only a segment in the entire test or monitoring system and is closely related to the subsequent segments. In an entire monitoring system, the sensor is always the first segment, and the accuracy and reliability of the sensor directly affect the system working conditions. Many monitoring systems can't work normally, where the main reason is that the outputs of the sensor are inaccurate due to improper selections. Therefore, it's significantly important to research the principles, structures, and installation methods of the sensor for equipment condition monitoring.

Common monitoring approaches for processing conditions are physical methods, where direct measurement and indirect measurement techniques are mainly used

[13, 14]. The direct measurement method is to directly measure the actual value of the variable, such as the tool wear state with camera images, radioisotopes, laser beams, and resistances. The direct measurement method has a high accuracy rate, but many direct measurement methods are inconvenient to operate because of the actual processing conditions and can only be applied in the laboratory. The actual variables in the indirect measurement method can be calculated based on the corresponding relationship based on experience. Indirect measurement methods are not as accurate as direct measurement methods but are relatively simple and more suitable for practical applications. The advantage is that the sensor can be used to continuously monitor the processing process, which helps to ensure the improvement of processing performance or provide information for the optimization of the processing process.

The physical variables that reflect the processing conditions include cutting force, vibration, acoustic emission, noise, temperature, surface roughness, etc. The signals of these process variables are collected by physical sensors. The prediction and judgment of processing conditions are realized by analyzing these variable signals. Based on these conclusions, corrections or other appropriate control operations are completed. Therefore, the monitoring process of processing conditions can be simplified as monitoring variables (environment)—sensors (input)—data processing and feature extraction (brain)—cognitive decision (brain)—action behavior. Monitoring cutting forces in processing can be used to monitor the tool wear conditions, detect tool failure, etc. Because the force signal is sensitive to the processing conditions and responds quickly, the monitoring method based on the cutting force is widely used to protect the tool and the electromechanical system of the machine tool.

Tool wear is an inevitable phenomenon in the machining process, and it is the focus of machining condition monitoring. Common monitoring methods for tool wear conditions are shown in Table 4.1.

Force is one of the most important signals in milling process monitoring. It has been reported by researchers that cutting forces contain reliable information on cutting conditions and most effective for tool wear monitoring [15–19].

Vibration signals contain various information of working conditions, and the advantages include high sensitivity, low cost, convenient installation, easy testing, and simple test instruments. Therefore, vibration signals are commonly used in actual engineering applications [20–24].

In the machine tool environment, AE signals are repeatedly reflected by the inner surface of the structure where the sensor is installed. This extension is recorded by the sensor. The installation of the AE sensor requires that there is no coupling medium between the sensor and the surface of the material, and the surface of the material must not have dust, paint, and other obstacles that may affect the acoustic coupling. The farther the sensor is from the AE source, the more the signal attenuates. It seriously affects the measurement of AE signals during processing. If the AE sensor is placed at one position of the workpiece, the varying distance between the sensor and the sound source during processing is a factor to be considered. Therefore, the installation location of the AE sensor is a difficult point, and its monitoring system requires higher equipment and higher cost [25–30].

Table 4.1 Common monitoring methods for machining conditions

Monitoring methods		Sensor	Working principles	Features
Direct methods	Optical image	Optical fiber, optical sensor, camera	Change of reflected light intensity on worn surface, image processing	The result is clear, and it is not easy to realize real-time monitoring due to cutting environment interference
	Contact	Probe, magnetic gap sensor	Detection of cutting-edge position	Simple, easy to be affected by cutting and temperature, and cannot be detected in real-time
	Radioactive technology	Radioactive elements	Detection of cutting radioactivity	Not affected by the process environment; the real-time performance is poor; protection problem; application is limited
Indirect methods	Cutting temperature	Thermo-couple	Temperature change between tool and workpiece	Used for turning, low sensitivity; can't be applied to the machining with coolant
	Ultrasound	Ultrasonic transducer	Detection changes of ultrasonic signals	Real-time detection, but affected by cutting vibration and environmental noise
	Surface roughness	Laser sensor	Detection changes of surface roughness	Used for turning, milling, drilling, etc., the sensitivity is not high
	Vibration	Accelerometer, vibration sensor	Detection changes of vibration	Sensitive, real-time detection, wide application; interference of self-excited vibration and environmental noise

(continued)

Table 4.1 (continued)

Monitoring methods		Sensor	Working principles	Features
	Cutting force	Piezoelectric, strain force sensor	Measuring changes of cutting forces and cutting component forces	Sensitive and widely used, but it is difficult to achieve dynamic test power in practice
	Power	Power sensor, transformer, power meter	Detection of motor power or current changes	Used in turning, milling, drilling, etc., low cost, easy to use, real-time detection; good application prospects
	Acoustic emission	Acoustic emission sensor	Detection of acoustic emission signals	Used in turning, milling, drilling, etc., sensitive, real-time, easy to use, moderate cost; great application prospects

The motor power/current signal has a close relationship with the tool wear and cutting force in processing. The main advantage is that the measuring instrument will not hinder the machining processing, monitoring in real-time, and economical and convenient [12]. The advantage of optical images is that the tool wear conditions can be directly obtained through image processing. Compared with other indirect methods, the results are more accurate. The disadvantage is that it is difficult to photograph the moving tool or workpiece in processing, which is not suitable for real-time monitoring. Affected by surface oil, cutting fluid, and light, the accuracy of the results is indelible.

Reliable processing conditions monitoring is impossible to achieve with only a single type of signal feature. The indirect method is convenient for real-time monitoring in processing. The direct measurement method has a high accuracy rate. Through the analysis and comparison of the above monitoring methods and combining the advantages, indirect monitoring based on vibration signals and power signals is chosen as the measurement method, as well as the machine vision and online detection. The measurement method provides real-time monitoring, which reflects the processing conditions and processing quality, especially the tool wear conditions. The signal sources of these indirect and direct monitoring methods, as well as the information feature extraction and fusion calculation related to processing conditions, affect the correct identification of working conditions, which are key issues that need to be resolved.

References

1. Teti R, Jemielniak K, O'Donnel G, Dornfeld D (2010) Advanced monitoring of machining operations. CIRP Ann Manuf Techn 59(2):717–739
2. Zhou Z, Chen Y (1999) The monitoring and fault diagnosis of modern manufacturing systems. Huazhong University of Science and Technology Press
3. Grzesik W (2017) Advanced machining processes of metallic materials: theory, modelling and applications, 2nd edn. Elsevier
4. Brecher C, Esser M, Witt S (2009) Interaction of manufacturing process and machine tool. CIRP Ann Manuf Technol 58(2):588–607
5. Ljung L (1999) System identification: theory for the user. Prentice-Hall
6. Bendat JS (2010) Random data analysis and measurement procedures. Wiley
7. Manolakis DG, Ingle VK, Kogon SM (2000) Statistical and adaptive signal processing. McGraw-Hill Education
8. Shumway RH, Stoffer DS (2017) Time series analysis and its application, 4th edn. Springer
9. Altintas Y, Yellowley I (1989) The process detection of tool failure in milling using cutting force models. ASME J Eng Ind 111:149–157
10. Kumar SA, Ravindra HV, Srinivasa YG (1997) In-process tool wear monitoring through time series modeling and pattern recognition. Int J Prod Res 35(3):739–751
11. Gradisek J, Govekar E, Grabec I (1998) Time series analysis in metal cutting: chatter versus chatter-free cutting. Mech Syst Signal Process 12(6):839–854
12. Tönshoff HK (ed) (2001) Sensors in manufacturing, vol 1. Wiley-VCH
13. Snr D (2000) Sensor signals for tool-wear monitoring in metal cutting operations—a review of methods. Int J Mach Tools Manuf 40(8):1073–1098
14. Zhou Y, Xue W (2018) Review of tool condition monitoring methods in milling processes. Int J Adv Manuf Technol 96:2509–2523
15. Kuntoglu M, Saglam H (2020) Investigation of signal behaviors for sensor fusion with tool condition monitoring system in turning. Measurement 108582
16. Özel T, Nadgir A (2002) Prediction of flank wear by using back propagation neural network modeling when cutting hardened H-13 steel with chamfered and honed CBN tools. Int J Mach Tools Manuf 42:287–297
17. Bhattacharyya P, Sengupta D, Mukhopadhyay S (2007) Cutting force-based real-time estimation of tool wear in face milling using a combination of signal processing techniques. Mech Syst Signal Process 21(6):2665–2683
18. Brinksmeier E, Preuss W, Riemer O, Rentsch R (2017) Cutting forces, tool wear and surface finish in high speed diamond machining. Precis Eng 49:293–304
19. Zhu KP, Zhang Y (2019) A generic tool wear model and its application to force modeling and wear monitoring in high speed milling. Mech Syst Signal Process 115(15):147–161
20. Sevilla P, Robles J, Muñiz J, Lee F (2015) Tool failure detection method for high-speed milling using vibration signal and reconfigurable bandpass digital filtering. Int J Adv Manuf Technol 81(5–8):1–8
21. Zhou Y, Liu X, Li F, Sun B, Xue W (2015) An online damage identification approach for numerical control machine tools based on data fusion using vibration signals. J Vib Control 21(15):2925–2936
22. Aghdam B, Vahdati M, Sadeghi M (2015) Vibration-based estimation of tool major flank wear in a turning process using ARMA models. Int J Adv Manuf Technol 76:1631–1642
23. Dimla DE (2002) The correlation of vibration signal features to cutting tool wear in a metal turning operation. Int J Adv Manuf Technol 19:705–713
24. Kataoka R, Shamoto E (2019) Influence of vibration in cutting on tool flank wear: Fundamental study by conducting a cutting experiment with forced vibration in the depth-of-cut direction. Precis Eng 55:322–329
25. Bhuiyan M, Choudhury IA, Dahari M, Nukman Y, Dawal S (2016) Application of acoustic emission sensor to investigate the frequency of tool wear and plastic deformation in tool condition monitoring. Measurement 92:208–217

26. Chiou RY, Liang SY (2000) Analysis of acoustic emission in chatter vibration with tool wear effect in turning. Int J Mach Tools Manuf 40:927–941
27. Maia LHA, Abrao AM, Vasconcelos WL, Sales WF, Machado AR (2015) A new approach for detection of wear mechanisms and determination of tool life in turning using acoustic emission. Tribol Int 92:519–532
28. Wang C, Bao Z, Zhang P, Ming W, Chen M (2019) Tool wear evaluation under minimum quantity lubrication by clustering energy of acoustic emission burst signals. Measurement 138:256–265
29. Jemielniak K, Arrazola P (2008) application of AE and cutting force signals in tool conditionmonitoring in micro-milling. CIRP J Manuf Sci Technol 1:97–102
30. Pechenin V, Khaimovich A, Kondratiev A, Bolotov M (2017) Method of controlling cutting tool wear based on signal analysis of acoustic emission for milling. Procedia Eng 176:246–252

Chapter 5
The Smart Machining System Monitoring from Machine Learning View

5.1 The Condition Monitoring Methods

The fundamental task of machining system monitoring is to identify the condition of the process to be detected according to the state information. There are many identification methods, which can be divided into empirical analysis method, mechanism analysis method, data-driven method, and data-mechanism fusion method. The mechanism-based analysis method is mainly to describe the failure or defect mechanism of the equipment by building a mathematical model, which has been introduced in Chaps. 2, 3 and 4. The fusion method refers to the combination of mechanism analysis and the data-driven method. Although the fusion approaches can take the advantages of the both sides, the modeling process is complicated and only limited studies reported at present. Some specific examples of cyber-physical fusion system will be introduced in Chaps. 10, 11 and 12. This chapter introduces the data-driven intelligent modeling and analysis methods.

5.1.1 Empirical Analysis

The empirical analysis method is the simplest and rough analysis method. The key is to establish the criterion. As long as compare the actual value of the characteristic parameters directly with the established criterion, the condition of the equipment, even the defect position, can be identified immediately according to the difference [1]. For example, the stability of machine tool and tool wear can be identified according to the abnormal amplitude of vibration signal. The criteria of the characteristic parameter (reference value) used to identify the state of the equipment can be divided into the following three categories:

K. Zhu, *Smart Machining Systems*, Springer Series in Advanced Manufacturing,
https://doi.org/10.1007/978-3-030-87878-8_5

(1) **Absolute criterion**

The absolute criterion comes from a large number of statistical data and theoretical analysis of the monitored objects, which is usually promulgated by enterprises, industries, or countries. Such standards are objective, comprehensive, and reliable. Users should pay attention to the scope of application and monitoring methods of the criterion, and revise them according to the actual situation.

(2) **Relative criterion**

It is widely used, but only suitable for establishing criterions for a single equipment.

(3) **Analogy criterion**

According to the monitoring data of many types of equipment with the same specification under the same working conditions, the criterion of reference value under normal conditions is established. Since it is not sure that the reference value reflects the normal condition of the equipment, it should be very careful when adopting it.

The above three empirical analysis methods are widely used in enterprises. The analysis results are valuable to be used as a reference but need to be confirmed manually because the accuracy is not high and it is easy to form a wrong judgment.

5.1.2 *Statistical Method*

Generally, the monitoring signal of machining process is a random signal, which obeys the statistical law and can be described by a probability density function. The identification method based on statistical law is more scientific and reliable, but it requires certain prior knowledge and a large amount of calculation. In this sense, probability and statistics is also an empirical analysis method.

There are many methods to identify machining conditions according to statistical law, which can be summarized into two categories from the perspective of geometry: posterior probability criterion and distance criterion.

Posterior probability criteria: The state domain is divided into several disjoint areas corresponding to the various states of the equipment, according to the distribution of the parameters. The state of equipment is determined by the maxima posterior probability of the area where the check point falls into.

Distance criterion: N feature parameters describing the machining condition are regarded as a point (or an N dimensional vector) in the N dimensional state domain. The point corresponding to the aggregation center of a certain state is called as a reference point; the point corresponding to the actual monitoring state of the equipment is called the check point. Which reference point the check point is closer to, the equipment belongs to which state.

The above two criteria can also be expressed in functional form, called discriminant functions, which are multivariate functions with feature parameters as independent variables [2]. The distance criterion is based on the distance formula between

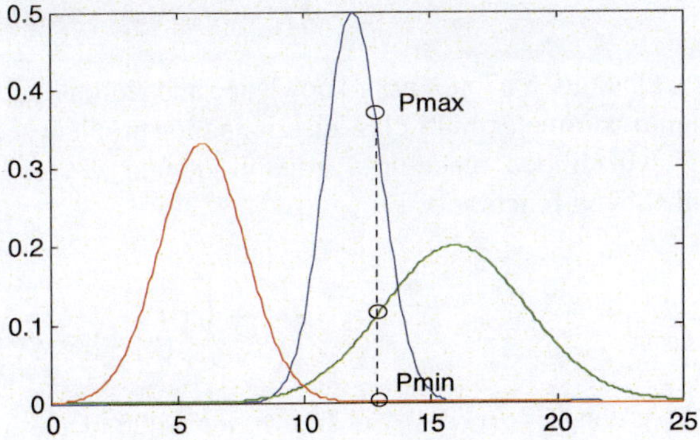

Fig. 5.1 Probability density of vibration amplitude

the check point and the reference point. The check point belongs to the state with the minimum value of the discriminant function. The posterior probability criterion takes the probability that the check point belongs to each state domains as the discriminant function (also known as likelihood discrimination) [3]. The check point belongs to the space with the highest possibility.

(1) **Posterior probability criterion**

The establishment of the posterior probability criterion requires complete prior knowledge and tedious calculations. For example, during machining, the vibration amplitude x of the machine tool is used to determine whether the machining condition is normal. The probability density function (conditional probability density function) $p(x/D_1)$ of vibration amplitude x of machine tool under normal condition D_1 and $p(x/D_2)$ under abnormal condition D_2 is statistically analyzed according to the historical data, as shown in Fig. 5.1.

In case the monitoring data x of the equipment detected is available, the probability $P(D_1/x)$ of state D_1 and the probability $P(D_2/x)$ of state D_2 can be calculated, which are called posterior probabilities. According to Bayes formula, the posterior probability [3] is:

$$P(D_i/x) = \frac{p(x/D_i)P(D_i)}{\sum_{i=1}^{2} p(x/D_i)P(D_i)} \; (i = 1, 2) \tag{5.1}$$

Then the higher posterior probability $P(D_i/x)$ determines the machining states.

The above example is the process to establish the state domain of single-parameter two-class discrimination according to the Bayes formula. For multi-parameter and multi-class discrimination, the process of establishing the state domain according to the Bayes formula is similar. The N states of the machine tool can be expressed as $D_i(i = 1, 2, ..., N)$ and the k features can be expressed as $x_j(j = 1, 2, ..., k)$. The conditional probability density function of the equipment state is $P(x_1, x_2, x_3, ..., x_k/D_i)$. The prior probability of equipment state is $P(D_i)$. According to Bayes formula, the posterior probability of the state can be calculated as

$$P(D_i/x_1, x_2, x_3, ..., x_k) \tag{5.2}$$

Obviously, a lot of prior knowledge and tedious calculation is needed to solve the multi-parameter multi-class discrimination problem according to the Bayes formula. The likelihood function is usually adopted to simplify the calculation [3]. The likelihood function is:

$$L_i = \prod_{j=1}^{k} P(x_j/D_i) \tag{5.3}$$

As long as $P(x_j/D_i)$ is known, the value of the likelihood function L_l can be calculated to represent the approximate value of the posterior probability. The state domain can be established to diagnose the condition of the machine tool through the above steps.

(2) **Distance criterion**

It can be seen from the above study, it is very difficult to establish regional discrimination criteria. This is not only because the prior knowledge is not clear, but also the calculation is complicated. In contrast, the distance criterion is more practical because the calculation is simple and the physical concept is clear. The disadvantage of the distance criterion is that it does not consider the probability of various states and the loss caused by misjudgment.

There are many ways to establish the reference point and calculate the distance between the reference point and the check point, so there are many distance discrimination criteria. The simplest and most direct is the Euclidean distance, which is the sum of the squares of the coordinate difference between the check point and the reference point.

Suppose the reference point is $X_0 = [x_{01}, x_{02}, ..., x_{0n}]^T$ and the check point is $X = [x_1, x_2, ..., x_n]^T$, the Euclidean distance can be expressed as:

$$D = \sum_{i=1}^{n} (x_i - x_{0i})^2 = (X - X_0)^T (X - X_0) \tag{5.4}$$

If the Euclidean distance criterion is taken to identify the equipment condition in Example 1, we only need to determine the positions of the two reference points according to the statistical average value of the features of the two states, so it is much easier to determine the state of the check point. In order to avoid the influence of the magnitude of features, the feature values should be normalized before the distance calculation. If the contribution of each feature is weighted before distance calculation, better results can be obtained.

Euclidean distance is one of the simplest geometric distance discriminant functions, including the extended Mahalanobis distance determined based on the statistical distribution of features and the information distance function constructed based on the information measurement of features, such as Kullback distance, etc.

5.1.3 *Intelligent Method*

The condition monitoring methods based on artificial intelligence mainly rely on pattern recognition and machine learning algorithms. Pattern recognition is an algorithm to describe, identify, classify and learn things or phenomena according to the characteristics of objects (equipment). It is suitable for the processes with obvious and easy-to-extract features, sufficient prior information, and requiring human control. In recent years, with the improvement of computing power, machine learning is in the ascendant. Different from pattern recognition, machine learning can actively extract the inherent laws for analysis and prediction, and directly build the problem model by training on the known samples. It is especially suitable for situations where the sample is huge, the problem is complicated and difficult to manually intervene. It can be considered that pattern recognition focus on extracting a set of features that can distinguish each pattern, while machine learning focuses on selecting an appropriate algorithm to build the current model. Although the research methods and strict definitions of pattern recognition and machine learning are different, there is no difference in practical application, so no distinction is made in this book. Machine learning generally includes the following steps:

(1) **Data preprocessing**: The data obtained from the actual monitoring often has strong randomness, more or less contains various types of noise generated by environment and monitoring [4, 5]. The noise increases the uncertainty of the model and the complexity of the problem. It is often necessary to carry out the noise filtering in advance to speed up the model convergence. In addition, pattern recognition is only effective under the premise that the data obey the specific distribution. However, if the premise hypothesis is not true, it is difficult to guarantee the predictive effect of the model or even the validity of the model. Therefore, data preprocessing is usually required to map the sample space to a feature space which is convenient for processing based on prior knowledge.

(2) **Feature extraction and selection**: The effects of feature extraction and selection are basically the same. Both of them are trying to transform the sample data into features that are easy to recognize, but their basic ideas are different. Feature extraction adopts the method of combination mapping, which maps the features or sample data of the original space to a new feature space through the transformation function. But feature selection is to select a new feature subset from the original feature set without changing the original feature space. The identification ability of the whole model is directly determined by the quality of the features. The specific implementation details will be given in this chapter.

(3) **Machine learning model (classifier)**: Learn and classify based on the extracted features. Machine learning can be divided into the following 5 categories according to learning methods [6]:

Supervised learning: It is suitable for situations where both sample data and labels are given. The label is the expected output of the algorithm which can be understood as the supervision information given by the designer. The essence of supervised learning is to fit a function to approximate the mapping relationship from sample data to label as much as possible. When the labeled data is sufficient, supervised learning is the simplest, most direct, and efficient machine learning algorithm, which is often used to solve regression and classification problems.

Unsupervised learning: It is suitable for situations where the sample data is not manually labeled. The goal of unsupervised learning is only a qualitative expected direction, which is vaguer than supervised learning, such as clustering. In order to quantitatively point out the direction of model training, it is generally necessary to give a discriminant function or evaluation function in advance to evaluate the quality of the model. Due to the diverse distribution of actual signals, it is difficult to accurately model real problems. The result of unsupervised learning is often determined by the selection of evaluation function.

Semi-supervised learning: This is a method between supervised learning and unsupervised learning. It is suitable for the case that the labeled samples are insufficient to make the network converge, caused by the high labeling cost or the infeasibility of large-scale labeling. Now researchers generally believe that unlabeled data cannot point out the optimization direction of the model, but implies the information of data distribution. The difficulty lies in how to make the model learn the knowledge of unlabeled data based on the information of the supervised data. It should be noted that semi-supervised learning is usually carried out under certain assumptions, such that the distribution of labeled samples and unlabeled samples are the same. If the assumption is not correct, the effect of semi-supervised learning is difficult to guarantee.

Reinforcement learning: The trained model is used to predict in the actual environment and is modified in real-time according to the actual environment at the same time.

5.2 Smart Machining System Monitoring (MSM) as a Machine Learning Problem

The problem of MSM can be considered as a typical pattern recognition problem. For pattern recognition, a given pattern is to be assigned to one of C categories $\omega_1, \omega_2, \ldots\omega_c$ based on a vector of d feature values $y = (y_1, y_2, \ldots, y_d)$. The features are assumed to have a probability density function conditioned on the pattern class. Thus, a pattern vector x belonging to class ω_i is viewed as an observation drawn randomly from the class-conditional probability function $p(y|\omega_i)$. We can formally specify the objectives of MSM to be a search for the most probable state ω_i given the extracted measurable signal feature y. This is a dynamic inference problem since we do not estimate the tool state only with prior knowledge, but also adapt to the current features. This is somewhat of Bayes inference [3]: assign input pattern y to class ω_i if

$$P(\omega_i|y) > P(\omega_j|y) \; for\, all\; j \neq i \tag{5.5}$$

Hence, the aim of MSM is to find,

$$\text{MSM} : \arg\max_{i} p(w_i|y) \tag{5.6a}$$

or in the physical form

$$\text{MSM} : \arg\max_{state\, i} p(machining\, state|signal\; features) \tag{5.6b}$$

Several state estimation approaches such as Bayes decision rule, neural networks, clustering, and hidden Markov models suitable for this purpose.

5.2.1 Feature

The feature is a quantitative description of a state of interest. It is usually arranged in the form of a feature vector as $y = [y_1, ...y_n]^T$, where y_1,y_2, ..., y_n are the features. Depending on the measurements of the class, features can be either discrete numbers or real continuous values. The requirement on features is that the features can reflect the characteristics of desired states and differ from those of other states to the largest extent.

5.2.2 State

State (or class) is a set of patterns that share some common properties. The feature vectors of the same type of a class will naturally form one set. Due to the diversity of the classes, the features extracted from the same type of classes are seldom identical. This can be interpreted as clusters of points in an n-dimensional space, which are called distributions of states. Since the purpose of pattern recognition is to classify these features, the distributions of classes are desired to be separable and not empty. Suppose we have K classes, in a mathematical form, the requirement is:

$$\omega_k \neq \phi\, k = 1, ...K; \tag{5.7}$$

and

$$\omega_k \cap \omega_l = \phi\, k \neq l \in \{1, ...K\} \tag{5.8}$$

5.2.3 Classifier

Classifier is a series of input–output functions $g_i(x, \Lambda_j)$, $i = 1, ..., K$, which are *discriminant* between states. In a discriminant function $g(x, \Lambda)$, x is the input vector and Λ is the parameter set of the class. Each discriminant function will output a value. Based on these values, the classifier then assigns x to one of the classes following the decision rule:

$$x \in Class\ i\ if\ g_i(x, \Lambda_i) = \max_{for\ all\ j \in K} g_i(x, \Lambda_j) \tag{5.9}$$

There are various classifiers developed in machine learning [6, 7], based on the cost function and classification strategies applied.

5.3 The MSM System Content

Pattern recognition involves mathematical and technical aspects of classifying different objects through their observable information, such as grey levels of pixels for an image, energy levels in the frequency domain for a waveform, and the percentage of certain contents in a product. The objective of an MSM system can be achieved in a three-step procedure, as shown in Fig. 5.2.

5.3.1 Signal Preprocessing

In today's big data era, the amount of process monitoring datasets is often extremely large. If directly feed the sensor signal into the model, the calculation amount tends to increase sharply and the model is difficult to converge. Moreover, the signal in the actual environment is often complex and changeable, which is not conducive to the unified processing of the model. Generally speaking, it is necessary to map the signal distribution to the desired distribution (normalization processing) through preprocessing first to reduce the computational difficulty of feature extraction. This issue will be discussed in the following chapters. The first step is signal preprocessing. The sensor signal $x(t)$ is firstly pre-processed to normalize the data and remove the noise. The purpose of signal pre-processing is to prepare the data for feature extraction.

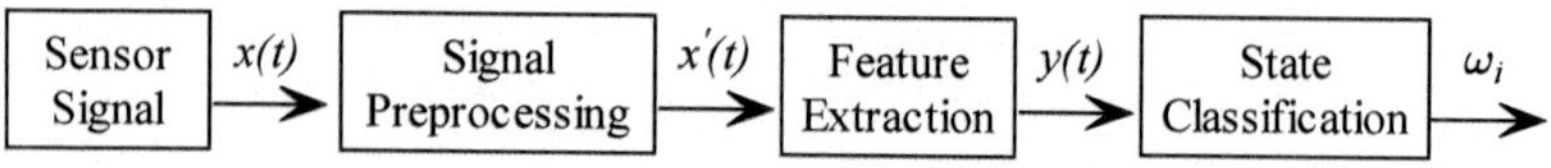

Fig. 5.2 Data flow in MSM system

Preprocessing generally requires completing the following two tasks:

(1) **Signal denoising**. If we have prior knowledge about noise, such as the frequency band of the noise, we can select a suitable filter to filter out these noises. If the frequency of the noise is much higher (or lower) than the frequency of the effective signal, then a low-pass (or high-pass) filter can be used to filter out the noise. For example, since the characteristic frequency of the acoustic emission signal is in a very high-frequency band, a high-pass filter can be used to filter the noise. But if the characteristic frequency of noise in some complex problems continues to change over time, an adaptive filter can be considered.

When the energy of the signal is very weak or the characteristics of the noise are very close to the signal, it is difficult for band filtering to play a role. For example, signals with small amplitudes such as vibration in the laser forming process are difficult to separate from noise. In this case, more advanced noise filtering methods are required. Independent component analysis (ICA) is a noise filtering algorithm commonly used in this situation [3]. It considers that the sampled signal is the instantaneous superposition of vibration signal and noise signal, so the denoising process is equivalent to separating noise source and force signal source, which is also called blind source separation (BSS). The advantage of blind source separation is that there is no need to assume a source distribution, which overcomes the limitation that traditional methods are only applicable to specific noise distribution of the signal.

(2) **Normalization.** As the machining status changes, the distribution range of features will change dynamically. The numerical difference between features is very large. Even though some features cannot reflect the state well, they have more influence than others in the discriminant function due to their high amplitude, which will affect the discriminant ability of the model. Normalization can map features to similar distribution spaces and avoid the influence of numerical differences of features themselves.

There are two ways of standardization. One is based on the mean and variance of the signal. In this method, the mean value of the signal is set to zero and the standard deviation is set to one [5], as shown in the following formula.

$$x_i' = \frac{x_i - \mu_i}{\sigma_i} \tag{5.10}$$

where x_i' is the normalized value, x_i represents the i-th data point of the original signal, μ_i is the mean value of the whole signal, and σ_i is the standard deviation of the whole signal.

The other normalization method is based on the amplitude of the signal, which simply divides all data points by the same value to limit all values to the range of $[-1, 1]$:

$$x = \frac{x}{\sqrt{\sum_{i=1}^{d} (x_i)^2}} \tag{5.11}$$

If the difference of signal amplitude is obvious, logarithmic transformation of the whole signal can be considered to reduce the dynamic range of the signal. If possible, signal preprocessing should be carried out before standardization as much as possible to eliminate singular values in the signal. When the dimension of the input signal is not higher than 3, the influence of singular points can be eliminated through human observation. However, when the signal is higher than 3 dimensions and the signal distribution cannot be represented by drawing, the singular points are difficult to remove. In this case, it is generally assumed that the data obeys Gaussian or approximate Gaussian distribution. The Mahalanobis distance [3] is used to measure the difference between each data point and the whole.

The simplest method is to set the corresponding Mahalanobis distance threshold of the sample points for each class. The points exceeding the threshold are regarded as singular points. The Mahalanobis distance of sample x belonging to class ω_j can be expressed as [3]:

$$M_j = (\mathrm{x} - \mu_j)^T \Sigma_j^{-1} (\mathrm{x} - \mu_j) \tag{5.12}$$

where μ_j represents the mean of class ω_j and Σ_j represents the variance of class ω_j.

(3) **Data compression**. In the process of developing any real-time monitoring system, the first consideration is the response speed, or to say the time required for control decisions, which mainly depends on the computational efficiency. In order to improve computational efficiency, data simplification or feature selection can be used to eliminate the feature components with the least relevant information to reduce the dimension of the data. But if want to speed up the calculation while avoiding information loss, data compression can be used. It should be noted that the main purpose of data compression is to reduce the dimension, not to reduce the misjudgment rate like feature extraction and selection. Therefore, data compression may not be the best transformation, or to say it can't achieve the best discrimination.

5.3.2 *Feature Extraction and Selection*

The next step is feature extraction and selection. Some interference information of the signal is filtered out, during preprocessing mentioned above, which is helpful for the feature extraction and feature selection. The process of feature extraction and feature selection to be described here is essentially a mathematical transformation of dimension reduction, which aims to search for the features with the best identification ability of the state (Fig. 5.3).

(1) **Feature extraction**

The purpose of feature extraction is to make the input data more suitable for pattern classifiers and/or reduce the dimensionality of the input data vectors. Information

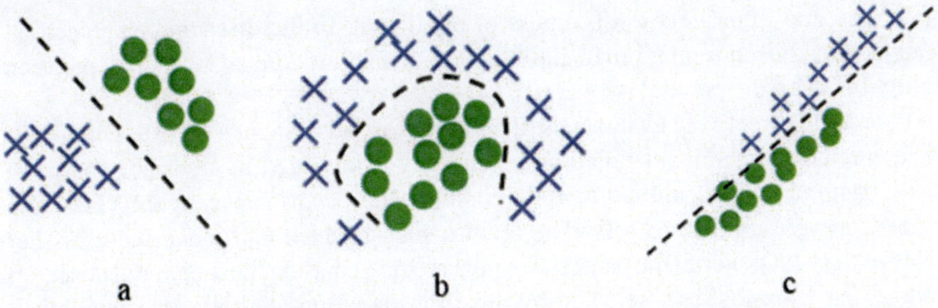

Fig. 5.3 Features with different characteristics, where **a** linear separable, **b** nonlinear separable, and **c** highly linear correlation

relevant to pattern classification is extracted from $x(t)$ to a feature vector $y(t)$ by a feature extractor. For example, the components of the signal that distinguish the various tool wear classes will be hidden in features that characterize the normal operating condition of the structure, particularly when the tool wear is not severe yet. The task of feature extraction is to enhance the characteristics of the various tool wear classes and suppress or filter off the normal background [8].

From the mathematical perspective, feature extraction is to reduce the dimension of the original feature. Namely, the pattern vector $\mathbf{U} = [\alpha_1, \alpha_2 ... \alpha_n, \beta_1, \beta_2 ... \beta_m, \sigma_t^2]^T$ is transformed to a pattern vector $\mathbf{V} = [\alpha_1^{'}, \alpha_2^{'} ... \alpha_p^{'}, \beta_1^{'}, \beta_2^{'} ... \beta_q^{'}, \sigma_t^{'2}]^T$, $(p < n, q < m)$ with lower dimension. Feature extraction should ensure that the pattern vector includes the dominant feature of the pattern vector. Hence, there exists certain kind of restriction on the transformation matrix **A**.

In the TCM [8], the task of feature extraction is to identify and enhance the characteristics of the various tool wear classes from the suitably processed sensor signals. Cutting forces are decomposed by the wavelet packet into five levels with 32 wavelet packet coefficients, and the features are represented by the mean energy of the respective packet node. The features from both wavelet packets and from the time domain, i.e. mean, standard deviation, skew, kurtosis, and dynamic component, are combined to form a 37-dimension feature vector. There are many feature extraction approaches with signal processing and pattern analysis techniques, such as PCA and linear discriminant analysis, which will be discussed in the later chapters.

(2) **Feature selection**

The feature vector is usually a high dimensional vector containing a lot of redundant or state-independent information, which greatly increases the decision-making calculation and system complexity. In order to reduce the computational cost, it is necessary to remove some features according to the performance indicators, which is often called feature selection. In many studies, the distance between classes is used as an index for feature selection, and the features are sorted accordingly. It is generally believed that the greater the distance between classes, the stronger the ability to distinguish features. Fisher linear discriminant analysis (FDA) and principal component analysis (PCA) are common algorithms that reduce the dimensionality of vectors and improve the robustness of prediction. In the discriminant selection, features are chosen to maximize class separation and are ranked by their separation ability between different classes.

In order to effectively evaluate the discriminative performance of different groups of features, a discriminant function is generally determined in advance. The most direct method of discrimination is to compare the performance of the classifiers trained on various feature sets. The feature subset which can make the classifier achieve the best generalization performance is selected as the best feature subset. Of course, the feature subset selected by this discrimination method corresponds to a specific problem and classifier. If the classifier or dataset changes, the best feature subset may also change. For example, using neural networks or decision trees as classifiers may result in different feature subsets. Even adjusting the structure of the neural network may affect the features selected.

5.3.3 State Classification

The final stage is state classification. The classifiers decide the machining state on the basis of the given feature vector based on a classification measure. Feature vector y is assigned to one of the K classes, ω_1, ω_2, ... ω_K, by the classifier based on a certain type of classification criteria, such as distance, likelihood and Bayesian, over class models. The details of classification approaches, such as neural networks, support vector machine and deep networks are discussed in the later sections.

For example, in the TCM [9], the Continuous Hidden Markov models (HMMs) are adapted for stochastic modeling of the tool wear process in micro-milling, and estimation of the tool wear state based on the output probability of the HMMs given the cutting force features. The framework of HMMs for MSM is illustrated in Fig. 5.4. The features from both time and wavelet domains are extracted and selected to train the HMMs. An HMM is built for each tool state for classification, and each HMM is trained with its own feature set. Once the models are trained, the $\boldsymbol{\lambda} = (\boldsymbol{\pi}, \boldsymbol{A}, \boldsymbol{c}, \boldsymbol{\mu}, \boldsymbol{\Sigma})$ is obtained that represent different tool wear states. In the tool state recognition state, the extracted features from an unknown state are extracted and input to the HMMs to match the trained states. Different tool wear states are modeled as separate HMMs. The tool wear state is then classified by the HMM that has the maximum probability to indicate the test features.

5.4 Feature Selection Method

The machining state feature vectors extracted from various sensor signal models are often high-dimensional eigenvectors, if this high-dimensional eigenvector is directly identified, it will not only take a lot of time but also the interference and noise in each component of the eigenvector will affect the recognition effect. Therefore, the

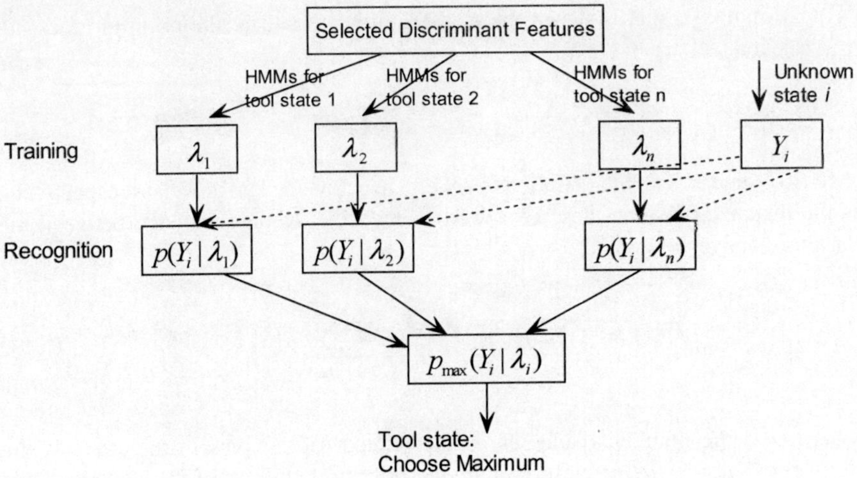

Fig. 5.4 Framework of Hidden Markov models for tool wear classification

acquired high-dimensional eigenvectors must be optimized to obtain the optimal feature group, and the dimension of the eigenvector can be reduced.

5.4.1 Effective Criteria for Monitoring Features

In the process of selecting the best feature group from a large number of processing state features, there must be a criterion to measure the validity of the feature group for recognition, the criterion shall meet the following requirements:

(1) It has a monotonous relationship with the error probability (or the upper and lower limits of the error probability), so that the effect of taking the maximum value of the judgment is generally less error probability.

(2) It is additive when features are independent, which is:

$$J_{ij}(x_1, x_2, \cdots, x_d) = \sum_{k=1}^{d} J_{ij}(x_k) \tag{5.13}$$

In the formula, J_{ij} is the separable criterion function of the i-th class and the j-th class, the larger J_{ij}, the greater the degree of separation of the two categories; $x_1, x_2, \cdots, x_d$ are random variables of the corresponding feature of a certain category.

(3) Measurement characteristics, which are: $J_{ij} > 0$, when $i \neq j$; $J_{ij} = 0$, when $i = j$; $J_{ij} = J_{ji}$.

(4) Monotonic, that is, when the new features are added, the criterion does not decrease,

$$J_{ij}(x_1, x_2, \cdots, x_d) \leq J_{ij}(x_1, x_2, \cdots, x_d, x_{d+1})$$

Let $x_k^{(i)}$ and $x_l^{(j)}$ be the D-dimensional eigenvectors in class and class respectively, is the distance between these two vectors, then the average distance between the various eigenvectors is

$$J_d(x) = \frac{1}{2}\sum_{i=1}^{c} P_i \sum_{j=1}^{c} P_j \frac{1}{n_i n_j} \sum_{k=1}^{n_i}\sum_{l=1}^{n_j} \delta\left(x_k^{(i)}, x_l^{(j)}\right) \tag{5.14}$$

where C is the number of classes, is the number of samples in the class, is the number of samples in the class, and are the prior probabilities of the corresponding category. There are many distance measurements between the two vectors in the multidimensional space. In the Euclidean distance, there are

$$\delta\left(x_k^{(i)}, x_l^{(j)}\right) = \left(x_k^{(i)} - x_l^{(j)}\right)^T \left(x_k^{(i)} - x_l^{(j)}\right) \tag{5.15}$$

The mean vector of the i-th sample set is denoted by m_i,

$$m_i = \frac{1}{n_i}\sum_{k=1}^{n_i} x_k^{(i)} \tag{5.16}$$

Use m to represent all kinds of sample lumped mean vectors

$$m = \sum_{i=1}^{c} P_i m_i \tag{5.17}$$

The following formula can be obtained

$$J_d(x) = \sum_{i=1}^{c} P_i \left[\frac{1}{n}\sum_{k=1}^{n_i}\left(x_k^{(i)} - m_i\right)^T\left(x_k^{(i)} - m_i\right) + (m_i - m)^T(m_i - m)\right] \tag{5.18}$$

The mean square distance representing the mean value vector can be obtained by weighted averaging of the prior probability

$$\sum_{i=1}^{c} P_i (m_i - m)^T (m_i - m) = \frac{1}{2}\sum_{i=1}^{c} P_i \sum_{j=1}^{c} P_j (m_i - m_j)^T (m_i - m_j) \tag{5.19}$$

Let

$$S_b = \sum_{i=1}^{c} P_i (m_i - m)(m_i - m)^T \tag{5.20}$$

As well as

$$S_w = \frac{1}{2} \sum_{i=1}^{c} P_i \frac{1}{n_i} \sum_{k=1}^{n_i} \left(x_k^{(i)} - m_i\right)\left(x_k^{(i)} - m_i\right)^T \tag{5.21}$$

Then

$$J_d(x) = tr(S_w + S_b) \tag{5.22}$$

The optimal monitoring feature is chosen by selecting the d features x^* from the acquired D-dimensional feature space so that the average distance $J(x)$ between the samples of the C categories is the largest, which is

$$J\left(x^*\right) = \max_x J(x) \tag{5.23}$$

$$J(x) = \frac{1}{2} \sum_{i=1}^{c} P_i \sum_{j=1}^{c} P_j \frac{1}{n_i n_j} \sum_{k=1}^{n_i} \sum_{l=1}^{n_j} \delta\left(x_k^{(i)}, x_l^{(j)}\right) \tag{5.24}$$

where n_i is the number of training samples for class C_i in design set S; P_i is the prior probability of class i. When these prior probabilities are unknown, the number of training samples can also be used to estimate, which is:

$$\hat{P}_i = \frac{n_i}{n} \tag{5.25}$$

where n is the total number of samples.

In the various distance measures, since Euclidean distance is easy to analyze and calculate in many cases, Euclidean distance measurement is the main method of feature selection. The formula is:

$$\delta(x_k, x_l) = \left[\sum_{j=1}^{d} \left(x_{kj} - x_{lj}\right)^2\right]^{\frac{1}{2}} = \left[(x_k - x_l)^T (x_k - x_l)\right]^{\frac{1}{2}} \tag{5.26}$$

When the expected value is used instead of the sample mean, a distance measure is obtained:

$$J_1(x) = tr(S_w + S_b) \tag{5.27}$$

where S_b is the inter-class dispersion matrix and S_w is the intra-class dispersion matrix. The ideal monitoring feature should make the degree of dispersion as large as possible, the intra-class dispersion as small as possible. In addition to $J_1(x)$, there are usually the following criteria:

$$J_2 = tr\left(S_w^{-1} S_b\right) \tag{5.28}$$

$$J_3 = \ln\left[\frac{|S_b|}{|S_w|}\right] \tag{5.29}$$

$$J_4 = tr\,S_b \big/ tr\,S_w \tag{5.30}$$

$$J_5 = S_w + |S_b| \big/ |S_w| \tag{5.31}$$

5.4.2 Optimal Monitoring Feature Group Selection

A set of optimal features of d ($D > d$) is selected from a set of features of D, and all possible combinations are

$$n = C_D^d = \frac{D!}{(D-d)!d!}$$

If you use the exhaustive method to compare the various possible combinations of J are obtained to compare, to find the optimal characteristics of the group, it is prone to a combination of explosion, resulting in too much computation cannot be achieved. To effectively reduce the amount of computation, the usual search technique is to add or remove some feature from the existing feature until the number of features is equal to d. If the number of features from zero gradually increased, known as the bottom down method; On the contrary, if the number of features from the D began to gradually reduce, known as the top-down method. Specific algorithms are:

(1) **Separate optimal feature combination method**. The simplest way is to calculate the criteria for the use of the individual values and to sort, take the former d as a result of the selection, but even if each feature are statistically independent, this result is not necessarily the best results, only when the separability criterion J can be written as follows

$$J(x) = \sum_{i=1}^{D} J(x_i)$$

This method can choose a set of optimal features.

(2) **Sequential Forward Selection (SFS)**. This is the easiest top-down search method, each time a feature is selected from the candidate's feature, the J value obtained when it is combined with the selected feature is maximized until the number of selected features increases to d. Suppose that k features have been chosen to form a feature group X_k of size k, and the unselected *D-k* features x_j, $j = 1, 2, \cdots, D - k$ are arranged in the same order as the J value of the selected feature, if:

$$J(X_k + x_1) \geq J(X_k + x_2) \geq \cdots \geq J(X_k + x_{D-k})$$

Then the next step of the feature group is $X_{k+1} = X_k + x_1$. At the beginning of time $X_0 = \Phi$, up to $k = d$.

The SFS method takes into account the correlation between the selected feature and the selected feature, in general, it is better than the choice of the maximum value of J when used, and the main drawback is that once a feature is selected, it cannot be removed by the feature that is added afterward.

(3) **Sequential Backward Selection (SBS)**. This is a top-down approach, from the beginning of the whole feature to remove one each time, and the feature removed should maximize the J value of the feature group that remains. For example, *k* has been removed from the feature, the remaining feature group is $\overline{X}_k$, $\overline{X}_k$ will be the feature x_j by *j* size of the order, $j = 1, 2, \cdots, D - k$. If

$$J\left(\overline{X}_k + x_1\right) \geq J\left(\overline{X}_k - x_2\right) \geq \cdots \geq J\left(\overline{X}_k - x_{D-k}\right)$$

Then

$$\overline{X}_{k+1} = \overline{X}_k - x_1$$

Compared with the sequential forward selection method, the sequential backward selection method has two characteristics:

(1) In the calculation process, it is possible to estimate the decrease in separability caused by the removal of a feature;
(2) Since the calculation of the sequential backward selection method is carried out in the high dimensional space, the computational complexity is larger than that of the sequential forward selection method. The main drawback of the sequential backward selection method is that once a feature is removed, it will no longer be considered in subsequent searches.

5.4.3 The Bidirectional Search Algorithm for Feature Selection

This method combines the advantages of the sequential forward selection method and sequential backward selection method. The so-called bidirectional search is in the process of selecting the optimal feature group, not only from the candidate feature set to select the most effective features, but also consider the selected feature set to remove the secondary features. In the bidirectional search algorithm, the feature selection and feature elimination process are repeated until no new features are selected and no secondary features are removed. The key to the algorithm is to determine the separability criteria for the state class, as well as the feature selection threshold and the culling threshold. The specific algorithm process is as follows:

(1) Determine the class separability criteria. For example, distance-based separability criterion J_3 (or J_4, J_5).

(2) Select a new feature. Suppose that k features have been chosen to form a feature group x_k of size k, sorting the candidate *D-k* features according to the J value of the selected feature, if:

$$J(X_k + x_1) \geq J(X_k + x_2) \geq \cdots \geq J(X_k + x_{D-k})$$

Then

$$J(X_{k+1})_{\max} = J(X_k + x_1)$$

And

$$J(X_{k+1})_{\max} - J_k > S_1$$

Then

$$X_{k+1} = X_k + x_1$$

where S_1 is the feature selected threshold.

(3) Remove an extra feature. From the selected feature group, sort the features by x_j by the value of J, $j = 1, 2, \cdots, k+1$,

If:

$$J(X_{k+1} - x_1) \geq J(X_{k+1} - x_2) \geq \cdots \geq J(X_{k+1} - x_{k+1})$$

Then

$$J(X_k)_{\max} = J(X_{k+1} - x_1)$$

And

$$J(X_{k+1}) - J(X_k)_{\max} < S_2$$

Then

$$X_k = X_{k+1} - x_1$$

where S2 is the feature culling threshold.

Repeat (2), (3) until the selected feature group vector X_k no longer changes, and finally output the results.

5.5 Machine Learning Method

5.5.1 *Bayesian Classifier*

Bayesian statistics allows one to use the distribution of the features for each class in determining the probability that a test feature belongs to a particular class. Applying the Bayes theorem, we can calculate from this the probability of the feature that belongs to that class. This probability can be obtained for each class. Then it can be decided that the test belongs to the class for which that feature gives the highest probability, which is similar to the posterior probability approach introduced in the Sect. 5.1.1.

Assume the prior probabilities $P(\omega_i)$ for each class, and $P(\omega_1|x)$ and $P(\omega_2|x)$ denote the conditional class densities of a feature x belonging to either class ω_1 or ω_2 respectively. The classification criterion can now be described as:

$$if\ P(\omega_1|x) > P(\omega_2|x) \text{ then decide } \omega_1, \tag{5.32}$$

$$if\ P(\omega_2|x) > P(\omega_1|x) \text{ then decide } \omega_2 \tag{5.33}$$

Bayes laws can be applied to these conditional probabilities to redefine them in terms of their density functions, which are denoted by $f(x|\omega_1)$ and $f(x|\omega_2)$. The derivation of the new classification criterion, now in terms of the conditional density functions $f(x|\omega_1)$ and $f(x|\omega_2)$ states that,

$$P(\omega_i|x) = \frac{f(x|\omega_i)P(\omega_i)}{P(x)} \tag{5.34}$$

Bayes decision rule is obtained by eliminating the denominator and is as follows:

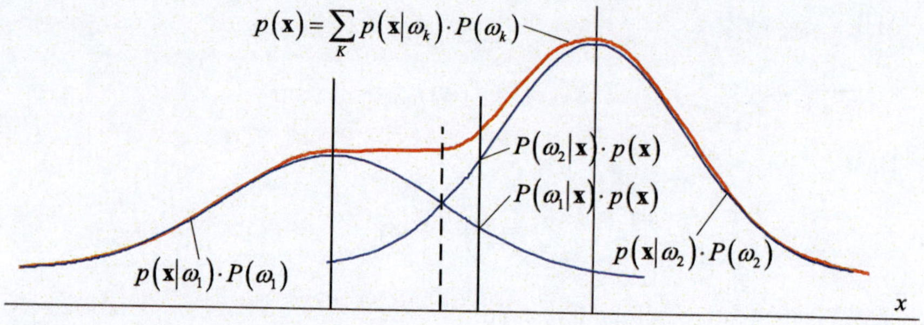

Fig. 5.5 Classification based on Bayesian inference

$$if\ P(\omega_1)f(x|\omega_1) \geq P(\omega_2)f(x|\omega_2) \text{ then } \omega_1 \tag{5.35}$$

$$if\ P(\omega_1)f(\overline{x}|\omega_1) < P(\omega_2)f(\overline{x}|\omega_2) \text{ then } \omega_2 \tag{5.36}$$

Figure 5.5 illustrated this decision-making, based on the likelihood comparison. This criterion can be generalized to situations involving more than two classes and multiple dimensional feature spaces. Let k be the number of classes involved and using the respective conditional density functions $f(x|\omega_i)$. The Bayesian classification can now be written as follows,

$$if\ f(x|\omega_i)P(\omega_k) = \underset{k=1,k}{Max}\{f(x|\omega_k)P(\omega_k)\}\ then\ select\ \omega_i \tag{5.37}$$

5.5.2 Fisher Linear Discriminant

The goal of linear discriminant analysis is to use a linear transformation to project the set of raw testing data vectors onto a vector space of lower dimension such that some metric of class discriminant is maximized [3]. The technique attempts to maximize the between-class covariance S_b and minimize the within-class covariance S_w for a set of features. The metric most often used is the ratio of the between-class scatter to the within-class scatter and that can be expressed as,

$$trace\{\frac{S_b}{S_w}\} \tag{5.38}$$

The LDA transform can be truncated to select only the n largest eigenvalues, i.e., the transformed features with the largest ratio of between-class covariance to the within-class covariance. By discarding the lower order LDA components, the dimensionality of the feature vector can be reduced.

5.5.3 Principal Components Analysis

The feature extraction step consists of mapping the input vector of observations $x \in R^n$ onto the new feature description $z \in R^m$ which is more suitable for classification. PCA is widely used for process monitoring for its excellent ability to handle high-dimensional, noisy, and highly correlated data. By applying a classifier to the lower-dimensional features produced by PCA, faults can be detected and diagnosed with greater proficiency.

The basic idea in PCA is to find the components $S_1, S_2, \ldots S_n$ so that they explain the maximum amount of variance possible by n linearly transformed components [10]. Define the direction of the first principal component, say w_1, by

$$w_1 = \arg\max E\{w^T x)^2\},\ \|w\| = 1 \tag{5.39}$$

where w_1 is of the same dimension m as the random data vector x. Thus the first principal component is the projection on the direction in which the variance of the projection is maximized. Having determined the first $k-1$ principal components, the $k-th$ principal component is determined as the principal component of the residual:

$$w_k = \arg\max E\{[w^T(x - \sum\nolimits_{i=1}^{k} w_i w_i^T x)^2]\},\ \|w\| = 1 \tag{5.40}$$

The principal components are then given by $s_i = w_i^T x$. In practice, the computation of the w_i can be simply accomplished using the (sample) covariance matrix $E\{xx^T\} = C$. The w_i is the eigenvectors of C that correspond to the n largest eigenvalues of C.

5.5.4 Kernel Principal Components Analysis

In complicated cases as machining processes with particularly nonlinear characteristics, PCA performs poorly due to its assumption that the process data are linear. In PCA the reduced dimension representation is generated by linear projections, and this can severely limit the usefulness of the approach. A new nonlinear PCA technique for tackling the nonlinear problem, called kernel PCA (KPCA), has been in development in recent years by SchJolkopf et al. [11] and Mika et al. [12]. The basic idea of KPCA is to first map the input space into a feature space via nonlinear mapping and then compute the PCs in that feature space. For any given algorithm that can be expressed solely in terms of dot products, this kernel method enables the construction of different nonlinear versions of the original algorithm [13].

Kernel PCA relies on the kernel trick: suppose we have an algorithm that depends only on dot products of the data. Consider using the same algorithm on transformed data:

$$x \rightarrow \Phi(x) \in F,$$

where F is a vector space.

Operating in F, the algorithm depends only on the dot product $\Phi(x_i) \cdot \Phi(x_j)$. Suppose there exists a kernel function $k(x_i, x_j)$ such that for all $x_i, x_j \in R^d$,

$$k(x_i, x_j) = \Phi(x_i) \cdot \Phi(x_j) \tag{5.41}$$

The eigenvectors of $\tilde{C}$ lie in the span of the mapped data $\Phi(x_i)$, if $\tilde{e}$ is an eigenvector of $\tilde{C}$ in F with eigenvalue λ, then

$$\tilde{C}\tilde{e} = \frac{1}{m}\sum_i \Phi(x_i) \cdot (\Phi(x_i) \cdot \tilde{e}) = \lambda\tilde{e} \tag{5.42}$$

Thus the eigenvalue equation can be replaced by m equations

$$\Phi(x_i)(\tilde{C} \cdot \tilde{e}) = \lambda\Phi(x_i) \cdot \tilde{e},\ i = 1, 2, \ldots m \tag{5.43}$$

In taking this step, the requirement that $\tilde{e}$ is an eigenvector of $\tilde{C}$ has been written entirely on terms of dot products of the mapped data, and so can be expressed in terms of the kernel matrix $k_{ij} = k(x_i, x_j)$; then expanding $\tilde{e} = \sum_{i=1}^{m} \alpha_i \Phi(x_i)$ results in eigenvector equation

$$\frac{1}{m}K\alpha = \lambda\alpha \tag{5.44}$$

Kernel PCA applies the idea to performing PCA in F. It's striking that, since projections are being performed in a space whose dimension can be much larger than d, the number of useful projections can exceed d, so kernel PCA is aimed more at feature extraction than dimensional reduction as PCA normally does.

The main idea of kernel methods can be illustrated in Fig. 5.6. The left figure of Fig. 5.6 shows a two-dimensional input space defined by attributes $\mathbf{x} = (x_1; x_2)$, with positive examples (y = +1) inside a circular region and negative examples (y = −1) outside. Clearly, there is no linear separator for this problem. Now, suppose we re-express the input data using some computed features, i.e., we map each input vector **x** to a new vector of feature values, $F(\mathbf{x})$. In particular, let us use the three features,

$$f_1 = x_1^2,\ f_2 = x_2^2, and\ f_3 = \sqrt{2}x_1x_2$$

The right figure shows the data in the new, three-dimensional space defined by the three features; the data are *linearly separable* in this space. This phenomenon is fairly general: if data are mapped into a space of sufficiently high dimension, then they will always be linearly separable.

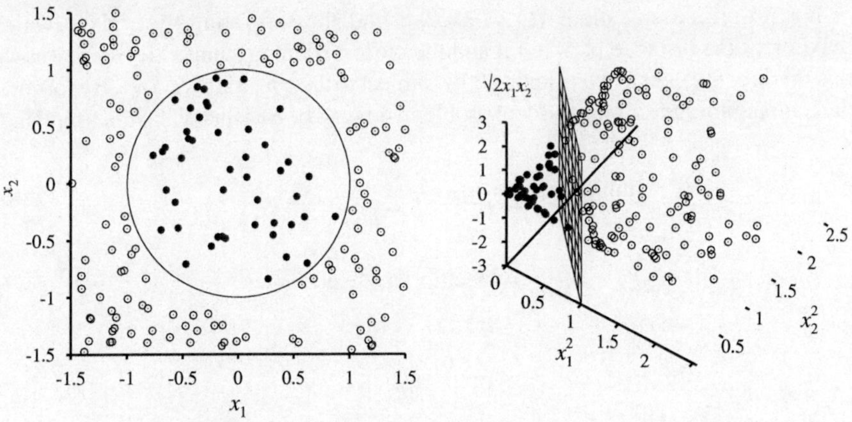

Fig. 5.6 A two-dimensional training with positive examples as black circles and negative examples as white circles. The same data after mapping into a three-dimensional input space and becomes a linear decision boundary in three dimensions

Compared to other nonlinear methods, the main advantage of KPCA is that it does not involve nonlinear optimization [11]; it essentially requires only linear algebra, making it as simple as standard PCA. KPCA requires only the solution of an eigenvalue problem, and due to its ability to use different kernels, it can handle a wide range of nonlinearities. In addition, KPCA does not require that the number of components to be extracted be specified before modeling. Due to these merits, KPCA has shown better performance than linear PCA in feature extraction and classification in nonlinear systems [11, 12].

5.5.5 Support Vector Machines

A support vector machine (SVM) is a supervised learning method that analyzes data and recognizes patterns for classification. Given a set of training examples, each marked as belonging to one of the categories, many hyperplanes might classify the data. The SVM chooses the hyperplane that represents the largest separation between the classes. So it chooses the hyperplane so that the distance from it to the nearest data point on each side is maximized [14].

Given a set of training samples $\mathrm{x}_1, \mathrm{x}_2, \ldots, \mathrm{x}_m$ with labels $y_1, y_2, \ldots, y_m$ we aim to learn the hyperplane $\mathrm{w}.\mathrm{x} + b$ which separates the data into classes such that:

$$\begin{aligned} &\mathrm{w}.\mathrm{x}_i + b \geq 1 - \xi_i \text{ if } y_i = 1 \\ &\mathrm{w}.\mathrm{x}_i + b \leq \xi_i - 1 \text{ if } y_i = -1 \\ &\xi_i \geq 0 \ \ \forall i. \end{aligned} \tag{5.45}$$

For no misclassifications ($\xi_i = 0$), we find the separating hyperplane which maximizes the distance between it and the closest training sample. It can be shown that this is equivalent to maximizing 2/|**w**| subject to the constraints above. By forming the Lagrangian and solving the dual problem, this can be translated into the following:

$$\text{minimize}: \sum_i \alpha_i - \frac{1}{2}\sum_{i,j} \alpha_i \alpha_j y_i y_j \mathrm{x}_i \cdot \mathrm{x}_j \tag{5.46}$$

$$\text{subject to}: \alpha_i \geq 0 \tag{5.47}$$

$$\sum_{\mathrm{i}} \alpha_{\mathrm{i}} y_i = 0.$$

where α_i is a Lagrange multiplier. There is one Lagrange multiplier for each training sample. The training samples for which the Lagrange multipliers are non-zero are called *support vectors*. Samples for which the corresponding Lagrange multiplier is zero can be removed from the training set without affecting the position of the final hyperplane. The above formulation is a well-understood quadratic programming problem for which solutions exist. The solution may be non-trivial though in cases where the training set is large.

The classification framework outlined above is limited to linear separating hyperplanes. It is possible however to use a non-linear hyperplane by first mapping the sample points into a higher dimensional space using a non-linear mapping. That is, we choose a map $\varphi : R^n \rightarrow N$ where the dimension of N is greater than n. We then seek a separating hyperplane in the higher dimensional space. This is equivalent to a non-linear separating surface in R^n.

As discussed above, the data only ever appears in our training problem in the form of dot products, so in the higher dimensional space, the data appears in the form of $\varphi(\mathrm{x}_i) \cdot \varphi(\mathrm{x}_j)$. If the dimension of N is very large, this product could be difficult or expensive to compute. However, by introducing a *kernel function* as in Eq. 5.41, which is $K(\mathrm{x}_i, \mathrm{x}_j) = \varphi(\mathrm{x}_i) \cdot \varphi(\mathrm{x}_j)$, we can use the $K(\mathrm{x}_i, \mathrm{x}_j)$ in place of $\mathrm{x}_i \cdot \mathrm{x}_j$ everywhere in the optimization problem and never need to know explicitly what φ is. Some examples of kernel functions are the polynomial kernel $K(\mathrm{x}, \mathrm{y}) = (\mathrm{x.y} + 1)^p$ and the Gaussian radial basis function (RBF) kernel $K(\mathrm{x}, \mathrm{y}) = e^{|\mathrm{x}-\mathrm{y}|^2/2\sigma^2}$.

After solving for **w** and b we determine which class a test vector $\mathrm{x}_\mathrm{t}$belongs to by evaluating $\mathrm{w} \cdot \mathrm{x}_t + b$ or $\mathrm{w} \cdot \varphi(\mathrm{x}_t) + b$ if a transform to a higher dimensional space has been used. It can be shown that the solution for **w** is given by $\overline{w} = \sum_i \alpha_i y_i \mathrm{x}_i$. Therefore,$\mathrm{w} \cdot \varphi(\mathrm{x}_t) + b$ can be rewritten,

$$\begin{aligned} \mathrm{w} \cdot \varphi(\mathrm{x}_t) + b &= \sum_i \alpha_i y_i \varphi(\mathrm{x}_i) \cdot \varphi(\mathrm{x}_\mathrm{t}) + b \\ &= \sum_i \alpha_i y_i K(\mathrm{x}_i, \mathrm{x}_t) + b \end{aligned} \tag{5.48}$$

Thus we again can use the Kernel function rather than transforming higher dimensional space since the data appears only in dot product form.

5.5.6 *Artificial Neural Network (ANN)*

Artificial neural networks are learning algorithms that have largely been motivated and inspired by the design of human neural systems. The key idea is to start with relatively simple computational units, the artificial neurons, and to connect them in a network that mimics the massive parallel computations performed in the brain [6]. A neural network thus has at least two basic ingredients: (1) the type of computational unit deployed, and (2) the topology of connectivity between the units.

For a neural network with just one layer of transformations between the input variables x and the final transformation f yielding the output y (one *hidden* layer), we have,

$$y = \sum_k w_k^{(2)} f_k^{(2)} \left(\sum_j w_j^{(1)} x_j \right) \tag{5.49}$$

Here the w are the weights in the linear combinations and the f are the nonlinear transformations.

The network functions as follows: Each neuron receives a signal from the neurons in the previous layer, and each of those signals is multiplied by a separate weight value. The weighted inputs are summed and passed through a limiting function, which scales the output to a fixed range of values. The output of the limiter is then broadcast to all of the neurons in the next layer. So, to use the network to solve a problem, one applies the input values to the inputs of the first layer, allows the signals to propagate through the network, and reads the output values (Fig. 5.7).

In the NNs there is no limit to the number of layers that can be used, though it can be proven that a single hidden layer (with enough nodes in that layer) is sufficient to model any continuous functions. Of course, the practicality of this will depend on the available data, and it might be convenient for other reasons (such as interpretability) to use more than one hidden layer.

The trained knowledge of the NNs is stored by the weights between neurons, and methods are needed to adjust the weights to solve a particular problem. The most common learning algorithm is called Back Propagation (BP). A BP network learns by example, that is, one must provide a learning set that consists of some input examples and the known-correct output for each case. So, these input–output examples are used to show the network what type of behavior is expected, and the BP algorithm allows the network to adapt. Other NN structures such as Self-Organizing Map (SOM) [15], Radial Basis Functions (RBF) [16], and Adaptive Resonance Theory (ART) [17, 18] have also been widely studied.

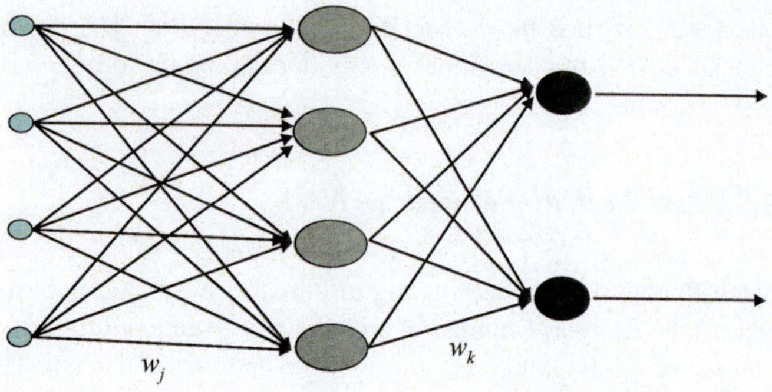

Fig. 5.7 A 3-layer neural network: signal/feature is applied to the inputs of the first layer, and signals propagate through the hidden layer to the output layer. Each link between neurons has a unique weighting value

5.5.7 *K-Nearest Neighbor (KNN)*

One of the most intuitive and historically one of the earliest approaches to classification is the k-nearest neighbor (KNN) method. The KNN is a very intuitive method that classifies unlabeled examples based on their similarity to examples in the training set. The classification of a new test pattern x given a training sample set $\Omega = \{(x_1, y_1), ..., (x_n, y_n)\}$ is performed according to the nearest neighbor rule:

$$f(x) = y_N(x, \Omega) \text{ where } N(x, \Omega) = \arg\min_i^n \|x - x_i\| \tag{5.50}$$

It means the label of a test point is given by the label of the closest training pattern—the nearest neighbor. The definition of the nearest neighbor depends on a suitable distance function or metric. In the above definition, we have simply used the Euclidean metric in Closeness depends on pattern space.

The KNN classifier is a generalization of the nearest neighbor rule, where the label at a test point will depend on the labels of the K closest training points, the K nearest neighbors. For a given unlabeled example $Xu \in R^D$, find the k closest labeled examples in the training data set and assign Xu to the class that appears most frequently within the k-subset. A neighbor is nearest if it has the smallest distance in feature space. For the calculation of the distance, metrics (i.e. Euclidean distance and Mahalanobis Distance) are needed (Fig. 5.8).

In the example below assume that we have a dataset with N examples, N_i from class ω_i and that we are interested in classifying an unknown sample Xu. We have three classes and the goal is to find a class label for the unknown example, Xu. In this case, we use the Euclidean distance and a value of $k = 5$ neighbors. Of the 5

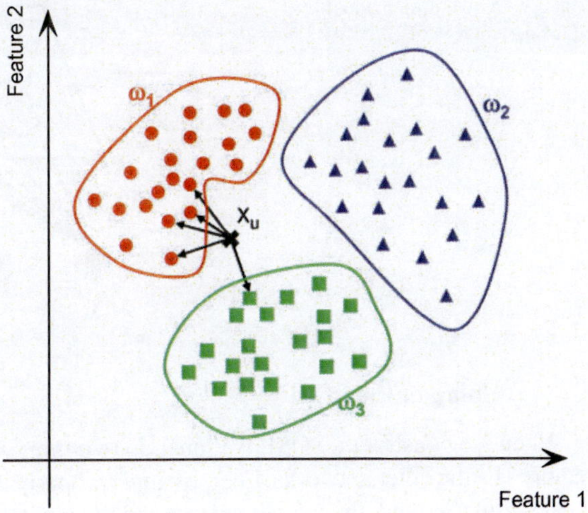

Fig. 5.8 The KNN classifier

closest neighbors, 4 belong to ω_1 and 1 belongs to ω_3, so *Xu* is assigned to ω_1, the Predominant class.

The main advantages of the KNN method are that it is easy to implement and it leads to a very simple approximation of the (optimal) Bayes classifier. If one can contrive to increase *k* at a suitable rate, the misclassification rate of the nearest neighbor rule will converge to a value related to the Bayes error rate. A potential drawback of nearest neighbor methods is that they do not build a model, relying instead on retaining all of the design set points. If the design set is large, searching through them to find the *k* nearest could be a time-consuming process.

5.5.8 Case Study: MSM with Self-Organizing Map (SOM)

(1) **Self Organizing Map (SOM)**

The most famous neural network model geared towards the type of unsupervised learning is the Self Organizing Map (SOM) [19]. Unlike back propagation NN, the inputs and outputs are not presented at the same time to the SOM it relies on a type of learning called competitive learning, where neurons compete for the privilege of learning, and the correct output is not known. A SOM is not a hierarchical system but consists of a fully interconnected array of neurons. The output of each neuron is an input to all other inputs in the network including itself. Each neuron has two sets of weights: one set is utilized to calculate the sum of weighted external inputs, and another one to control the interactions between different neurons in the network. The weights on the input pattern are adjustable, while the weights between neurons are fixed. A block diagram of a simple SOM with N neurons is shown in Fig. 5.9.

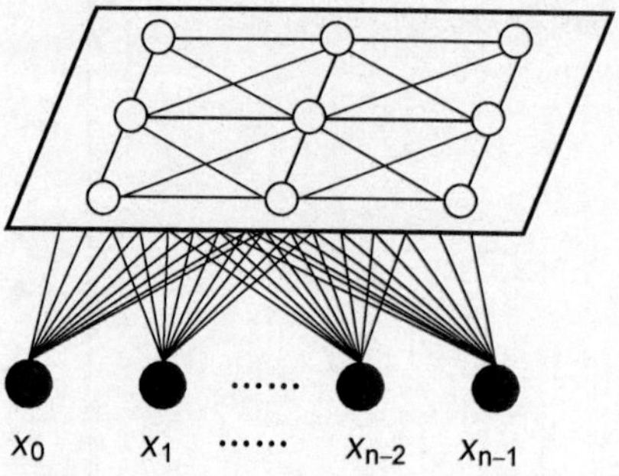

Fig. 5.9 A two dimensional SOM network

(2) **Training of the SOM**

When we construct a SOM, we must do two things that have not been generally required by the other networks. First, we must properly initialize the weight vectors of the neurons. Second, the weight vectors and the input vectors should be normalized. These two steps are vital to the success of the SOM network as follows:

1. Normalize the randomly selected weights W_i.
2. Present an input pattern vector x to the network. All neurons in the SOM layer receive this input vector.
3. Choose the winning neuron as the one with the largest similarity measure between all weight vectors W_i and the input vector x. If the shortest Euclidean distance is selected as similarity measure within a cluster, then the winning unit m satisfies the following equation:

$$\|x - W_m\| = \min_i\{\|x - w_i\|\} \tag{5.51}$$

where *m* is referred to as the winning unit.

4. Decrease the radius of the Nm region as the training progress, where Nm denotes, as a set of the index associated with the winning neighborhood around the winner unit C. The radius of the Nm region can be fairly large as the learning starts and is slowly reduced to include only the winner and possibly its immediate neighbors.
5. The weight of the winner unit and its neighborhood units are obtained as follows:

$$(W_i)_{new} = (W_i)_{old} + \alpha\left[x - (W_i)_{old}\right] \tag{5.52}$$

where Wi is the weight vector, x is the input pattern vector and a is the leaning rate ($0<\alpha<1$) Since a depends on the size of the neighborhood function, Eq. 5.52 can be rewritten as

$$(W_i)_{new} = (W_i)_{old} + \alpha\, N_{ci}\left[x - (W_i)_{old}\right] \tag{5.53}$$

where the function N_c can be chosen appropriately such as a Gaussian function or a Mexican hat function.

6. Present the next input vector. Repeat steps 3–5 until the training phase is completed for all inputs.

To achieve a good convergence for the above procedure, the learning rate α, as well as the size of neighborhood Nc should be decreased gradually with each iteration. As was mentioned before, at the beginning of the training phase, the selected region around the winner unit might be fairly large. Therefore, a substantial portion of the network can learn each pattern. As the training proceeds, the size of the neighborhood slowly decreases, so fewer and fewer neurons learn with each iteration. Finally, the winner itself will adjust its weights. After the completion of this procedure, the network is trained for the next input vector similarly.

(2) **Tool wear Estimation with SOM**

The average force and the standard deviation over one rotation have been found to have a strong correlation with flank wear [20]. These two features are extracted in each machining pass, as pass-by-pass features. A one-dimensional SOM is constructed with 15 units. In training, the majority of the data in the *k*-th pass are unlabeled (only the two force features) and used in phase one. A small portion of the data is labeled (the two force features and their corresponding wear increment value) according to the overall wear increment during the whole pass. The labeled data are used in phase two. The trained SOM is then applied on each data in the $(k + 1)$-th pass to estimate flank wear increment at different rotations.

The output of SOM is flank wear increment concerning the wear value at the beginning of this $(k + 1)$-th pass. This wear value has been made available employing an automatic vision-based image processing technique. Thus, the absolute wear obtained each time is the summation of the wear value plus the wear increment value estimated by SOM.

Figure 5.10a shows the wear estimation result throughout the entire test (till the end of the pass). (b) shows the breakage result inspected by vision, with the gray-level image of the insert on the left and the corresponding binary image on the upper right while the extracted breakage zone is shown on the lower right. (c) shows the force and force features, i.e. residual error and peak rate against time, superimposed with the thresholds (outlined by two horizontal lines) being equal to the maximum amplitudes of the features extracted in the previous pass. It is observed that the flank wear estimation result is good in association with the in-cycle measurement by vision, especially at the linear wear stage, although there is deviation at the initial wear stage.

For the ML based approaches, it is noted that these shallow ML approaches are originated from statistical data analysis. More details and in-depth study of these techniques can be found in multivariate statistical analysis [21–23].

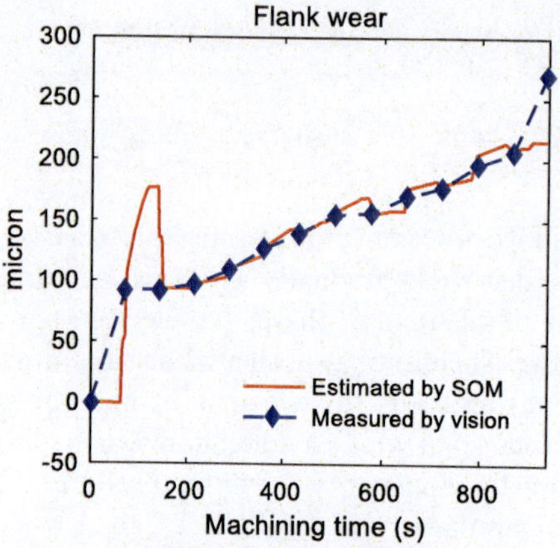

(a) Wear estimation result

(b) Breakage by vision after pass 13

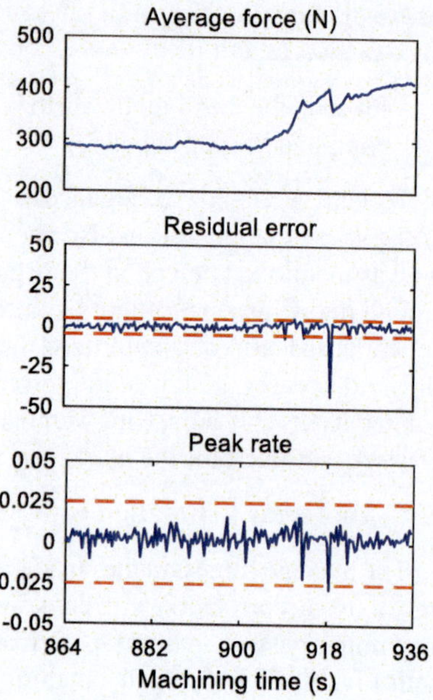

(c) Force features in pass 13

Fig. 5.10 On-line MSM result for test 5

5.6 Deep Learning

5.6.1 Introduction to Deep Learning

Due to the rapid development of sensor and network technology in recent years, it has become a reality to use monitoring systems in industrial systems to sample multiple machining data in real-time. With the improvement of hardware computing speed, more complex monitoring algorithms can be used to deal with massive data and make correct decisions quickly.

Deep learning is such an algorithm that can realize feature extraction, feature abstraction, and model learning independently [24]. It has obvious advantages compared with traditional algorithms in parallel computing, self-adaptive, self-learning, and nonlinear mapping, so it has been widely used in pattern recognition, function approximation, and other fields [25, 26]. In addition, with good fault tolerance and associative memory function, there is no need for excessive manual intervention in the whole signal process, which provides a new way to solve the problem of condition monitoring.

After the signal is transferred into the input layer, the feature extraction will be carried out to map the monitoring signal from the original input space to the feature space which is easy to process. This process can greatly reduce the pressure of the decision-making layer. Then the extracted features are sent to a fully connected layer to output the expected result. In contrast, the calculations performed by other algorithms in the feature extraction process are often insufficient, so the functional relationship between the extracted features and the monitoring state is not direct, such as feature compression. This is not enough to reduce the complexity of features. For example, if using statistical analysis, Fourier analysis, and wavelet analysis, the decision-making layer needs to deal with the time-domain, frequency-domain, and energy-domain information to complete the whole monitoring task. The complexity of the decision-making layer becomes the bottleneck of the whole algorithm.

In dealing with the problem of multi-source heterogeneous data, the advantage of deep learning is more obvious than the traditional algorithm. Compared with the signal collected by a single sensor, the sampling space of multi-source heterogeneous data is greatly different. In the fusion process, it is inevitable to map different sampling spaces to the same feature space. However, for data with dimensionality spanning like vectors and images, the mapping function will be very complicated. The existing algorithms are not sufficient to solve such fusion problems. Compared with traditional machine learning which can only learn task models, deep learning has the inherent ability of feature learning and feature abstraction. Through the reasonable design of network structure, deep learning can fit the mapping relationship between various feature spaces, which makes cross-dimensional data fusion possible. What's more, deep learning will make the whole processing flow more concise.

As a new paradigm originated from neural network research, deep learning is powerful in processing industrial big data and meets the challenges of complex and big volume monitoring data analytics for condition monitoring. Deep learning generally uses multiple-layer NNs to learn the data, with which the depth of layers in the learned model correspond to the levels of abstraction. It has shown much better performances than shallow ML approaches in problem adaptability, network extension, complex feature representation, etc. After a decade's exploration, deep learning has been successfully applied in many fields, such as image recognition [27], speech recognition [28], and fault diagnosis [29]. In the whole development process of deep learning, deep belief network (DBN) [30, 31], deep auto-encoder (AE) [30, 32], convolutional neural network (CNN) [33, 34], recurrent neural network (RNN) [35], constitute the fundamental models. Most follow-up studies are based on these or the improved models. Different networks are developed according to the different definitions of structures, layers, weight sharing, and edge characteristics. It should be noted that deep learning has two folds, and the discussion in this paper uses the deep structure to process signals and information, rather than the deep understanding of them, although in some cases these two aspects may be similar.

The main steps of the deep learning-based method for process monitoring are feature representation and defect classification. The input original signal is processed layer by layer, and the initial sample features which are not closely related to the quality state are transformed into the feature expression closer to the quality state. In this way, the low level measurement is gradually transformed into the high level feature expression, and the feature extraction and defect detection are integrated with the network to directly establish the high level feature and the quality relationship.

This high dimensional nonlinear mapping between defects and samples enables the learned model to complete process monitoring tasks.

Table 5.1 shows the network structure, training algorithm, and problem-solving of the typical deep learning algorithms, and gives a brief summary of the existing problems. Most of the deep learning algorithms are embodied in the deep structure of spatial dimension and belong to the feedforward neural network. According to the degree of dependence on annotation data, DBN, AE, and their derivatives are mainly unsupervised learning or semi-supervised learning, while CNN, RNN, and their derivatives are mainly supervised learning.

5.6.2 *Sparse Autoencoder (AE)*

The auto-encoder mainly consists of two parts: encoder and decoder, as shown in Fig. 5.11. The encoder maps the input data from high-dimensional space to low-dimensional space to obtain the feature representation, while the decoder reverses this process to reproduce the input. These two processes are called encoding and decoding respectively. The auto-encoder can reduce the dimension of data and effectively filter out redundant information, which has great advantages in image recognition and pedestrian detection [30, 32]. To make full use of the implicit information such as data categories and patterns, it can be improved by adding a logic regression (LR) layer at the top of the model. In addition, the BP algorithm, Newton interpolation [38] and other methods can fine tune the parameters. To increase the robustness of SAE, the denoising auto-encoder (DAE) [38] was also developed.

The encoding and decoding processes of an unlabeled observation sample set $S = \{x_i\}_{i=1}^n$ can be expressed as

$$h = f_\theta(x) = s_f(Wx + b) \tag{5.54}$$

$$\hat{x} = g_{\theta'}(h) = s_g\left(W'h + b'\right) \tag{5.55}$$

where h and $\hat{x}$ represent the encoding and decoding vector, f_θ and $g_{\theta'}$ are the encoding and decoding function respectively. The s_f and s_g represent the activation functions of the network, which are usually taken as sigmoid functions. $\theta = \{W, b\}$, W is the weight matrix and b is the bias vector.

The pre-training of AE is essentially a process of updating parameters θ based on the sample set S. The degree of closeness can be characterized by the error of reconstruction function $L\left(x, \hat{x}\right)$, which can be defined as:

$$L\left(x, \hat{x}\right) = \sum_{i=1}^{n} \left[x_i \ln\left(\hat{x}\right) + (1 - x_i)\ln\left(1 - \hat{x}\right)\right] \tag{5.56}$$

Table 5.1 Deep learning models and their algorithms

DL approach	Network topology	Learning algorithms	Model properties	References	Comments
Deep belief network (DBN)	Multi-layer, full connection with/without directed edge	Greedy layer by layer training; BP	The stack of RBM; Using unsupervised learning parameters as the initial value of supervised learning	[30]	Visual layers receive binary values; optimization difficult
Deep Boltzmann Machine (DBM)	Multi-layer, undirected edge, full connection	BP	Special form of BM; bottom-up generation structure; reduce error caused by propagation	[31]	Low learning efficiency
Deep Auto-encoder (DAE)	Multi-layer	Greedy layer by layer training; BP	Unsupervised greedy pre-training of the hidden layer; BP fine tuning; Dimension reduction	[30]	Increase of hidden layers and neurons results in Gradient vanishing or explosion
Sparse Auto-encoder (SAE)	3-layer/ Multi-layer	Gradient descent; BP	Dimension reduction and sparse feature representation	[32]	As above
convolutional neural network (CNN)	Multi-layer, undirected edge, local connection, shared weights	Gradient descent; BP	Contains convolution layer and subsampling layer; Strong distortion robustness for image recognition	[33, 34]	Requiring large labeled data training, high computational power
Recurrent neural network (RNN)	Multi-layer	BPTT; Gradient descent	Deep structure in multiple time dimensions; Strong processing of sequential data	[35]	Gradient vanishing or explosion

(continued)

Table 5.1 (continued)

DL approach	Network topology	Learning algorithms	Model properties	References	Comments
Long Short-Term Memory (LSTM)	Multi-layer	BPTT; Gradient descent	Gradient vanishing or explosion solved; capable of processing complex sequence data	[36]	High training complexity, decoding required
Generative adversarial network (GAN)	Multi-layer, undirected edge, local connection, shared weights	BP; Dropout	Combination of feedforward network and convolution network	[37]	Difficult to train; unstable training process

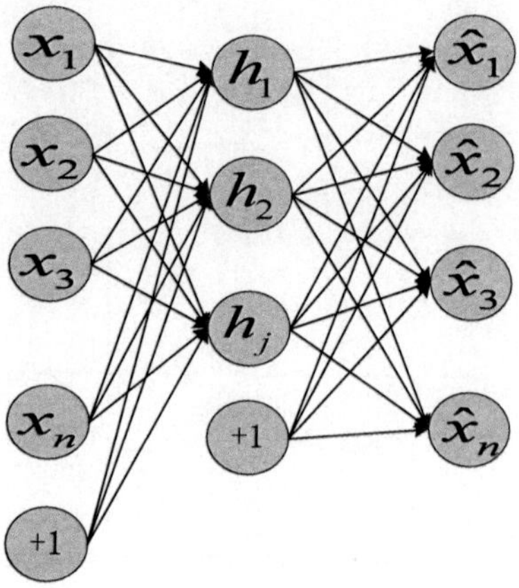

Fig. 5.11 The model structure of AE

The process of obtaining the parameters of this layer by minimizing the cost function $J_1(\theta_{AE})$ can be defined as:

$$J_1(\theta_{AE}) = \frac{1}{n}\sum_{x\in S} L[x, g_{\theta'}(f_\theta(x))] \tag{5.57}$$

This operation to minimize the cost function only preserves the information of x, which is often insufficient to obtain an effective feature representation of the signal. Considering the robustness and generality of sparse representation, sparsity constraints can be added to the cost function, which is also known as sparse autoencoder (SAE). The loss function with weight attenuation can be expressed as

$$J_2(\theta_{SAE}) = \frac{1}{2n}\sum_{i=1}^{n}\left\|x_i - \hat{x}_i\right\|^2 + \lambda_1\left(\|W\|_F^2 + \left\|W'\right\|_F^2\right) + \lambda_2\sum_{j=1}^{m} KL(\rho\|\hat{\rho}_j) \tag{5.58}$$

where λ_1 represents the weight attenuation term coefficient; λ_2 represents the weight coefficient that controls the sparsity penalty term; ρ represents the sparsity coefficient; $\hat{\rho}_j$ represents the average activation of the *j-th* neuron in the hidden layer on the training set S when the input is x_i. The expression of $KL(\rho\|\hat{\rho}_j)$ can be expressed as

$$KL(\rho\|\hat{\rho}_j) = \rho\log\frac{\rho}{\hat{\rho}_j} + (1-\rho)\log\frac{1-\rho}{1-\hat{\rho}_j} \tag{5.59}$$

(1) **Stacked sparse auto-encoder**

The sparse auto-encoder is a 3-layer neural network structure, which has limited ability to map and generalize complex functions, so it is restricted in the application of complex classification problems. The stacked sparse auto-encoder, a multi-layer stacked deep network trained layer-by-layer, can overcome this weakness, as shown in Fig. 5.12. Since the training of deep neural networks is very difficult, the initialization method of layer-wise pre-training can be used. In this pre-training method, only a single hidden layer of the network is trained each time. Only after the training of this layer is finished, the training of the next layer is started. 'Layer-by-layer initialization' can local the initial value of the model parameters in a position closer to the global optimum, making the model easier to obtain better results.

(2) **Softmax classifier**

The stacked auto-encoder is a deep neural network with only feature learning ability, and cannot classify. Therefore, we need to insert a layer of network with classification

Fig. 5.12 The stacked auto-encoder network

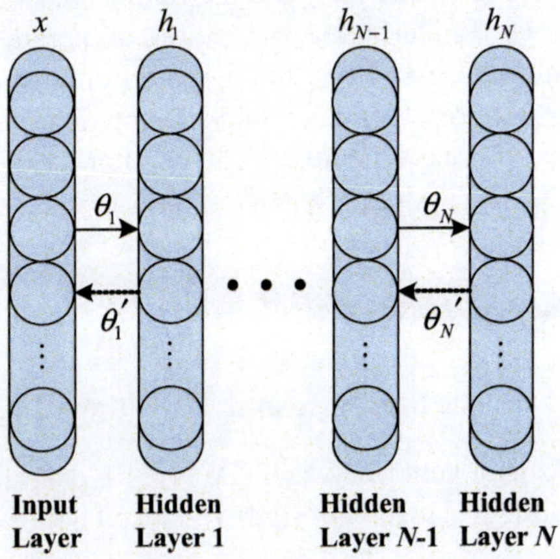

and prediction capabilities behind the stacked auto-encoder, such as the softmax classifier.

Softmax classifier is a generalization of the logistic regression model in classification problems, which can solve multi-class pattern classification. The function can be expressed as

$$h_\theta(x_i) = \begin{bmatrix} p(y_i = 1|x_i; \theta) \\ p(y_i = 2|x_i; \theta) \\ \vdots \\ p(y_i = k|x_i; \theta) \end{bmatrix} = \frac{1}{\sum_{j=1}^{k} e^{\theta_j^T x_i}} \begin{bmatrix} e^{\theta_1^T x_i} \\ e^{\theta_2^T x_i} \\ \vdots \\ e^{\theta_k^T x_i} \end{bmatrix} \tag{5.60}$$

where $\{(x_i, y_i); i = 1, 2, \cdots, n\}$, is the dataset, x represents the input signal and represents the probability that x belongs to class j estimated by the hypothesis function.

The cost function of the softmax regression model can be expressed as

$$J_3(\theta) = -\frac{1}{m}\left[\sum_{i=1}^{m}\sum_{j=1}^{k} 1\{y_i = j\} \lg \frac{e^{\theta_j^T x_i}}{\sum_{l=1}^{k} e^{\theta_l^T x_i}}\right] \tag{5.61}$$

where $1\{\cdot\}$ is an indicative function. when the value in the curly bracket is true, the result of the function is 1, otherwise, the result is 0.

(3) **Fine-tuning algorithm of multi-layer auto-encoder**

In the step of unsupervised pre-training, the parameters of other layers are fixed during the training of each layer of parameters. In order to further improve the prediction accuracy and generalization ability of the network, the entire model will also be supervised trained based on the backpropagation algorithm to further optimize the parameters of each layer after the above-mentioned pre-training process is over, also known as fine-tuning. Fine-tuning treats all layers of the stacked auto-encoder as a whole. During each iteration, all parameters of the model will be optimized and adjusted. After fine-tuning based on a small amount of labeled data, the ability of feature representation and state prediction can be obtained, and the performance of the entire deep network can be improved.

5.6.3 Deep Belief Neural Network (DBN)

(1) **The Structure and Algorithm of Basic DBN**

A deep belief network (DBN) is a deep network composed of multiple restricted Boltzman machines (RBM) stacked with a classification or regression layer. RBM is

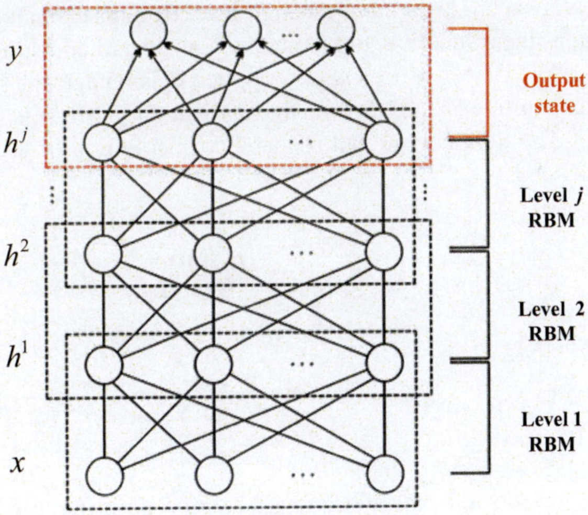

Fig. 5.13 The architecture and training process of deep belief network

a symmetric coupled stochastic binary network [31]. The DBN cannot only extract the features of observation data from low to high levels but also discover the distributed features of input data [30, 31]. The set of visible and hidden units,h, $v \in [0, 1][0, 1]$, and the connection between restricted units consist of the hierarchy of RBMs (Fig. 5.13). The energy function of joint configuration determines the probability distribution by connecting the weight and bias of visible and hidden units in the binary RBM level, which is defined as:

$$E(\mathbf{v}, \mathbf{h}) = -\sum_{\substack{i=1 \\ i \in visible}}^{n} a_i v_i - \sum_{\substack{j=1 \\ j \in hidden}}^{m} b_j h_j - \sum_{i=1}^{n} \sum_{j=1}^{m} v_i h_j w_{ij} \tag{5.62}$$

where w_{ij} is the symmetric interaction term of visible unit i and hidden unit j, a_i and b_j represent the bias of visible unit and hidden unit respectively.

Figure 5.13 shows a 3-layer DBN stacked by 3 RBMs, where the random vector input to the network is denoted as x, the node of the i-th hidden layer is denoted as $h^i, i = 1, 2, 3$, and the output is denoted as y. There are two layers composed of each RBM. The upper layer is the hidden unit and the lower layer is the visible unit. The hidden unit of the former RBM is used as the visible unit of the next RBM, which is stacked in turn. Finally, an output layer is added to output the classification results.

The variable set $\theta = (\mathbf{w}, \mathbf{a}, \mathbf{b})$ is the parameter to determine the RBM model, and the goal of RBM training is to find the optimal value θ^* representing the training sample. In general, DBN trains the model through the combination of greedy forward learning layer by layer and backward fine-tuning of gradient descent. Stacked RBM mainly completes the feature extraction of input data, usually pre-selects the contrast divergence algorithm to determine the network parameters, and the classification (regression) layer maps the high-level features extracted in front of the label through non-linear function, to facilitate the backward adjustment of DBN parameters.

The DBN has been widely applied in language and speech processing, and image recognition [22, 23], while the application to MSM is much less studied.

The joint conditional probability distribution of visible and hidden units can be defined as

$$p(\mathbf{v}, \mathbf{h}) = \frac{e^{-E(\mathbf{v},\mathbf{h})}}{Z} \tag{5.63}$$

where

$$Z = \sum\nolimits_{\mathbf{v},\mathbf{h}} e^{-E(\mathbf{v},\mathbf{h})} \tag{5.64}$$

represents the normalization factor or partition function which is obtained by summing all possible visible and hidden vectors. The probability value that the network assigns to the visible vector v is obtained by the summation of all possible hidden vectors:

$$p(\mathbf{v}) = \frac{\sum_{\mathbf{h}} e^{-E(\mathbf{v},\mathbf{h})}}{Z} \tag{5.65}$$

The probability distribution that the network assigns to the training image can be increased by adjusting the weights and biases, to reduce the energy of the image and increase the energy of other images, especially those images with lower energy, to make a greater contribution to the partition function.

The derivative of the logarithmic probability of the training vector can be expressed as

$$\frac{\partial \log p(v)}{\partial w_{ij}} = \langle v_i h_j \rangle_{data} - \langle v_i h_j \rangle_{\text{model}} \tag{5.66}$$

where angle brackets are used to indicate the expected value under the distribution of the specified subscript. One of the simplest learning rules is to select the direction with the most obvious difference in expected values as the direction of weight updating, which can be expressed as:

$$\Delta w_{ij} = \varepsilon\left(\langle v_i h_j \rangle_{\text{data}} - \langle v_i h_j \rangle_{\text{model}}\right) \tag{5.67}$$

where ε represents the learning rate.

Since there is no direct relationship between hidden units in RBM, it is easy to get an unbiased sample $\langle v_i h_j \rangle_{data}$.

Given the randomly selected training image v, the probability of the binary state of hidden unit j $h_j = 1$ is:

$$p(h_j = 1|\mathbf{v}) = g(b_j + \sum\nolimits_i v_i w_{ij}) \tag{5.68}$$

where $g(x)$ represents the sigmoid function $1/(1+\exp(-x))$, and $v_i h_j$ is an unbiased sample.

Because there is no direct connection between visible units in RBM, it is very easy to obtain unbiased samples of the state of visible units. Given a hidden vector h, the probability of binary state of hidden unit j $h_j = 1$ is

$$p(v_i = 1|\mathbf{h}) = g(a_i + \sum\nolimits_j h_j w_{ij}) \tag{5.69}$$

However, it is much more difficult to obtain an unbiased sample $\langle v_i h_j \rangle_{\mathrm{mod}\,el}$ that needs to start from any random state of the visible unit and complete by alternate Gibbs sampling. The single iteration of alternating Gibbs sampling can be expressed as two steps: first, update all hidden units in parallel according to Eq. 5.68, and then update all visible units in parallel according to Eq. 5.69.

Hinton [39] proposed a faster learning process. First, set the state of the visual unit to the training vector. Then use Eq. 5.68 to calculate the binary state of the hidden unit in parallel. Once the hidden unit has selected the binary state, all v can be set to 1 and the probability is given by Eq. 5.70 can be used for reconstruction. The weight update can be expressed as

$$\Delta w_{ij} = \varepsilon\left(\langle v_i h_j \rangle_{data} - \langle v_i h_j \rangle_{recon}\right) \tag{5.70}$$

There is also a simplified version of the learning rule, which calculates the deviation only by the state of a single unit instead of the pairwise product.

Although this kind of learning only roughly estimates the logarithmic probability gradient of training data, the calculated gradient is closer to the gradient of another objective function called contrast divergence [39] which can also achieve good results. This approach is simpler because it ignores a tricky term in the objective function and does not even need to follow the gradient, which is also the difference between the two Kullback-Liebler divergences. In fact, Sutskever and Tieleman have proved that this method does not follow any functional gradient [40]. But it's enough to succeed in many important applications. If more alternating Gibbs sampling steps are used before the statistical data collection of the second item in the learning rule, RBMs will usually learn better parameters. In this book, n complete Gibbs alternate sampling is called CDn. The update of model parameters $\theta = (\mathbf{w}, \mathbf{a}, \mathbf{b})$ can be expressed as

$$w_{ij} = \eta w_{ij} + \varepsilon\left(\langle v_i h_j \rangle_{data} - \langle v_i h_j \rangle_{recon}\right) \tag{5.71}$$

$$a_i = a_i + \varepsilon(\langle v_i \rangle_{data} - \langle v_i \rangle_{\text{model}}) \tag{5.72}$$

$$b_j = b_j + \varepsilon\left(\langle h_j \rangle_{data} - \langle h_j \rangle_{\text{mod } el}\right) \tag{5.73}$$

In a multi-layer RBM classification algorithm, it is usually necessary to insert a softmax layer at the top to realize the classification function.

(2) **Gaussian visible unit**

Logical units often perform poorly in processing data such as images and Mel-Cepstrum coefficients of speech. One solution is to replace the binary visible unit with a linear unit with independent Gaussian noise. The energy function can be described as

$$E(\mathbf{v}, \mathbf{h}) = -\sum_{\substack{i=1 \\ i \in visible}}^{n} \frac{(v_i - a_i)^2}{2\sigma_i^2} - \sum_{\substack{j=1 \\ j \in hidden}}^{m} b_j h_j - \sum_{i=1}^{n} \sum_{j=1}^{m} \frac{v_i}{\sigma_i} h_j w_{ij} \tag{5.74}$$

where σ_i is the standard deviation of the Gaussian noise of the visible unit i.

Using the CD1 approach [39], it is difficult to learn the noise variance of each visible unit. In many applications, it is much easier to first normalize each component of the data to zero mean and unit variance, and then use noise-free reconstruction and set the variance to 1. Then, the reconstruction value of the Gaussian visible unit is equal to the top-down input of its binary hidden unit plus its deviation.

The learning rate of the network composed of Gaussian visible units is generally one to two orders of magnitude smaller than that of the network composed of binary visible units. The failure cases reported in some kinds of literature may be due to the high learning rate used. This is because there is no upper limit on the size of artifacts in refactoring. If the weight of a component is very large, its update span will be large. In contrast, binary hidden and visible unit weight updates must be between -1 and 1, so they are more stable. This is because there is no upper limit of the component in the reconstruction. If the weight of a component is very large, the update span will be large. In contrast, the weight update of the binary hidden and visible units must be between -1 and 1, so they are more stable.

5.6.4 *Convolution Neural Network (CNN)*

CNN is a neural network that realizes feature extraction through convolution, which is especially suitable for classification and prediction tasks of two-dimensional images [33]. LeNet first applied CNN to the MNIST handwritten dataset digit recognition task, which greatly improved the accuracy of the image recognition task. Subsequently, networks such as InceptionNet, VGGNet, and ResNet appeared one after

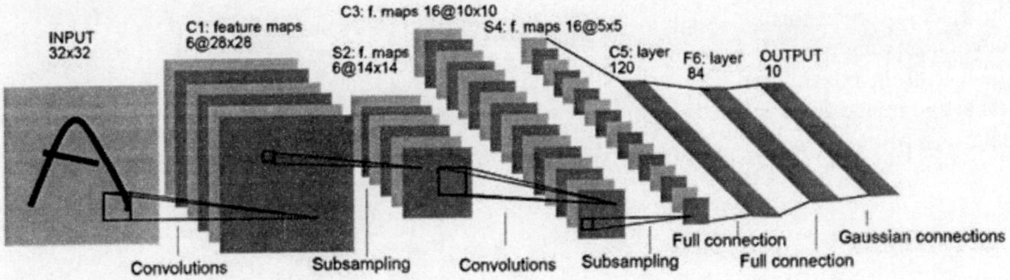

Fig. 5.14 The network architecture of LeNet-5 [33]

another [24], using CNN for more complex tasks and winning prizes in multiple pattern recognition competitions. CNN has become the most widely used and the most mature architecture in deep networks.

Compared with a fully connected neural network that realizes nonlinear function mapping through the stack of fully connected layers, CNN extracts two-dimensional features through the convolutional layer and pooling layer. The learning process is alternately carried out by convolution and subsampling. The extracted features are sent to the fully connected layer for the classification decision. For example, feature extraction of LeNet is achieved through two convolutional layers and two down sampling layers to reduce the data dimension, as shown in Fig. 5.14. The mathematical model can be summarized as

$$x_j^l = f(\sum_{i \in M_j} x_i^{l-1} \times k_{ij}^l + b_j^l) \tag{5.75}$$

where k is the convolution kernel of the l-th layer, x_i^{l-1} and are the input and output of the *l-th* layer, b is the network bias,M_j is the input feature patch. The subsampling layer is a scaling mapping of the data graph of the previous layer to reduce the data dimension. The extracted features have scaling invariance and can also prevent overfitting. The calculation method of its neurons can be expressed as,

$$x_j^{l+1} = f(\beta_j^l down(x_i^l) + b_j^l) \tag{5.76}$$

where $down(\cdot)$ is the sub-sampling function,β network bias multiplier. The output layer of CNN generally uses the cascaded feature map of the previous layer for linear full connection and generally uses the softmax classifier. Its advantage is that it can solve the problem of multi-classification.

$$O = f(b_o + w_o f_v) \tag{5.77}$$

where f_v is the feature vector, and b_o, w_o are bias vector and weight matrix. The training of the CNN is similar to that of the traditional supervised NN method, i.e.,

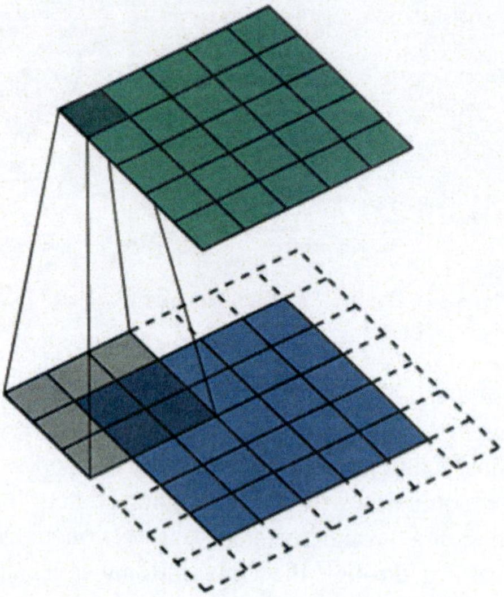

Fig. 5.15 2-dimensional convolution diagram [30]. The size of the convolution kernel and image are 3 × 3 and 5 × 5 respectively

the BP algorithm is often used to transfer the error layer by layer, and the gradient descent method is applied to adjust the parameters between each layer. With its special network structure and learning algorithms, the CNN can extract the local features of input data, and combine and abstract the high-level features layer by layer, which can effectively achieve state monitoring [41–43] and defect identification [44, 45].

The convolution layer essentially realizes the two-dimensional convolution between convolution kernels and the image, as shown in Fig. 5.15. The convolution kernels pass through all the positions of the image and multiply with the pixels in the receptive field in turn during convolution. For example, if the kernel is a Gaussian kernel, the function which convolution realizes is fuzzy denoising. The parameters of the convolution kernel are modified iteratively based on the training set during training. The two-dimensional convolution diagram is shown in Fig. 5.15.

CNN is very suitable for image processing because of local connection (local receptive field), weight sharing, and translation invariance.

(1) **Local connection**: The convolution kernel only multiplies with the local image each time, so each point of the output matrix is only related to the corresponding area. This method can greatly reduce the calculation. Taking the first layer of LeNet as an example, if using a fully connected layer, the number of multiply–add required is $32 \times 32 \times 28 \times 28 = 802{,}816$. But the number of multiply–add required by the first convolution layer of LeNet is only 5×5 (convolution kernel size) $\times$ 6 (number of convolution kernel) $\times$ 28 $\times$ 28 (image size) = 117,600 which only accounts for 15% of the calculation of the fully connected layer under the same conditions. When the input image is a color image with a higher resolution, the computational advantage of the convolution layer will be more obvious. In addition, CNN is very suitable for the parallel operation mechanism of GPU, so the computation speed of convolution is faster in actually.

(2) **Weight sharing**: Although only the features of the local area can be extracted during each convolution iteration, but the features of the whole image can be extracted after many iterations. The kernel used in each convolution is the same, or to say, the weight of the convolution kernel is shared, which greatly reduces the number of parameters learned. Still taking the first layer of LeNet as an example, the number of weight parameters that a fully connected layer needs to learn is $32 \times 32 \times 28 \times 28 + 1 = 802{,}817$. But only 5×5 (kernel size) $\times$ 6 (number of convolution kernel) + 1 (bias) = 151 parameters need to learn if use convolution layer. The small number of learning parameters reduces the difficulty of model training and eases the overfitting phenomenon.

(3) **Translation invariance**: In a sense, each convolution operation is to monitor whether there is a corresponding feature in this receptive field, then the convolution layer is equivalent to scoring the possibility of the feature in all positions. Intuitively speaking, the natural translation invariance of convolution matches the geometric characteristics of the image very well. Taking the image of the molten pool as an example, the position and angle of the camera will cause the difference in the position of the molten pool on the image, but the position is not significantly related to the monitoring results. In this case, the translation invariance of the convolution kernel can ensure that the extracted features are not affected by the position and angle, so a more general result can be obtained.

5.6.5 Recurrent Neural Network (RNN)

Ordinary neural networks cannot process temporal data, that is, there is no direct connection between this input and the next input, and there is no sequential connection between the same input. To solve this problem, RNN, a feedforward neural network used to model the temporal dependence of sequence signals [36] is proposed, as shown in Fig. 5.16. It can be found that the difference between RNN and other neural networks is that the output of other neural networks o_t is only based on the current signal x_t, while the output of RNN o_t is based on the current input x_t and the state of

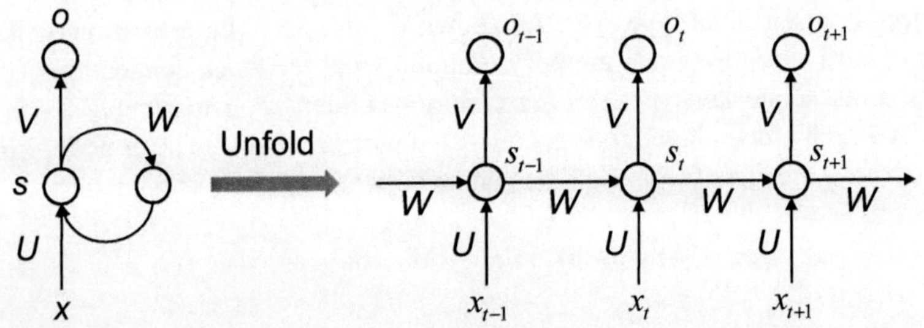

Fig. 5.16 Schematic diagram of recurrent neural network [31]. The left is a schematic diagram of folding; the right is a schematic diagram of expansion

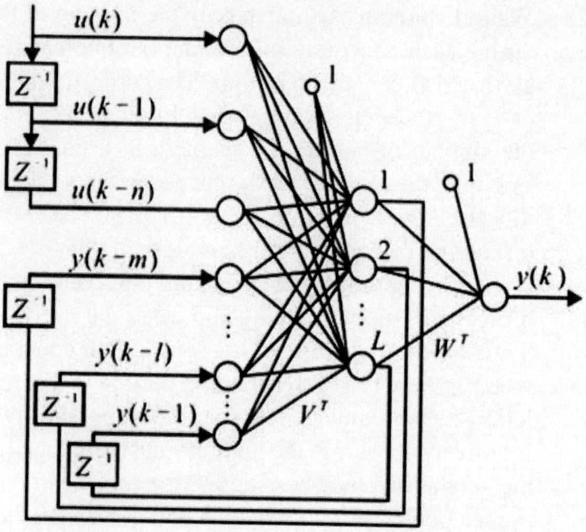

Fig. 5.17 The architecture of double-layer Elman recurrent neural network [6]

the previous iteration s_{t-1}. At the same time, a new state of RNN s_t is generated based on the current input x_t and previous state s_{t-1} which retains useful information of previous periods. Therefore, each output of RNN depends on previous information. Note that the RNN units in each period are reused, that is, the weight parameters of RNN are shared, which is very similar to CNN in a sense. However, CNN is suitable for extracting spatial features of images at different positions, while RNN is suitable for extracting temporal features at different times. In addition, compared with processing temporal signal through a fully connected layer, the size of the dataset required, the number of parameters to be learned, and the computation load can be further reduced by RNN. Therefore, RNN can alleviate the overfitting phenomenon.

The essential feature of RNN is that there are both internal feedback connections and feedforward connections between its processing units (Fig. 5.17). The internal feedback connection can reserve the state of hidden layer nodes and provide a memory mode for the network. The output of the network not only depends on the current input but also relates to the previous internal state of the network, reflecting better dynamic characteristics [46, 47]. According to different feedback paths, RNN can be the Jordan and Elman types [6]. As shown in Fig. 5.17, the Elman network is a feedforward NN with a local memory unit and local feedback connection. There is a special associate layer besides the common hidden layer to receive feedback signals, and each hidden layer node is connected with an associate layer node, which is equivalent to the state feedback. The Elman network can be formulated as,

$$x(k) = f\left(W'x_c(k) + W^2u(k-1)\right) \tag{5.78}$$

$$x_c(k) = ax_{c-1}(k) + x(k-1) \tag{5.79}$$

$$y(k) = g\left(W^3 x(k)\right) \tag{5.80}$$

where, $u(k - 1)$, $y(k)$, $x(k)$, x_c (k) are the external input, network output, hidden layer output, and associate layer output of the network, respectively. W_1, W_2, W_3, are the connection weight matrix from context layer to the hidden layer, input layer to hidden layer, and hidden layer to the output layer, respectively. $f(\cdot)$, $g(\cdot)$are the nonlinear excitation functions and is the self-connection feedback gain factor. $f(\cdot)$,

However, due to the vanishing and exploding gradient problems, the traditional RNN usually meet problems when processing long-time sequence, and there will be a large prediction error. To solve this problem scholar have developed long short-term memory(LSTM) [36] and the Gated Recurrent Unit (GRU) [48] formulations to train long sequences of data. LSTM mainly includes two parts: (1) long-term state and short-term state; (2) three control gates (forge gate, input gate, and output gate). The model parameters can be learned from the sample by the back propagation through the time (BPTT) algorithm. Due to the unique structure of LSTM, gate structure can determine the information characteristics under the optimal conditions. Guo et al. [49] trained the LSTM with features from the time–frequency domain, effectively overcome the gradient disappearance or explosion, and avoided the manual threshold selection for bearing fault prediction. In order to solve the uncertainty caused by operation disturbance, Zhang et al. [50] developed a fault prediction method based on a bidirectional LSTM. It calculated the forward and backward path independently and connected the output in series, which could effectively smooth the degradation path and prediction results.

Because of its special network structure, RNN has a strong advantage in the field of complex dynamic system modeling with excellent learning and model stability for condition monitoring [47]. On the other hand, RNN mainly deals with time-series data and is difficult to model multidimensional signals (i.e. image), which usually needs to be combined with other methods for practical applications in MSM.

In summary, RNN is essentially an RNN unit that is reused many times. According to the different unit structures of RNN, it can be divided into many categories. Among them, the earliest and most popular units are the Long-short term memory model (LSTM) and the Gated Recurrent Unit (GRU) model.

(1) **Long-short term memory model (LSTM)**

LSTM is first proposed by Hochreiter et al. [36] which is mainly used to solve the problem that RNN does not have long-term effective memory (forgetting). Other RNN units generally have only one state transmission channel, but LSTM units have two as shown in Fig. 5.18. The function of the hidden state h is similar to that of the ordinary RNN unit used to transmit short-term temporal information. But this is enough. Hochreiter S et al. believe that the reason why RNN has no long-term memory is that the information transmission channel h is easily modified and not easy to retain for a long time. Moreover, if the output o_t and state h_t of RNN are the same, it is necessary to weigh the accuracy of the output and the effectiveness of the information retained, such as GRU. This

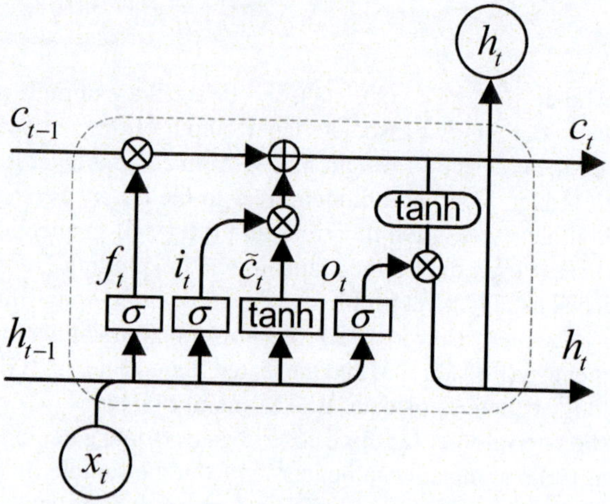

Fig. 5.18 The structure of the LSTM unit [33]. The black lines indicate the transmission path related to the cell state

is often fatal in the sequence-to-sequence RNN. In order to enable the network to process long-term signals, the cell state c is introduced to LSTM.

It can be seen intuitively from Fig. 5.18 that cell state c is only going through two steps in one iteration - forget and update. The cell state at the last moment c_{t-1} will be selected and forgotten based on the forget gate f_t to discard useless information. Then update c_t according to the input gate i_t, input x_t and state h_{t-1}. Compared with h_t which is obtained by the recalculation of h_{t-1} and x_t in each iteration, c_t is less likely to be changed undoubtedly, so c_t is easier to remember long-term information.

(2) **Gated Recurrent Unit (GRU)**

The complete workflow of LSTM can be found in [48]. For more detailed derivation, please refer to Hochreiter et al. [36].

GRU is a simplified form of LSTM proposed by Cho et al. [48]. Although later than LSTM, it is widely used because of its simple structure. The structure of GRU is shown in Fig. 5.19. The cell state c_t is omitted in GRU. And the original three gates are simplified into two gates: update gate z_t and reset gate r_t. The reset gate r_t is used to determine how much information in h_t needs to be forgotten, while the update gate z_t is used to determine which information is useful and how much needs to be transferred to the current hidden state h_t.

In each iteration, the GRU unit determines the gate values of z_t and r_t according to h_{t-1} and x_t, and output the result h_t which also is sent to the GRU unit in the next iteration. In many cases, GRU and LSTM have the same excellent results, but GRU is not easy to overfit and training is easier because of fewer parameters.

The popular variants of the RNN unit have been compared on a large number of datasets in [51] which points out that there is no significant difference between the units in the results, but when the calculation is insufficient (This is usually the case), the choice of the structure of RNN unit is still meaningful. It is generally believed that

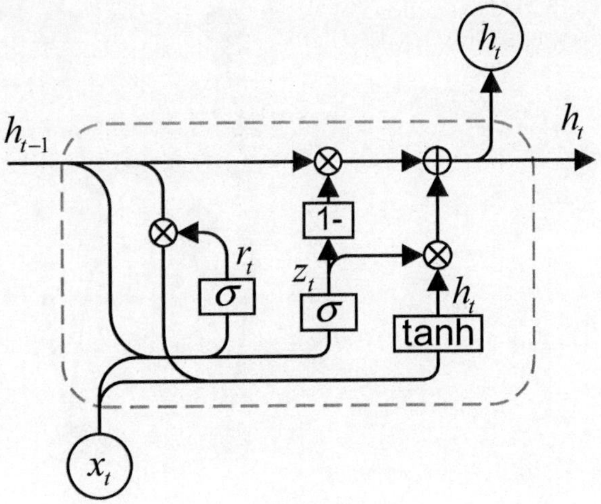

Fig. 5.19 The structure of GRU [48]

GRU and its variants only require a small dataset and relatively simple calculation, but have the capacity of short-term memory. On the contrary, the calculation of LSTM is more complex and the dataset required by LSTM is relatively large, but it is easier to remember long-term information. Therefore, LSTM is more advantageous in the case of a large dataset, affordable calculation, long-term signal, and large information span.

The disadvantage of RNN is that it can only analyze the current information, and it needs to propagate signals and errors between input and output iteratively. The number of iterations required for RNN increases with the number of sampling points. This leads to the forgetting of long-term information and the slow computation speed of RNN. Moreover, the problem of gradient explosion and gradient disappearance is more serious in training, so the formation of information processing channels is difficult in training.

In recent years, bidirectional RNN (biRNN) has been widely used in signal processing. Unlike ordinary RNN that can only transmit information in one direction, bidirectional RNN can transmit information forward and backward, as shown in Fig. 5.20. BiRNN combines the forward features and the backward features extracted at each time point as the temporal feature extracted at that time point. Considering the long-term forgetting characteristics of RNN, the information represented by the features extracted by biRNN at each time point is closer to the temporal information near the time point. Although the working mode is different from that of machine scanning and human reading, biRNN shows better performance than ordinary RNN in many studies.

In order to extract more complex and practical temporal features, the structure often used in practical applications is multi-layer RNN inspired by deep networks. Figure 5.21 shows a schematic diagram of multi-layer biRNN. The temporal features extracted from each RNN layer are sent to the next RNN layer. Compared with

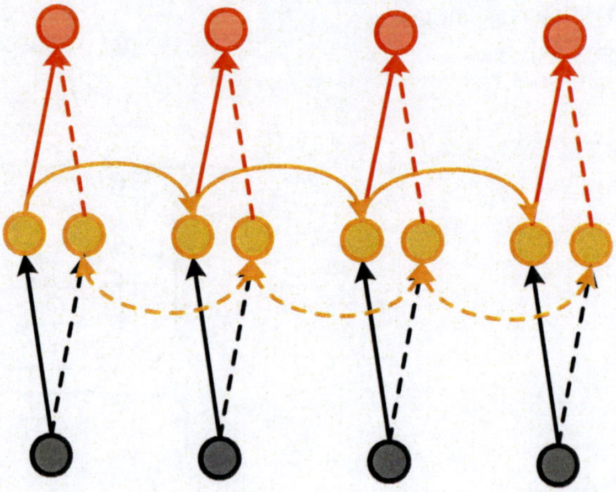

Fig. 5.20 The schematic diagram of biRNN [52]

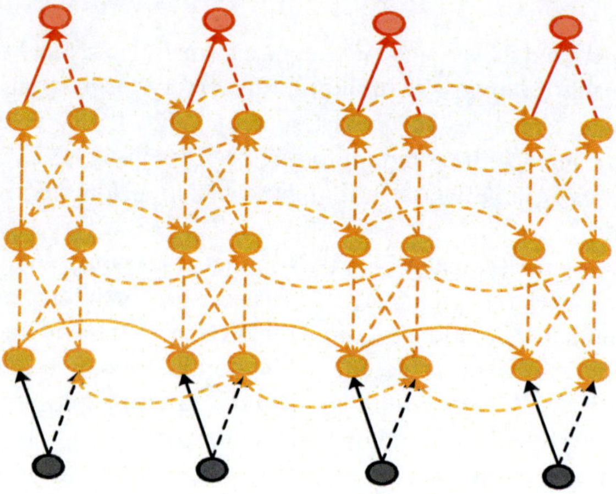

Fig. 5.21 The multi-layer biRNN [52]

the single-layer RNN network, the multi-layer network can realize more complex functions and extract richer features.

It should be noted that in the field of condition monitoring, the input signal is usually a sequence containing many sampling points, and the output is generally the monitoring state. The network used for condition monitoring is usually a multi-input single-output RNN. It is generally believed that the output of the feature by the RNN unit at the last moment summarizes all the valid information of the whole signal, and can make decisions accordingly. However, the information represented by the RNN unit at other times h_t and c_t are not complete which only needs to be updated but not used as output. The temporal features extracted from the multi-layer RNN are usually sent to fully connected layers to complete the prediction or classification. This is because RNN has extracted the temporal information from the signal, the

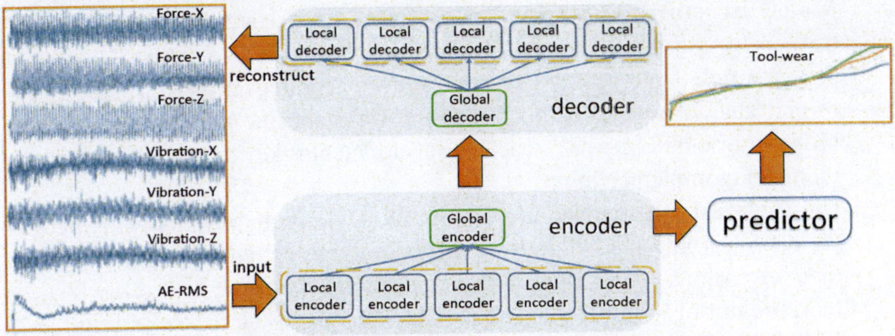

Fig. 5.22 The framework of pyramid LSTM auto-encoder. [53]

subsequent network only needs to realize the mapping to the output on this basis. Obviously, the fully connected network is more capable of this task.

Due to the harsh working conditions, time-varying, and non-stationary signals in the highly dynamical high-speed milling process, the milling condition monitoring is difficult to apply to practical production. It is necessary to find a robust and powerful tool wear feature extraction model. Guo et al. [53] proposed a pyramid LSTM auto-encoder based on the cascade nesting of milling signal to monitor tool wear, as shown in Fig. 5.22. The signal with a long duration is firstly decomposed into several short-term signals with a fixed length according to the periodic distribution of the Fourier transform. Then the underlying local encoder extracts the short-term features in turn. The upper global encoder extracts global features from these short-term features in the same way. The output of the global encoder in the last cycle is taken as the features extracted. The extracted features are sent into the fully connected network to realize the tool wear prediction.

By simplifying the units, reducing the number of units and connections, the efficiency of long-term signal processing is greatly improved. The introduction of a pyramid decoder can further improve the generality under varied milling conditions through unsupervised learning. Moreover, the inherent ability to process variable length signals also enhances the monitoring flexibility in practical milling. Compared with the classic multi-layer LSTM, pyramid LSTM auto-encoder has advantages in accuracy, efficiency, and stability. In addition, the visual analysis is carried out on each layer to reveal the evolution of features intuitively.

5.6.6 Challenges of Deep Learning Approaches in MSM Process Monitoring

Deep learning has shown good performances and great potential in MSM process monitoring. However, there are still many key issues to further investigate for its practical applications.

(1) **Model structure innovation.** Most of the deep learning models are still formed by the superposition of simple models (such as SAE, RBM, etc.) or the simple superposition of several deep learning models. This may fail to play the advantages of deep learning for a specific monitoring task. The problem-oriented network structure development is key to a successful quality monitoring implementation.

(2) **Network training method improvement**. At present, the training methods of the deep learning model are mainly random gradient descent algorithms and their extensions. The model with too many layers of the network will meet the problem of gradient vanishing. The strategies such as pre-training [54], ReLU [55], dropout, and batch normalization [56], may alleviate this problem. Another challenge is the small samples training problem, as in practice only limited samples are labeled for quality monitoring.

(3) **The selection of model hyper-parameters**. For practical problems, there are many hyper-parameters to select in deep learning, including data encoding parameters, network structures, model-updating rules. This is nontrivial and tedious. The Google DeepMind approach, which used Learning to Learn algorithm [57] and another network to adjust the learning rate, would speed up the convergence. Meanwhile, the interpretability of the model is also to be further studied for results explanation.

References

1. Ahmed H, Nandi AK (2020) Condition monitoring with vibration signals. Wiley
2. Farrar CR, Worden K (2013) Structural health monitoring: a machine learning perspective. Wiley
3. Duda RO, Hart PE, Stork DG (2001) Pattern classification. Wiley, US
4. Mallat SG (1998) A wavelet tour of signal processing, 2nd edn. Academic Press
5. Bendat JS, Piersol AG (2010) Random data: analysis and measurement procedures, 4 edn. Wiley
6. Haykin SO (2006) Neural networks and learning machines, 3rd edn. Prentice-Hall
7. Bishop (2006) Pattern recognition and machine learning. Springer
8. Zhu KP, Hong GS (2008) Feature selection for HMM in micro-milling. Mach Sci Technol
9. Zhu KP et al (2009) Multi-category tool wear classification HMM. Mechanical system and signal processing
10. Jolliffe IT, Anderson RE (2010) Multivariate data analysis with readings. J R Stat Soc Ser A (Stat Soc) 151(3):558
11. Scholkopf B, Smola A, Muller K-R (1998) Nonlinear component analysis as a kernel eigenvalue problem. Neural Comput 10(5):1299–1319
12. Mika S, Scholkopf B, Muller AK-R, Scholz M, Ratsch G (1999) Kernel PCA and denoising in feature spaces. Advances Neural Inform Process Syst 11
13. Christianini N, Shawe-Taylor J (2000) An introduction to support vector machines and other kernel-based learning methods. Cambridge University Press, Cambridge
14. Vaplink (2003) The nature of statistical learning. Springer
15. Kohonen T. Self-organizating maps, 3rd edn. Springer, New York
16. Buhmann MD (2003) Radial basis functions: theory and implementations. Cambridge University

17. Principe, J.C, Euliano NR, Lefebvre WC (1999) Neural and adaptive systems: fundamentals through simulations. Wiley
18. Ripley BD (1996) Pattern recognition and neural networks. Oxford University Press
19. Wang WH et al (2007) Sensor fusion for online tool condition monitoring in milling. Int J Prod Res
20. Dimla E, Lister PM (2000) On-line metal cutting tool condition monitoring. I: force and vibration analyses. Int J Mach Tools Manuf 40(5):739–768
21. Multivariate analysis. Academic Press (1976)
22. Johnson R, Wichern D (2006) Applied multivariate statistical analyses, 5th edn. Prentice-Hall
23. Izenman AJ (2007) Modern multivariate statistical techniques. Springer
24. Goodfellow I, Bengio Y, Courville A (2016) Deep learning. MIT press
25. LeCun Y, Bengio Y, Hinton G (2015) Deep learning. Nature 521: 436–444
26. Poggio T, Banburski A, Liao Q (2020) Theoretical issues in deep networks. PNAS
27. Zhao Z-Qi, Zheng P, Xu S-t, Wu X (2020) Object detection with deep learning: a review. IEEE Trans Neural Netw Learning Syst
28. Hinton G, Deng L, Yu D, Dahl GE, Mohamed A-R, Jaitly N, Senior A, Vanhoucke V, Nguyen P, Sainath TN et al (2012) Deep neural networks for acoustic modeling in speech recognition: the shared views of four research groups. IEEE Signal Process Mag 29(6):82–97
29. Jia F, Lei Y, Lin J et al (2016) Deep neural networks: s promising tool for fault characteristic mining and intelligent diagnosis of rotating machinery with massive data. Mech Syst Signal Process S72–73:303–315
30. Hinton GE, Salakhutdinov RR (2006) Reducing the dimensionality of data with neural networks. Science 313(5786):504–507
31. Salakhutdinov R, Hinton G (2009) Deep Boltzmann machines. J Mach Learn Res 5(2):1997–2006
32. Ranzato MA, Boureau YL, Lecun Y (2007) Sparse feature learning for deep belief networks. Adv Neural Inf Process Syst 20:1185–1192
33. Lecun Y, Bottou L, Bengio Y et al (1998) Gradient-based learning applied to document recognition. Proc IEEE 86(11):2278–2324
34. Krizhevsky A, Sutskever I, Hinton GE (2012) ImageNet classification with deep convolutional neural networks. Adv Neural Inf Process Syst 25(2):2012
35. Hihi SE, Hc-J MQ, Bengio Y (1995) Hierarchical recurrent neural networks for long-term dependencies. Adv Neural Inf Process Syst 8:493–499
36. Hochreiter S, Schmidhuber J (1997) Long short-term memory. Neural Comput 9(8):1735–1780
37. Goodfellow I, Pougetabadie J, Mirza M et al (2004) Generative adversarial Nets. Advances in neural information processing systems, pp 2672–2680
38. Vincent P, Larochelle H, Lajoie I, Bengio Y, Manzagol P (2010) Stacked denoising autoencoders: learning useful representations in a deep network with a local denoising criterion. J Mach Learn Res 11:3371–3408
39. Hinton GE (2002) Training products of experts by minimizing contrastive divergence. Neural Comput 14(8):1711–1800
40. Sutskever I, Tieleman (2010) On the convergence properties of contrastive divergence. In: Proceedings of the 13th international conference on artificial intelligence and statistics (AISTATS), Sardinia, Italy
41. Cheng C, Li J, Liu Y, Nie M, Wang W (2019) Deep convolutional neural network-based in-process tool condition monitoring in abrasive belt grinding. Comput Ind 106:1–13
42. Sun H, Zhang J, Mo R, Zhang X (2020) In-process tool condition forecasting based on a deep learning method. Robot Comput Integr Manuf 64(2020):101924
43. Wang P, Liu Z, Gao RX, Guo Y (2019) Heterogeneous data-driven hybrid machine learning for tool condition prognosis. CIRP Ann Manuf Technol 68:455–458
44. Shevchik SA, Kenel C, Leinenbach C, Wasmer K (2018) Acoustic emission for in situ quality monitoring in additive manufacturing using spectral convolutional neural networks. Addit Manuf 21:598–604

45. Caggiano A, Zhang J, Alfieri V, Caiazzo F, Gao R, Teti R (2019) Machine learning-based image processing for on-line defect recognition in additive manufacturing. CIRP Ann Manuf Technol 68:451–454
46. Zhang B, Zhang S, Li W (2019) Bearing performance degradation assessment using long short-term memory recurrent network. Comput Ind 106:14–29
47. Chow TWS, Fang Y (1998) A recurrent neural-network based real-time learning control strategy applying to nonlinear systems with unknown dynamics. IEEE Trans Ind Electron 45(1):151–161
48. Cho K, Merriënboer V, Gulcehre C, Bahdanau D, Bougares F, Schwenk H, Bengio Y (2014) Learning phrase representations using RNN encoder-decoder for statistical machine translation. arXiv preprint. arXiv:1406.1078
49. Guo L, Li N, Jia F et al (2017) A recurrent neural network based health indicator for remaining useful life prediction of bearings. Neurocomputing 240(C):98–109
50. Zhang JJ, Wang P, Yan RQ et al (2018) Long short-term memory for machine remaining life prediction. J Manuf Syst 48:78–86
51. Greff K, Srivastava RK , Koutník J et al (2015) LSTM: a search space Odyssey. IEEE Trans Neural Networks Learn 28(10):2222–2232
52. Schuster M, Paliwal KK (1997) Bidirectional recurrent neural networks. EEE Trans Signal Process 45(11):2673–2681
53. Guo H, Zhu KP (2020) Pyramid LSTM auto-encoder for tool wear monitoring. In: Proceedings of CASE, Hong Kong
54. Hinton GE, Osindero S, Teh YW (2006) A fast learning algorithm for deep belief nets. Neural Comput 18(7):1527–1554
55. Glorot X, Bordes A, Bengio Y (2011) Deep sparse rectifier neural networks. In: Proceedings of the 14th international conference on artificial intelligence and statistics (AISTATS), pp 315–323
56. Ioffe S, Szegedy C (2015) Batch normalization: accelerating deep network training by reducing internal covariate shift. Proceedings of the 32nd international conference on machine learning, vol 37, pp 448–456
57. Andrychowicz M, Denil M, Gomez S et al (2016) Learning to learn by gradient descent by gradient descent. NIPS 2016 3988–3996

Chapter 6
Signal Processing for Machining Process Modeling and Condition Monitoring

6.1 Signal Processing in Condition Monitoring

Signal processing plays an important role in manufacturing automation and industrial control. In CNC machining, by utilizing the monitored information from the platform to direct the further actions, the signal processing bridges the gap of human instruction and full automation. Effective signal processing techniques are prerequisites for smart machining system monitoring. Fourier transform is the basis of modern signal processing and has been the main signal processing method. It associates the signal from the time domain to the frequency domain and provides useful information. Wavelet analysis is inherited from Fourier analysis and is most suitable for analyzing nonstationary signal processing in machining. This chapter serves as an introduction to the conventional signal analysis methods and some recently developed approaches, i.e. wavelet analysis and sparse decomposition, regarding their applications to signal modeling and condition monitoring.

6.1.1 Overview of Condition Monitoring

As introduced in the Sect. 5.2, the smart machining monitoring system can be regarded as a in a typical machine learning problem (Fig. 6.1). Under this framework, the objective of condition monitoring is to search the most probable state C_i given the extracted measurable signal feature $y(t)$ at time t. The sensor signal $x(t)$ is firstly pre-processed to remove the noise and prepare the data for feature extraction. Then the information relevant to pattern classification is extracted from $x(t)$ to a feature vector $y(t)$. The task of feature extraction is to enhance the characteristics of the various tool wear classes and suppress or filter off the normal background. The final stage is state classification. Feature vector $y(t)$ is assigned to one of the K tools wear states, C_1, C_2, … C_K, by the classifier based on a certain type of classification criteria. Practically indirect measurement methods such as force, acoustic emission, and vibration

K. Zhu, *Smart Machining Systems*, Springer Series in Advanced Manufacturing,
https://doi.org/10.1007/978-3-030-87878-8_6

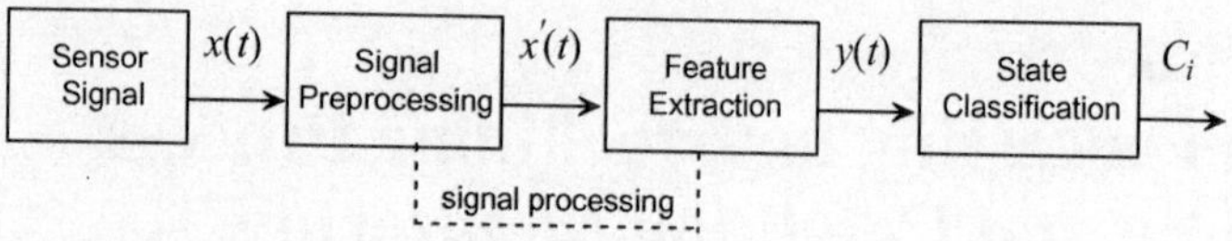

Fig. 6.1 Basic condition monitoring scheme

are used to capture machining signals. After the signal's pre-processing and feature extraction, approaches such as autoregressive models, and Hidden Markov models (HMMs) are then applied to estimate the machining states.

6.1.2 Signal Processing Issues in Condition Monitoring

Although some monitoring signals can directly provide some useful information for the monitoring and detection, their discriminative and robust capabilities are often very limited. Generally, the features that reflect the condition of the equipment are extracted from the monitoring signal through signal processing and analysis. The feature parameters extracted by different signal analysis methods are different, but the purpose is the same: remove or weaken the interference information in the signal, and extract feature parameters that reflect the condition of the equipment from different aspects of the signal.

The task of signal processing is to understand the intrinsic characteristics of signals in the objective world. Understanding and analyzing signals from different aspects helps to understand the intrinsic characteristics of signals. The main methods are the time domain, frequency domain, time–frequency domain, and model-based analysis method [1–5]. The signal was originally expressed in the form of time (space). For example, the signal time-domain analysis method is a process that directly performs various operations and processing on the monitoring signal (time history function) and extracts various time-domain feature parameters that reflect the condition of the equipment.

Besides time, frequency is the most important way to express signal characteristics. The frequency domain analysis is the process that the monitoring signal is first converted to the frequency domain function, and then perform operations to extract various frequency domain features that reflect the condition of the system. The frequency representation method is based on Fourier transform [6, 7]. On the other side, as the Fourier transform is a global transformation, either completely in the time domain or completely in the frequency domain, it is impossible to express the time–frequency local properties of the signal. However, the local properties of time–frequency are exactly the most basic and critical properties of non-stationary signals. To analyse and process non-stationary signals, based on the Fourier analysis theory, a series of new signal analysis theories will be discussed separately in this chapter: short-time Fourier transform, Gabor transform, wavelet transform, etc. [8–10].

These signal processing methods, with their advanced mathematical theory, are promising that could extract useful information for condition monitoring [10–20]. Unfortunately, the signal processing works to date are predominantly oriented toward applications with strong ties to the communication field, or toward theoretical issues which are far beyond the scopes of mechanical engineering. The purpose of this chapter is to bring the academic researchers and the industrial users together to identify the promising signal processing technologies that meet the demands of machining system condition monitoring, which is ultimately applied for the process control and optimization. However, an exhaustive presentation of these vast topics would be beyond the scope and possibilities of a single chapter. The signal processing approaches discussed in this chapter are expressed with linear transforms, from Fourier transform to sparse decomposition, focusing on wavelet transform. The outline of the chapter is the following: First, some basic concepts of signal processing are reviewed including signal space, the Fourier transform, and sampling. Then, classical spectral analysis methods based on the Fourier transform and their limitations will be discussed. In the following, the theory of wavelet analysis is introduced and discussed in more detail. The theory of sparse signal representation is then introduced. In the last part of the chapter, some of the approaches' applications to tool condition monitoring (TCM) are presented.

6.2 Signal Space, Linear System, and Fourier Transform

For a fundamental understanding of signal theory, the basic knowledge of functional analysis is needed, such as the definition of Hilbert space, metric space, frame, and operator theory. These concepts are far beyond the limit of this chapter, and we only introduce the theory related to Fourier and wavelet analysis. The theories presented herein are not mathematical rigorous but aimed for easy understanding. The theory of signal analysis can be referred in [21] and [22] from a vector space of view.

6.2.1 Signal Spaces and Inner Product

In a broad sense, a signal $f(t)$ is defined as a time-varying function of time variable t,

$$\text{Signal} = \text{function } f(t), t \in R \tag{6.1}$$

A signal space is simply a collection of signals (functions) that satisfies a certain mathematical structure. The signal spaces with finite energy and finite power structures are particularly interested in signal processing. Finite energy space $L_2(R)$ is the vector space of finite energy functions (or say, square integral) such that,

$$\int_{-\infty}^{\infty} |f(t)|^2 dt < +\infty \tag{6.2}$$

The signal in this space is called energy signal. For a signal $f(t)$, if the energy is infinite, but the mean power is finite in the time period (−T/2, T/2) in finite,

$$\frac{1}{T}\int_{-T/2}^{T/2} |f(t)|^2 dt < +\infty \tag{6.3}$$

It is called power signal. Most signals encountered in technical applications belong to these two classes.

If $f(t)$ is an energy signal, then we can consider $f(t)$ belonging to the Hilbert space, $L_2(R)$. The reader unfamiliar with Hilbert space may think of it as a generalization of the three-dimensional Euclidean space, in which the concepts of distance and angle are extended as the norm and inner product respectively. Any signal in the Hilbert space has the basic operation properties that can be added, subtracted, and multiplied by constants.

The inner product of two signals $x(t)$ and $y(t)$ is defined as,

$$\text{inner product} = \langle x(t), y(t)\rangle = \int_{-\infty}^{\infty} x(t)y^*(t)dt \tag{6.4}$$

where $y^*(t)$ is the complex conjugate of $y(t)$, and if $y(t)$ is real then $y^*(t) = y(t)$. The inner product is a generalization of the dot product in the signal (vector) space.

The $x(t)$ and $y(t)$ are said to be orthogonal if their inner product $\langle x(t), y(t)\rangle = 0$, and if also each vector is a unit vector, the vector set is called orthonormal. The collection of these vectors can span completely any vector in the space. They are called the orthogonal (orthonormal) basis. The inner product can be roughly understood as the correlation of $x(t)$ and $y(t)$: The higher their correlation, the larger the absolute value. If they are uncorrelated, then $\langle x(t), y(t)\rangle = 0$, $x(t)$ and $y(t)$ are orthogonal.

In the Hilbert space, we can always find a set of functions $\{f_i(t)\}_{i=1}^{\infty}$ in $L_2(R)$ that are orthonormal, i.e., $\langle f_i(t), f_j(t)\rangle = 0, \, when\, i \neq j$, such that any $s(t) \in L_2(R)$, we have,

$$s(t) = \sum_{i=1}^{\infty} s_i f_i(t) \tag{6.5}$$

The set $\{f_i(t)\}_{i=1}^{\infty}$ is called a complete basis for $L_2(R)$. This leads to the signal's linear decompositions with the basis for the later parts of this chapter.

6.2.2 Fourier Transform

For any signal $f(t)$, the Fourier Transform $\hat{f}(\omega)$ is obtained by the inner product of $f(t)$ with a sinusoidal wave $e^{j\omega t}$,

$$\hat{f}(\omega) = \left\langle f(t), e^{j\omega t} \right\rangle = \int_{-\infty}^{\infty} f(t) e^{-j\omega t} dt \tag{6.6}$$

The FT transforms the signal $f(t)$ from the time domain into the frequency domain and results in the amplitudes of the frequency ω.

Another equation reverses the transform and takes the signal from the frequency domain back to the time domain,

$$f(t) = \frac{1}{2\pi} \int_{-\infty}^{\infty} \hat{f}(\omega) e^{j\omega t} d\omega \tag{6.7}$$

Equations 6.6 and 6.7 can be used on any signal, periodic or non-periodic. The FT lays down the basis of modern signal processing. It was extensively applied to machining process monitoring, for example, Dimla and Lister [12], and Elbestawi et al. [13].

6.2.3 Linear System, Sampling Theorem, and Convolution

The Fourier transform is powerful and dominates signal processing because the sinusoidal waves $e^{-j\omega t}$ are the eigenvectors of the time-invariant linear system. A system or transform maps an input signal $x(t)$ into an output signal $y(t)$,

$$y(t) = T[x(t)] \tag{6.8}$$

where T denotes the system or transforms, a function from input signals to output signals.

Systems come in a wide variety of types. One important class is known as linear systems. A linear system obeys rules of linearity are homogeneity and additivity. Specifically, a linear system T that maps a vector space X into a vector space Y is said to be linear if for every scalars α, β we have,

$$T(\alpha x + \beta y) = \alpha T(x) + \beta T(y) \tag{6.9}$$

Please be noted that the linear system here refers to the signal processing system, not to claim that our machining system is linear. Linear signal processing brings a lot of good properties and favorable with signal analysis. Fourier analysis, wavelet analysis, and sparse decomposition are all linear. The nonlinear time–frequency approaches such as Wigner-Ville distribution [4] are not discussed in this paper.

In practice, the continuous signals are sampled at the discrete-time to process. It can be completely represented by a set of equally spaced samples if the samples occur at more than twice the frequency of the highest frequency component of the signal. This sampling frequency is called the Nyquist frequency [22]. If we sample the signal $x(t)$ with sampling frequency f_s (sampling interval $T_s = 1/f_s$), then we generate the sequence $\{\ldots, x(-nT_s), \ldots, x(-T_s), x(0), x(T_s), \ldots, x(nT_s), \ldots\}$,

$$y(t) = \sum_{n=-\infty}^{\infty} x(nT_s) \cdot \delta(t - nT_s) = x(t) \cdot \sum_{k=-\infty}^{\infty} \delta(t - nT_s) \tag{6.10}$$

where $\delta(t - nT_s) = 1$ at $t - nT_s$ and equals to 0 elsewhere. The maximum seeable frequency in the $X(\omega)$ by DFT is $f_s/2$ according to this theory.

The Nyquist frequency has dominated the sampling principles in signal processing, but it has been discovered lately Candès et al. [23] and Donoho [24] that this constraint is not necessary when the signal is sparse in some domain (either time, frequency, or else). This finding leads to the newly developed theory named Sparse Decomposition (or compressive sensing/sampling in information theory), which is to be discussed in Sect. 6.4.

Convolution is another important concept in signal processing. For a linear system, let $h(t)$ be the impulse response (the output of the system to the input of delta function) of the system, then for any input $x(t)$, the output $y(t)$ is obtained by convolution,

$$y(t) = Lx(t) = h(t) * x(t) = \int_{-\infty}^{\infty} h(\tau)x(t - \tau)d\tau \tag{6.11}$$

then the system is called linear.

The convolution has a clear physical meaning that: At any time, the system's output $y(t)$ is obtained by the input $x(t)$ and the convolution of the impulse response of the system $h(t)$. And if $y(t - t_0) = Lx(t - t_0)$, any input delay will result in an output delay, the system is called linear time-invariant system (LTI).

6.3 Spectrum Analysis of Machining Signals

6.3.1 The Spectrum of Machining Signals

The signals in engineering applications are generally non-periodic signals with periodic components. When the period of the periodic signal tends to infinity, the original periodic signal can be treated as a non-periodic signal. At this time, the interval between adjacent spectral lines of the signal tends to be infinitesimal, and the spectral lines become denser and denser, eventually becoming a continuous spectrum. Although the amplitude of each frequency component also tends to be correspondingly infinitesimal, these components still maintain a certain proportional relationship.

For non-periodic signals, Fourier transform is needed to calculate its frequency spectrum. The sufficient condition for the existence of Fourier transform of non-periodic function a is that $x(t)$ is integral in the interval $-\infty$, $+\infty$, namely as shown in Eq. 6.18 [22].

$$\int_{-\infty}^{\infty} |x(t)|\mathrm{d}t < \infty \tag{6.12}$$

The Fourier transform $X(\omega)$ of signal $x(t)$ is defined as Eq. 6.13.

$$X(\omega) = \int_{-\infty}^{\infty} x(t)\mathrm{e}^{-\mathrm{j}\omega t}\mathrm{d}t \tag{6.13}$$

The corresponding inverse Fourier transform is as follows.

$$x(t) = \frac{1}{2\pi}\int_{-\infty}^{\infty} X(\omega)\mathrm{e}^{\mathrm{j}\omega t}\mathrm{d}\omega \tag{6.14}$$

Or, as,

$$X(f) = \int_{-\infty}^{\infty} x(t)\mathrm{e}^{-\mathrm{j}2\pi f t}\mathrm{d}t \tag{6.15}$$

$$x(t) = \int_{-\infty}^{\infty} X(f)\mathrm{e}^{\mathrm{j}2\pi f t}\mathrm{d}f \tag{6.16}$$

A non-periodic function can be decomposed into the superposition of harmonics with continuously changing frequency f, where the $X(f)\mathrm{d}f$ is the coefficients of the harmonics $\mathrm{e}^{\mathrm{j}2\pi ft}$, which determines the amplitude and phase of the signal. Because $\mathrm{d}f$ in the $X(f)\mathrm{d}f$ is same for different frequency f and only $X(f)$ can reflect the variation of amplitude and phase of different harmonic components, so the $X(f)$ is called the continuous spectrum of $x(t)$. Since $X(f)$ is generally a complex function of the real variable f, it can be written as Eq. 6.17.

$$X(f) = |X(f)|\mathrm{e}^{\mathrm{j}\varphi(f)} \tag{6.17}$$

where the $|X(f)|$ is called the amplitude spectrum of the non-periodic signal and the $\varphi(f)$ is called phase spectrum.

It should be noted that although the amplitude spectrum $|X(f)|$ of the non-periodic signal is the same as the amplitude spectrum $|C_n|$ of the periodic signal in name, but $|X(f)|$ is continuous while $|C_n|$ is discrete.

In addition, the two are also different in dimension. $|C_n|$ is consistent with the dimension of the signal amplitude, while the dimension of b is inconsistent with the dimension of the signal. The dimensions of $x(t)$ and $X(f)\mathrm{d}f$ are the same, and $X(f)$ is the amplitude per unit bandwidth. Therefore, strictly speaking, $|X(f)|$ is spectral density function.

The Fourier transform establishes the relationship between the time domain and the frequency domain of signal, and frequency is one of the physical natures of the signal. The Fourier transform defined by the above formula is essentially an integral calculation, which is characterized by continuity, and the signals are obtained by discretized sampling in practical applications. To sample signal through discretization and effectively use a computer to realize the calculation of the Fourier transform, it is necessary to realize efficient and high-precision discretization of the Fourier transform pairs. For this reason, discrete Fourier transform (DFT) is needed.

Given a set of orthogonal basis (vector):

$$\mathbf{\Phi}_k = \left\{1, \mathrm{e}^{\frac{2k\pi}{N}}, \mathrm{e}^{\frac{2k\cdot 2\pi}{N}}, \ldots, \mathrm{e}^{\frac{2k\cdot(N-1)\pi}{N}}\right\}, k = 0, 1, 2, \ldots, N-1.$$

The basis satisfies the inner product relationship directly: $\langle \mathbf{\Phi}_k, \mathbf{\Phi}_l \rangle = N\delta_{k,l}\mathbf{I}_N$, where the I_N is N order unit matrix.

$$\delta_{k,l} = \begin{cases} 1, k = l \\ 0, k \neq l \end{cases} \tag{6.18}$$

$$ck = \sum_{n=0}^{N-1} x_n W_N^{nk} \tag{6.19}$$

where $k = 0, 1, 2, \cdots, N-1$; $W_N = e^{-\frac{2\pi}{N}\mathrm{i}}$

$$X_l = \frac{1}{N\pi}\sum_{k=0}^{N-1} c_k \int_{-\pi}^{\pi} e^{i(k-l)x}dx = \frac{2\pi}{N}\sum_{n=0}^{N-1} x_n W_N^{nl} \tag{6.20}$$

In addition to the constant 2π, then Eq. 6.20 is the discrete Fourier transform (DFT) in common sense. Where the input x_n and output X_l are the discrete information in the time domain and frequency domain of the signal, respectively.

In particular, if another orthogonal basis is used, discrete orthogonal transformations with different meanings can be obtained by using least square approximation, such as discrete cosine transform (DCT, 4 types in total), discrete sine transform (DST, 4 types in total), discrete Hartley transforms (DHT) and discrete Walsh transform (including Discrete Hadmard transform), etc. Interesting readers can refer to other relevant literature [1, 8, 22].

6.3.2 Spectrum Characteristics of Stochastic Signals

The amplitude and phase of the stochastic signal change irregularly, unknowingly, and randomly over time. This time it was this waveform, and the next time it might be a different waveform. The stochastic signal cannot be expressed by a definite time function, but can only be expressed by its statistical characteristics over time or by its amplitude. These statistical characteristic values are as follows:

(1) Mean, describing the average value of a stochastic signal.
(2) Variance, describing the intensity of amplitude variation of stochastic signal.
(3) Probability density function, describing the probability of the amplitude of the signal.
(4) Correlation function, a function of time interval, describing the correlation degree between each two amplitude values of a stochastic signal with a certain time interval.
(5) Power spectral density, describing the power spectral characteristics of stochastic signals in the average meaning.

These statistical characteristics are the main digital characteristics describing stochastic signals. The mathematical method of studying stochastic signals is stochastic process theory.

(1) **Power spectral density**

The power spectrum is the abbreviation of the power spectral density function, which is defined as the signal power in the unit frequency band, the unit is W/Hz. For power limited signals (energy limited signal using energy spectrum density), the signal power changes with frequency in the unit frequency band, that is, the distribution of signal power in the frequency domain.

The average power of the power signal $x(t)$ can be represented by the mean square value, namely as shown in Eq. 6.21.

$$\psi_x^2 = \lim_{T\to\infty} \frac{1}{T} \int_{-\frac{T}{2}}^{\frac{T}{2}} x^2(t)\mathrm{d}t \tag{6.21}$$

$$\lim_{T\to\infty} \frac{1}{T} \int_{-\infty}^{\infty} x^2(t)\mathrm{d}t = \frac{1}{2\pi} \int_{-\infty}^{\infty} \lim_{T\to\infty} \frac{|X(\omega)|^2}{T} \mathrm{d}\omega \tag{6.22}$$

Let $S_x(\omega) = \lim_{T\to\infty} \frac{|X(\omega)|^2}{T}$, then the average powder $\psi_x^2 = \frac{1}{2\pi} \int_{-\infty}^{\infty} S_x(\omega)\mathrm{d}\omega$. As $S_x(\omega)$ has the average power dimension of unit frequency, it is often called the power spectral density function, which describes the distribution of the average power of the signal concerning frequency.

According to the Wiener-Khintchine formula [8], the power spectral density $S_x(\omega)$ and autocorrelation function $R_x(\tau)$ of the stationary stochastic process are a pair of Fourier transforms, namely as shown in Eqs. 6.23, 6.24.

$$R_x(\tau) = \frac{1}{2\pi} \int_{-\infty}^{\infty} S_x(\omega)\mathrm{e}^{\mathrm{j}\omega\tau}\mathrm{d}\omega \tag{6.23}$$

$$S_x(\omega) = \int_{-\infty}^{\infty} R_x(\tau)\mathrm{e}^{-\mathrm{j}\omega\tau}\mathrm{d}\tau \tag{6.24}$$

The power spectral density function can be obtained by the Fourier transform of the autocorrelation function.

The power spectrum is often used for the expression and analysis of power signals (different from energy signals). Generally, the horizontal coordinate of the curve (the power spectrum curve) is frequency, and the vertical ordinate is power. The frequency spectrum of the periodic continuous signal $x(t)$ can be expressed as a discrete non-periodic sequence X_n, and the sequence formed by the square of its amplitude spectrum $|X_n|^2$ is called the "power spectrum" of the periodic signal. Since the power has no negative value, there is no negative vertical ordinate value on the power spectrum curve. The area covered by the power spectrum curve is numerically equal to the total power (energy) of the signal.

There are two calculation methods for power spectrum:

(1) The square of Fourier transform/length of the interval;
(2) Fourier transform of the autocorrelation function.

These two methods are called the direct method and correlation function method respectively.

Figure 6.2 is the power spectrum obtained by the direct method, and the result is theoretically the same as the correlation function method.

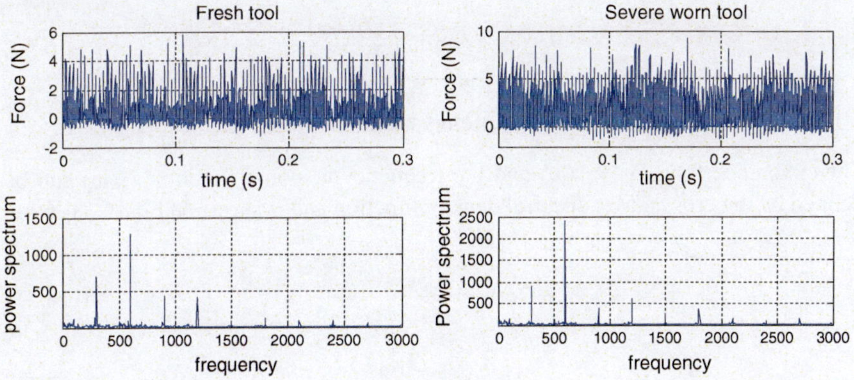

Fig. 6.2 Signal spectrum

(2) Cross power spectral density

The cross power spectral density function $S_{xy}(w)$ and the cross-correlation function $R_{xy}(\omega)$ between two stochastic signals $x(t)$ and $y(t)$ form a pair of Fourier transforms, namely as shown in Eqs. 6.25, 6.26.

$$S_{xy}(\omega) = \int_{-\infty}^{\infty} R_{xy}(\tau)\mathrm{e}^{-\mathrm{j}\omega\tau}\mathrm{d}\tau \tag{6.25}$$

$$R_{xy}(\tau) = \frac{1}{2\pi}\int_{-\infty}^{\infty} S_{xy}(\omega)\mathrm{e}^{\mathrm{j}\omega\tau}\mathrm{d}\omega \tag{6.26}$$

Unilateral cross power spectral density function is expressed in Eq. 6.27.

$$G_{xy}(\omega) = 2\int_{-\infty}^{\infty} R_{xy}(\tau)\mathrm{e}^{-\mathrm{j}\omega\tau}\mathrm{d}\tau, (0 <\omega<\infty) \tag{6.27}$$

Due to the cross-correlation function is a non-even function, the cross power spectral function is a plurality.

$$G_{xy}(\omega) = C_{xy}(\omega) - \mathrm{j}Q_{xy}(\omega) = \left|G_{xy}(\omega)\right|\mathrm{e}^{-\mathrm{j}\theta_{xy}(\omega)} \tag{6.28}$$

In practical applications, the amplitude and phase of the spectral density are commonly used to express the cross power spectral density function, namely is expressed in Eqs. 6.29, 6.30.

$$\left|G_{xy}(\omega)\right| = \sqrt{C_{xy}^2(\omega) + Q_{xy}^2(\omega)} \tag{6.29}$$

$$\theta_{xy}(\omega) = \arctan \frac{Q_{xy}(\omega)}{C_{xy}(\omega)} \tag{6.30}$$

(3) **Coherence function and frequency response function**

The coherence function $\gamma_{xy}(\omega)$ and the frequency response function $H(\omega)$ can be defined by the cross power spectral density function and is shown in Eqs. 6.31, 6.32.

$$\gamma_{xy}^2(\omega) = \frac{|G_{xy}(\omega)|^2}{G_x(\omega)G_y(\omega)} \tag{6.31}$$

$$H(\omega) = \frac{G_{xy}(\omega)}{G_x(\omega)} \tag{6.32}$$

The coherence function, also known as aggregate function, is a dimensionless proportional coefficient that describes the causal relationship between two signals in the frequency domain, which is a discriminant indicator to indicate whether the two signals are correlated in the frequency domain [2].

It links the cross power spectral density between the two measurement point signals with their respective auto-power spectral density to determine the frequency components of the output $y(t)$ and how much it comes from the input signal $x(t)$. Through that, the influence of input signal on output signal can be known, which is very useful in the identification of fault cause.

6.4 Correlation Analysis

The correlation function describes the correlative degree between the values of the signals $x(s)$ and $y(t)$ (the signals can be stochastic or deterministic) at any two different times s and t.

6.4.1 Autocorrelation Function

The autocorrelation function is used to describe the linear dependence between two values of a signal at a certain time interval. The larger the function value (correlation value), the stronger the correlation degree of the signal, and vice versa [2]. The autocorrelation function is defined as follows.

$$R_x(n) = E\left[x_t x_{t+n}\right] \tag{6.33}$$

Its discrete calculation formula is shown in Eq. 6.34:

$$R_x(n) = \frac{1}{N-n}\sum_{i=1}^{N-n} x_i x_{i+n} \tag{6.34}$$

where the n is called the delay number, which value is 0, 1, 2, ..., m, and $m << N$.

When $n = 0$, the autocorrelation function of the stochastic signal has the largest value (correlation value), the best correlation, and complete linear correlation. When n increases, the function value decreases rapidly and the correlation rapidly weakens. The autocorrelation function of a periodic signal is a periodic function of the same frequency.

6.4.2 Cross-Correlation Function

Autocorrelation is a special case of cross-correlation. The cross-correlation function describes the dependence of the values of the stochastic signal $x(t)$ and $y(t)$ at any two different moments i and $i + n$, which is defined as Eq. 6.35.

$$R_{xy}(n) = \frac{1}{N-n}\sum_{i=1}^{N-n} x_i y_{i+n} \tag{6.35}$$

In image processing, the definition of autocorrelation and cross-correlation function are as follows: Let the original function be $f(t)$, then the autocorrelation function is defined as $R(u) = f(t)f * (-t)$, where the * represents convolution. Let two functions be $f(t)$ and $g(t)$ respectively, then the cross-correlation function is as $R(u) = f(t)f * (-t)$, which reflects the degree of mutual matching between two functions in different relative positions.

Both correlation functions are the measurement of correlation, that is, similarity. Autocorrelation is the correlation between the function and the function itself. When there are periodic components in the function, the maximum value of the autocorrelation function can reflect this periodicity. Cross-correlation is the similarity between two functions. When both functions have the same periodic component, the maximum value of the cross-correlation function can also reflect this periodic component.

From the perspective of linear space, correlation operations are inner product operations, and the inner product of two vectors in linear space represents the projection of one vector to another vector, indicating the similarity of the two vectors, so correlation operations reflect this kind of similarity.

6.5 Common Signal Features in Time and Frequency Domain

There are two kinds of signal feature parameters widely used in condition monitoring and fault diagnosis: the characteristic parameters of state signals and the parameters of mathematical models of signals. There are many feature parameters in the time domain, such as signal amplitude and its statistics [2, 3]. We only introduce some common parameters in the time domain and frequency domain here. The parameters of the mathematical model, such as the coefficients and residuals of time series, are the description of signal parameters after signal modelling.

It is generally considered that the normal operation machining process is a stochastic process, which is stationary, normal, and ergodic. Stationary means that the statistical characteristic parameter sets of a random process do not change with time. Normality means every random variable in the random process follows the normal distribution. Ergodicity means that the statistical characteristic parameter sets of a random process are the same as those of any sample function (a long-time monitoring record). Therefore, the statistical characteristic parameter sets of the equipment operation process can be calculated based on a sample function (a state signal of the equipment) obtained by the monitoring method. This method not only saves plenty of computing time but also greatly reduces the workload of monitoring. Of course, in practical application, the sample length is always limited so that the statistical characteristic parameter values calculated according to the finite state signal are estimated values, which are not true values, and have certain errors.

6.5.1 Feature Parameters in the Time Domain

Signal feature values, which are usually divided into dimensional parameters and dimensionless parameters, are important indexes to represent the characteristics of signals. "Dimension" can be simply understood as "Units". A dimensional parameter has units, for example in the aspect of mean, the units of the average values of a temperature set (units is Centigrade) being still Centigrade (°C). A dimensionless parameter has no units. It is usually written as the product or ratio of dimensional parameters and the final dimensionless value can be gotten after the elimination of dimension with each other. For example, strain, which can be defined as the ratio of length difference to the original length, is a measurement of deformation. Feature values with dimension normally have intuitive physical meanings and they are the most commonly used characteristic indicator. Dimension feature values mainly include: maximum, minimum, peak-to-peak amplitude, mean, variance, standard deviation, mean square value, root mean square (RMS) value, mean square error (MSE), root mean square error (RMSE), square root amplitude, and so on.

If the time series of a state signal's feature is $x_1, x_2, \ldots, x_N$, the common signal feature parameters and their estimation equations are shown as follow:

(1) Mean: $\mu_x = \frac{1}{N}\sum_{i=1}^{N} x_i$, which is also the first moment, is the average value of the signal.
(2) Variance: $\sigma_x^2 = \frac{1}{N}\sum_{i=1}^{N}(x_i - \mu_x)^2$

Standard deviation: The positive root of the variance σ_x is the standard deviation or root-mean-square deviation.

The variance is the average of the squares of the differences between the mean of each sample and the mean of the whole sample. It represents the dynamic component of the signal energy; however, the square of mean is the static component. The variance shows the dispersion degree between data, which can be seen as the distance between the variable and the expectation. Meanwhile, it is also the second-order central moment.

(3) Mean Square (average power): $\psi_x^2 = \frac{1}{N}\sum_{i=1}^{N} x_i^2$.

Root Mean Square (RMS): The positive square root of the mean square is the root mean square x_{rms}. In physical terms, the mean square represents the average power of the signal; the expectation represents the direct current (DC) component; the variance represents the alternating current (AC) component.

Root Mean Square (RMS) is also known as the effective value. It can be gotten through summing all squared values, achieving its mean, and making the square root of the mean. It can be understood in the following physical terms: Let DC and AC pass through the same resistance respectively. If DC and AC consume the same amount of energy at the same time, the value of DC can be seen as the effective value of AC.

(4) Peak Amplitude: $x_{\max} = E\{\max|x_i|\}$.

In this equation, $\max|x_i|$ is the data with the largest absolute value in each equal segment of a time series.

(5) Skewness: $\alpha_x = \frac{1}{N}\sum_{i=1}^{N}(x_i - \mu_x)^3$.
(6) Kurtosis: $\beta_x = \frac{1}{N}\sum_{i=1}^{N}(x_i - \mu_x)^4$.

The mean is the static (stable) component of the signal and the distribution centre of the signal value; the variance and standard deviation reflect the dispersion degree of the signal value to the distribution centre; the standard deviation is a measure of the dynamic component of a signal; the mean square represents the signal power. When the equipment is running normally, its state is stable and it has small values in dynamic component, variance, and means square. However, all of these values increase when a failure occurs to the equipment. The peak amplitude, which is often used as the measure of velocity and acceleration, reflects the magnitude of the instantaneous amplitude. Whereas, the measure of displacement usually uses peak-to-peak amplitude (the average of the distance between peaks and troughs in a fluctuating signal) rather than peak amplitude.

Skewness reflects the degree to which the distribution of the data $x_1, x_2, ..., x_N$ deviates from the ideal centre. When the equipment is running normally, monitoring

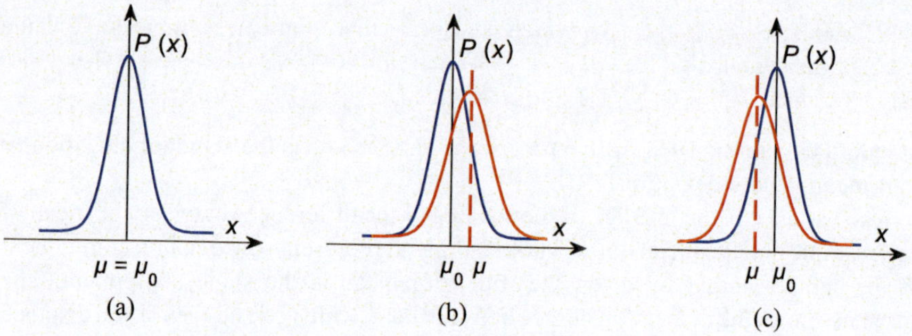

Fig. 6.3 The meaning of skewness. **a** Normal. **b** Positive. **c** Negative

data should follow normal distribution and the distribution centre is the ideal value μ_0, which is shown in Fig. 6.3a. If the mean of monitoring data is μ, and

$$\frac{1}{N}\sum_{i=1}^{N}(x_i - \mu)^3 = 0 \tag{6.36}$$

where the average deviation's cube between data x_i and mean μ equals zero, it can be concluded that $\mu = \mu_0$ and the data distribution is normal without skewness. If the equation is:

$$\frac{1}{N}\sum_{i=1}^{N}(x_i - \mu)^3 > 0 \tag{6.37}$$

It can be concluded that $\mu > \mu_0$ and the data distribution, which is shown in Fig. 6.3b, is abnormal and leans to the right. If the equation is:

$$\frac{1}{N}\sum_{i=1}^{N}(x_i - \mu)^3 < 0 \tag{6.38}$$

It can be concluded that $\mu < \mu_0$ and the data distribution, which is shown in Fig. 6.3c, is also abnormal and leans to the left. No matter which side the distribution leans to, this phenomenon is the sign of equipment abnormality. The larger the absolute value of skewness is, the more serious the degree of abnormality is.

Kurtosis reflects the convex peaks of the data distribution curve. Compared with a specified kurtosis parameter, which can be seen as zero kurtosis, data that is called negative kurtosis with the kurtosis parameter smaller than the specified one has a larger dispersion to the mean of data, a higher probability away from the mean, and a lower and flat peak; on the contrary, data that is called positive kurtosis with the kurtosis parameter larger than the specified one has a smaller dispersion to the mean of data, a lower probability away from the mean and a higher and sharp peak

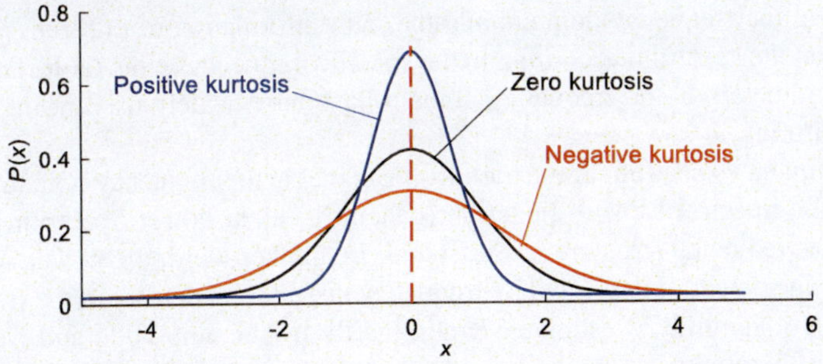

Fig. 6.4 The meaning of kurtosis

(Fig. 6.4). With the occurrence and development of fault, the value of kurtosis will increase gradually. This parameter is particularly sensitive to large amplitudes and is particularly effective for detecting whether a signal contains a pulse component.

The mean of a signal is the stable component so that it is useless for diagnosis. As a result, the mean is usually eliminated from raw data to calculate diagnosis parameters, which only uses the dynamic part of the signal for calculation.

Although the aforementioned feature parameters change with equipment state, they also change with working condition and measurement condition (load, rotate speed, instrument sensitivity, and so on), resulting in bad stability. To overcome these weaknesses, dimensionless parameters are used, and commonly used parameters are:

(7) Form Factor: $k = \frac{x_{rms}}{\overline{|x|}}$

(8) Crest Factor: $C = \frac{x_{\max}}{x_{rms}}$

(9) Pulse Factor: $I = \frac{x_{\max}}{\overline{|x|}}$, where $\overline{|x|}$ is the average of amplitude's absolute value, $\overline{|x|} = \frac{1}{N}\sum_{i=1}^{N}|x_i|$

(10) Margin Factor: $L = \frac{x_{\max}}{x_r}$, where x_r is the square of the average of the amplitude's root, $x_r = (\frac{1}{N}\sum_{i=1}^{N}\sqrt{|x_i|})^2$

(11) Coefficient of skewness: $S_v = \frac{\alpha_x}{\sigma_x^3}$

(12) Coefficient of kurtosis: $k_v = \frac{\beta_x}{\sigma_x^4}$.

6.5.2 Feature Parameters in the Frequency Domain

The feature parameters in the time domain can only give the information of whether the equipment (or the forming process) is normal or not, but hardly provide the important information of fault (defect) location. In addition, the monitoring signals in the time domain often contain a lot of interference information, especially in the early fault so that the useful information is often covered by interference information. However, in the frequency domain, the characteristic information of fault and interference information of each part of the equipment have their specific frequency and occupy different positions on the spectrum, so it is easy to avoid the interference influence in the frequency domain, and the fault information of different parts can be found on the spectrum according to the specific frequency of the fault. Therefore, feature parameters in the frequency domain have been widely used in the field of state monitoring.

The frequency spectrum of the monitoring signal in the frequency domain always refers to the Spectral Density Function (generally called Power Spectrum), which reflects the frequency structure of the signal. In another word, the power spectrum is the average power of the signal by frequency distribution density in the frequency domain. Its magnitude is estimated preliminarily by the amplitude and frequency spectrum of N monitoring data:

$$S_X(f) = \frac{1}{N}|X(f)|^2 \tag{6.39}$$

where $f = \frac{0}{T}, \frac{1}{T}, \frac{2}{T}, \ldots, \frac{N-1}{T}$, T is the interval of a sample.

The power spectrum contains a lot of useful information. Meanwhile, the power spectrum has a significant change when the equipment has a failure. As a result, it is an important basis of fault diagnosis in the frequency domain and is widely used.

Although it is comprehensive to use the complete power spectrum to judge the state of the equipment, the workload is large and the application is inconvenient.

(1) Peak frequency and amplitude: The frequency and amplitude of the peak component in the spectrum, which has a wide application, are the simplest and the most commonly used feature parameters. Many faults have their specific frequency. A clear judgment of the existence and extent of these faults can be made based on observing the corresponding peaks and analyzing the fluctuation of peaks in the spectrum.

(2) Mean magnitude of the frequency window: It is more stable and reliable than the former method to set a window to the frequency band which has the most sensitivity to changes and to take the average of the amplitudes in the window as the diagnostic parameter.

In addition, like the statistic feature parameters of a signal in the time domain, a signal also has statistic feature parameters in the frequency domain: spectral centre of the power spectrum, variance in the frequency domain, mean square frequency, and so on [2, 25].

Assuming that the power spectrum of a signal is shown in Fig. 6.5, several other features can be extracted:

(3) Spectral Centre: $F_C = \frac{\sum f_i S(f_i)}{\sum S(f_i)}$

(4) Mean Square Frequency: $MSF = \frac{\sum f_i^2 S(f_i)}{\sum S(f_i)}$

(5) Variance in Frequency Domain: $V_f = \frac{\sum (f_i - F_C)^2 S(f_i)}{\sum S(f_i)}$

F_C and MSF are parameters to describe the main frequency band position of the power spectrum, which also can be seen as the position of power concentration. When F_C and MSF have a small value, it means the energy of the power spectrum

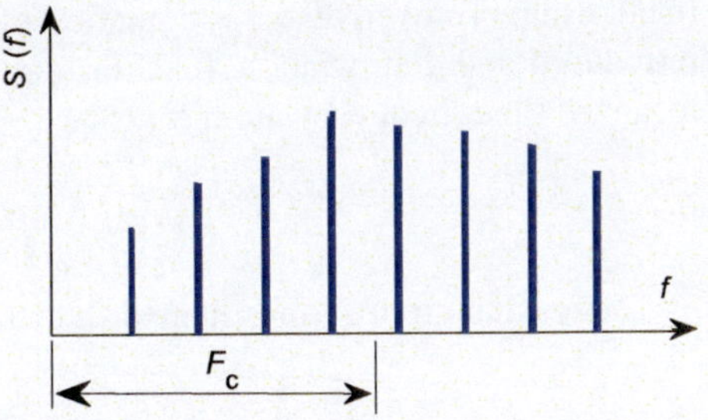

Fig. 6.5 The power spectrum of a signal

is mainly in the low-frequency band. V_f is the parameter to describe the degree of spectral dispersion. These parameters have certain diagnostic abilities and can be selected according to different monitoring purposes.

6.6 Wavelet Analysis

6.6.1 Limitation of Fourier Methods

Despite its popularity, the Fourier transform has certain serious theoretical drawbacks in processing signals, which limits its applications to TCM. This is because the support of $e^{j\omega t}$ covers the entire real dimension so that $\hat{f}(\omega)$ depends on the value of $f(t)$ for all times $t \in R$. This global inclusivity of information makes it difficult to analyze any local property of $f(t)$ from $\hat{f}(\omega)$ with WT. As can be seen from Fig. 6.6, it has infinite time resolution (upper Fig. 6.6a), and have infinite frequency resolution (Fig. 6.6b), but it doesn't have the information regarding the time–frequency

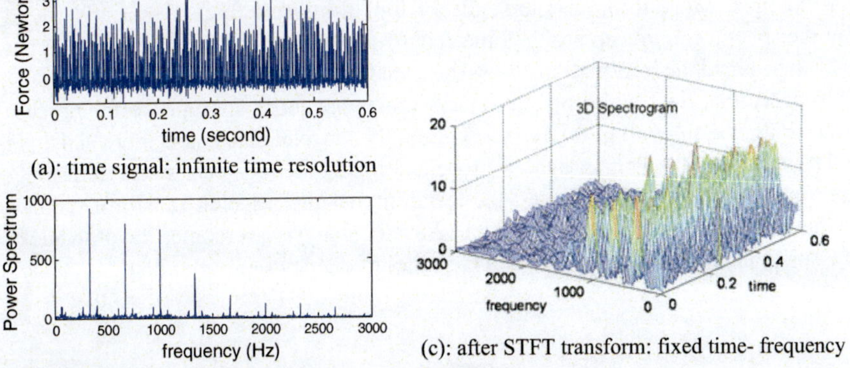

(a): time signal: infinite time resolution

(b) FFT transform: infinite frequency resolution

(c): after STFT transform: fixed time- frequency resolution

Fig. 6.6 Cutting forces, Fourier transform, and spectrogram relationship. To overcome the limitation of the lack of localization, Garbor (1946) introduced a sliding window function $g(t)$ to the Fourier transform and obtains a localized time–frequency atom ϕ [26]:

$$\phi_{u,\xi}(t) = e^{j\xi t} g(t-u) \tag{6.40}$$

The resultant transform is named short-time Fourier Transform (STFT).

$$(STFT)(f(t)) = \langle f(t), \phi_{u,\xi} \rangle = \int_{-\infty}^{\infty} f(t) g(t-u) e^{-j\xi t} dt \tag{6.41}$$

The corresponding energy density $|STFT(f(t))|^2$ is called a spectrogram, which is widely used for time–frequency analysis before wavelet dominated the area in the 1990s, and applied to TCM by Doweling [11]. Figure 6.6c is the spectrogram of the force from Fig. 6.6a. The power spectrum is localized in the time axis and the variation of spectrum energy can be detected in the spectrogram.

In the STFT analysis, however, once the window is chosen, the frequency resolution σ_ω^2 and time resolution σ_t^2 are constant for both time and frequency, and according to Heisenberg Uncertainty Principle [26, 27],

$$\sigma_t^2 \sigma_\omega^2 \geq \frac{1}{4} \tag{6.42}$$

Equality holds if $\phi(t)$ is a Gaussian.

The Eq. (6.42) states that Fourier methods cannot reach high resolution both in frequency and time. On the other hand, the above-mentioned approaches assume that the sensor signals are stationary. However, due to the nature of manufacturing processes, the signals are usually non-stationary [28]. Constant time and frequency resolutions of STFT are not suitable for the analysis of the non-stationary signal. For example, Mori et al. [29] took the FFT of the thrust force signal and found that the spectra cannot capture the time-localized aspect of the saw-tooth signal, but instead it spread the information across the transformed signal. The signal processing approaches that deal with non-stationary signals are more appropriate for process monitoring. Gong et al. [14] have shown that the wavelet analysis is more sensitive and reliable than the Fourier analysis for recognizing the tool wear states in turning. Yoon and Chin [30] also verified the reliability of the wavelet transform method compared to the spectra method of FFT. Wavelet analysis overcomes the drawbacks of Fourier methods and permits adaptive time–frequency representation.

6.6.2 Continuous Wavelet Analysis (CWT) and Its Time–Frequency Properties

6.6.2.1 Continuous Wavelet Analysis (CWT)

To meet the needs for adaptive time–frequency analysis in applied mathematics, physics, and engineering, the wavelet theory was developed in the late 1980s by Morlet, Meyer, Mallat, and Daubechies (see [26] and [8]). In wavelet transform (WT), a signal of interest is decomposed into a set of basic localized waveforms, called wavelets. The signal is analyzed by examining the coefficients of the wavelets. Wavelet transform is a convenient tool for processing nonstationary signals. It has been widely used to analyze machining signals for tool condition monitoring.

Let $\psi_{a,b}(t), a, b \in R, a \neq 0$, be a family of functions defined as translations and re-scales of a single function $\psi_{a,b}(t) \in L^2(R)$ [27],

$$\psi_{a,b}(t) = \frac{1}{\sqrt{|a|}}\psi\left(\frac{t-b}{a}\right) \tag{6.43}$$

with the following basic properties,

$$\int_R \psi_{a,b}(t)dt = 0 \tag{6.44}$$

$$\int_R \psi_{a,b}^2(t)dt = 1 \tag{6.45}$$

These properties indicate that the wavelet is a small wave: oscillating around zero (zero mean) and diminishing quickly (finite energy).

But not all $\psi_{a,b}(t) \in L^2(R)$ with the properties in Eqs. 6.44 and 6.45 can be regarded as a wavelet. The wavelet $\psi_{a,b}(t)$ has to meet the following admissibility condition for the transformation to be invertible (to reconstruct the signal after WT transform),

$$C_\psi = \int_{-\infty}^{\infty} \frac{\left|\hat{\Psi}(\omega)\right|^2}{\omega} d\omega < \infty \tag{6.46}$$

Figure 6.7 shows samples of Haar, Morlet, and Daubechies wavelets.

The continuous wavelet transform (CWT) of the signal $f(t) \in L^2(R)$ is defined as the inner product of $f(t)$ and $\psi_{a,b}(t)$,

$$W_\psi f(a,b) = \left\langle f(t), \psi\left(\frac{t-b}{a}\right)\right\rangle = \frac{1}{\sqrt{a}}\int_{-\infty}^{\infty} f(t)\psi\left(\frac{t-b}{a}\right)dt \tag{6.47}$$

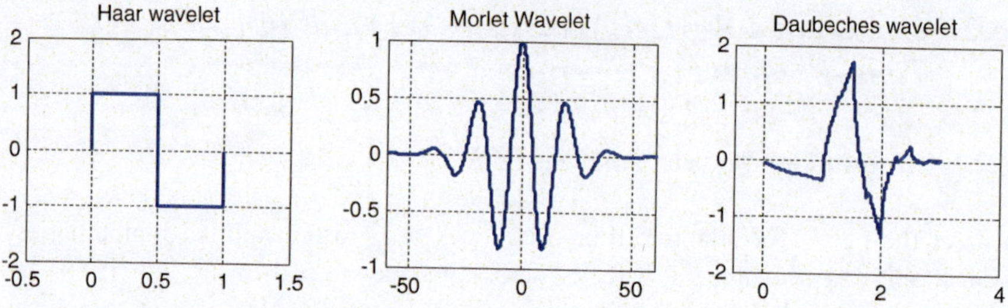

Fig. 6.7 Haar wavelet, Morlet wavelet, and Daubechies wavelet

For a fixed scale, the CWT is the convolution of the signal $f(t)$ with the time-reversed wavelet $\psi(-(t-b)/a)$.

The following Fig. 6.8 shows a signal $f(t)$ along with the Morelet wavelet at three scales and shifts. As increases, the wavelet stretches by a factor of two which analysis the signal at lower frequency; as b increases, the wavelet shifts right which analysis the signal at different localizations. The variations of a and b lead to the multiscale and localized analysis with wavelet.

In physical interpretation, the coefficients of the wavelet transform indicate the variation of energy of the signal with time and frequency. In engineering applications, the square of the coefficients of the CWT is often called a scalogram, defined as

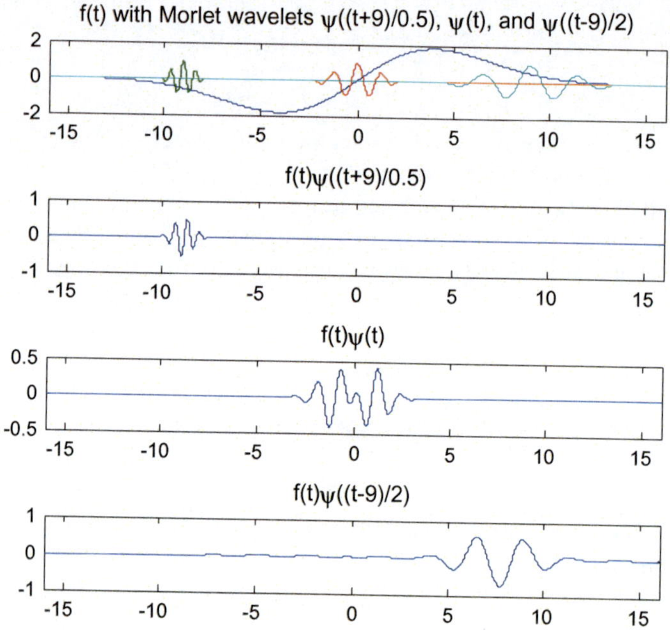

Fig. 6.8 Morlet wavelet transform at different scales and locations

Eq. 6.48, which has been widely used for machinery monitoring and TCM [31].

$$SC_t(a, b; \psi) = \left| W_\psi f(a, b) \right|^2 \tag{6.48}$$

6.6.2.2 Comparison of Time–Frequency Resolution of FT and WT

The time–frequency resolution of Fourier methods and wavelet analysis are illustrated in Fig. 6.9. Figure 6.9a is the T–F resolution of the time signal. As can be seen from the figure, it has infinite time resolution but has no frequency resolution. Figure 6.9b is the T–F resolution for FFT transformed signal. On the contrary, it has no time resolution but infinite frequency resolution. Figure 6.9c is the T–F resolution of STFT. Once the window is chosen, the resolution is fixed. Figure 6.9d is the T–F for wavelet analysis. The T–F boxes are adaptive. Figure 6.9a–c can be regarded as a special case of Fig. 6.9d. The signal is localized in the area with time width Δ_t:$[u_0 - \frac{1}{2}a_0\sigma_t, u_0 + \frac{1}{2}a_0\sigma_t]$ and frequency width Δ_ω:$[\frac{\eta}{a_0} - \frac{\sigma_\omega}{2a_0}, \frac{\eta}{a_0} + \frac{\sigma_\omega}{2a_0}]$. The area of Heisenberg box remains to,

$$\Delta_t \Delta_\omega = (a_0\sigma_t) \times (\sigma_\omega / a_0) = \sigma_t \sigma_\omega \tag{6.49}$$

Small a: CWT resolves events closely spaced in time.

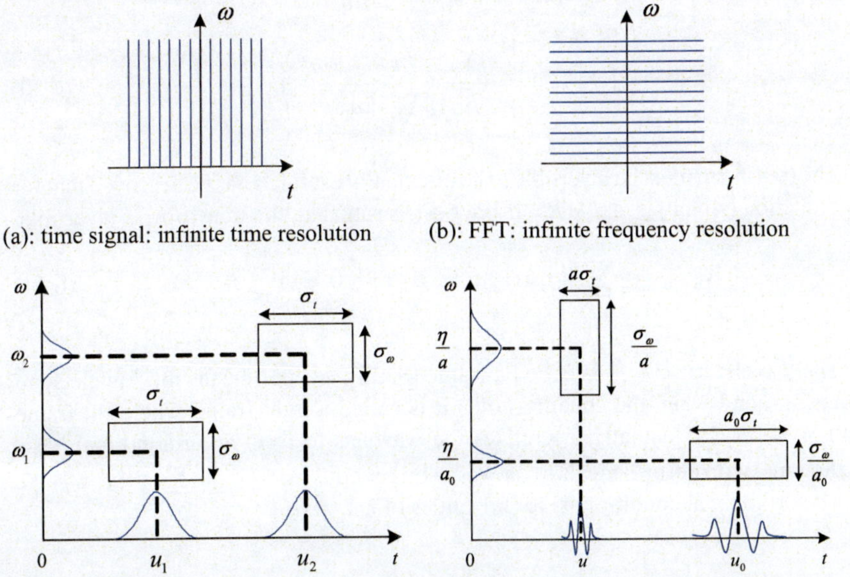

(a): time signal: infinite time resolution (b): FFT: infinite frequency resolution

(c): STFT: fixed time-frequency resolution (d) WT: adaptive time- frequency resolution

Fig. 6.9 Time–frequency resolution of and Fourier transform and wavelet transform

Large a: CWT resolves events closely spaced in frequency.

Compared with the STFT, whose time–frequency resolution is constant, the time–frequency resolution of the wavelet transform depends on the frequency of the signal. At high frequencies, the wavelet reaches at a high time resolution but at a low-frequency resolution; whereas at low frequencies, high-frequency resolution and low time resolution can be obtained. Such adaptability of time–frequency analysis reinforces the important status of the wavelet transform in the fault diagnostics field.

6.6.3 Discrete Wavelet Transform and Its Implementation

6.6.3.1 Discrete Wavelet Transform (DWT)

The CWT is recognized as an effective tool for both stationary and non-stationary signals. However, the CWT transforms the signal of one variable into a signal of two variables. It involves much redundant information and is computationally very slow. To "minimize" the transformation one can select discrete values of a and b and still have an invertible transformation.

The redundancy in the continuous wavelets is can be fixed. First, the continuous variables a and b are discretized so that the mother wavelet can only be scaled and translated in discrete steps. The discretized values b are made dependent on a in such a manner, that low frequency components are sampled less often. As a result of the discretization, we get the discretized wavelets [26],

$$\psi_{a,b}(t) = \frac{1}{\sqrt{a_0^j}} \psi\left(\frac{t - kb_0 a_0^j}{a_0^j}\right) \tag{6.50}$$

The transform must be invertible to be practically useful. It has been proven that for any function (signal) $f \in L_2(R)$, if Eq. 6.51 is satisfied, the transform is invertible,

$$A\|f\|^2 \leq \sum_{j,k} \left|\langle f, \psi_{j,k}\rangle\right|^2 \leq B\|f\|^2, 0 < A \leq B < \infty \tag{6.51}$$

The wavelet set $\{\psi_{j,k}\}$ is called a frame [8, 26]. $A < B$ means the wavelets are redundant, and when the equality holds it is called a tight frame. The tight frame with $A = B = 1$ actually discretizes the wavelets to form an orthonormal basis (orthogonal with norm 2 equals to one) as in this condition, $\|f\|^2 = \sum_{j,k} \left|\langle f, \psi_{j,k}\rangle\right|^2$.

The orthogonal condition is, for all values of j, k, m, n,

$$\int \psi_{j,k}(t)\psi_{n,m}^*(t)dt = \delta(j - n)\delta(k - m) \tag{6.52}$$

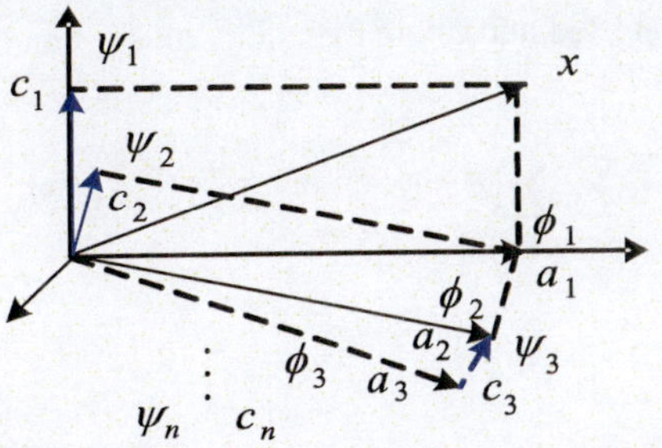

$$
\begin{aligned}
x &= c_1\psi_1 + a_1\phi_1 \\
&= c_1\psi_1 + c_2\psi_2 + a_2\phi_2 \\
&= c_1\psi_1 + c_2\psi_2 + c_3\psi_3 + a_3\phi_3 \\
&= c_1\psi_1 + c_2\psi_2 + c_3\psi_3 + \ldots + c_n\psi_n + a_n\phi_n
\end{aligned}
$$

Fig. 6.10 Geometry of MRA

The equation meets only if $j = n$ and $k = m$. Notice that orthogonality is not essential for discrete wavelets to work, and sometimes it is not required, as it reduces the flexibility of the wavelets.

So the key to wavelet analysis is to construct a wavelet function $\psi_{j,k}(t)$, so that the wavelets $\{2^{-j/2}\psi(2^{-j}t - k),\ i, k \in Z\}$ form an orthogonal basis in $L_2(R)$. Mallat [27] developed a framework for such construction under the idea of Multiresolution analysis (MRA) (please refer to Mallat [26, 27] for details). Basically, an MRA consists of a function space $\{V_j\}_{j\in Z}$ of $L^2(R)$ satisfying some important properties, such as scaling, approximation, nesting, and shifting properties. The function space has nesting property such that $V_0 \supset V_1 \cdots V_j \supset V_{j+1} \cdots$, and the function defined in the space $\{V_j\}_{j\in Z}$ with $\phi(t) \in V_j \Leftrightarrow \phi(\frac{t}{2}) \in V_{j+1}$, is called scaling function. By defining W_k as an orthogonal complement of V_k in V_{k-1}, i.e. $V_{j-1} = V_j \oplus W_j$ and $V_k \bot W_k$. Space V_j is the sums of W_j:

$$V_0 = W_1 \oplus W_2 \oplus \cdots \oplus W_j \oplus W_{j+1} \oplus V_{j+1} \tag{6.53}$$

This construction can be geometrically viewed through Fig. 6.10, the function x can be approximated by the basis ψ and ϕ, and the scaling basis ϕ could be further by approximated by finer wavelet and scaling basis.

According to Mallat's Multiresolution analysis (MRA), the discretizing of (a, b) that preserves all information about the decomposed function cannot be coarser than the critical sampling, and any coarse discretization will not give a unique inverse transformation. Under the critical sampling, a_0 is assigned a value of 2 and b_0 the value of one. By discretizing $(a, b) = (2^j, k * 2^j)$, we have the wavelet function

$$\psi_{j,k}(t) = 2^{-j/2}\psi(2^{-j}t - k), \tag{6.54}$$

and define an accompanying scaling function as

$$\phi_{j,k}(t) = 2^{-j/2}\phi(2^{-j}t - k) \tag{6.55}$$

at the level of detail j and shift k.

The signal $f(t)$ is then represented by shifted and dilated versions of $\psi(t)$ and $\phi(t)$ with Discrete Wavelet Transform (DWT),

$$f(t) = \sum_{k \in Z} c_{j_0,k}\phi_{j_0,k}(t) + \sum_{j \geq j_0}\sum_{k \in Z} d_{J,k}\psi_{J,k}(t) \tag{6.56}$$

where

$$c_{j,k} = 2^{-j/2}\int_R f(t)\phi(2^{-j}t - k)dt \tag{6.57}$$

is the scaling coefficient, which measures the local mean around the time $2^{-j}i$, and

$$d_{j,k} = 2^{-j/2}\int_R f(t)\psi(2^{-j}t - k)dt \tag{6.58}$$

is the wavelet coefficient, which characterizes the local variation around the time $2^{-j}i$ and the frequency $2^{-j}f_0$.

6.6.3.2 DWT Implementation with Filter Banks

Practically, the DWT is implemented using a filter bank structure above. The scaling function $\varphi(t)$ lies in the central space V_0, so $\varphi(t) \in V_0 \subset V_1$ since the spaces are nested. Also, the translations $\varphi(2t - k)$ of $\varphi(2t)$ form an orthonormal basis for V_1.

Since $\varphi(t) \in V_1$, we can write $\varphi(t)$ in terms of this basis, and relate it to the two-scale difference equation [32]:

$$\phi(t) = \sqrt{2}\sum_k h_0(k)\phi(2t - k) \tag{6.59}$$

where coefficient $h(k)$ must satisfy several conditions to make bias wavelet functions unique, orthonormal, and have a certain degree of regularity (1). The mother wavelet is related to the scaling function via [26, 32]:

$$\psi(t) = \sqrt{2}\sum_k h_1(k)\phi(2t - k) \tag{6.60}$$

where

$$h_1(k) = (-1)^k h(1 - k) \tag{6.61}$$

Note that $h_0(k)$ and $h_1(k)$ can be viewed as filter coefficients of half-band low pass and high pass filters, respectively. At this point, if valid $h_0(k)$ is available, one can obtain $h_1(k)$. The different $h(k)$ and scaling functions may lead to different qualified wavelets such as Haar, Morlet, and Daubechies.

6.6.4 Wavelet Basis Function

The wavelet basis function determines the efficiency and effect of the wavelet transform. The wavelet basis function can be chosen flexibly, and the basis function can be constructed according to the problems. The wavelet function is generated by the generating function, which must be a damped oscillation waveform with alternating positive and negative amplitude. Many functions meet this condition and some common continuous wavelet basis functions are shown as follow [26]:

(1) **Haar Function**

The earliest found and simplest function is Haar Function. It can be written as:

$$\psi(t) = \begin{cases} 1 & 0 \le t < 1/2 \\ -1 & 1/2 \le t < 1 \\ 0 & others \end{cases} \tag{6.62}$$

The class of functions generated by it is called the Haar wavelet function (shown in Fig. 6.11), and the Harr wavelet function whose parameters are discrete binary values is called the Haar Dyadic Wavelet and is expressed as:

$$\psi_{j,k}(t) = 2^{-j/2}\psi(2^{-j}t - k) = \begin{cases} 2^{-j/2} & k/2^{-j} \le t < (2k+1)/2^{-j+1} \\ -2^{-j/2} & (2k+1)/2^{-j+1} \le t < (k+1)/2^{-j} \\ 0 & others \end{cases} \tag{6.63}$$

where $j = 0, 1, 2, ..., k = 0, 1, 2, ...$

In Eq. 6.63, 2^{-j} is the scaling parameter, whose variation causes the wavelet to stretch or shorten to form the wavelet with different scale 'levels' or different frequency bands. The wavelet with a large scaling parameter 2^{-j}, which belongs

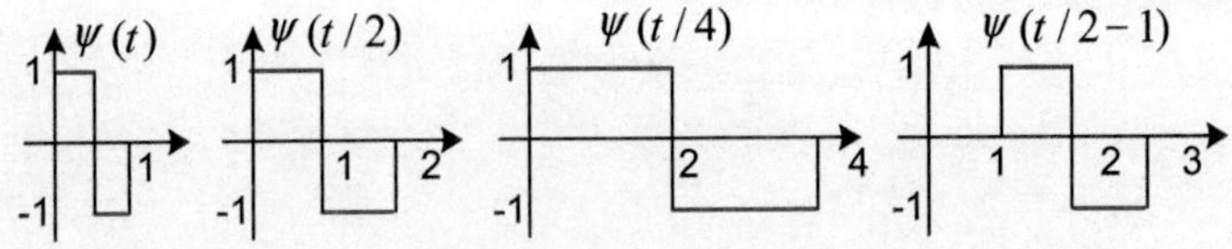

Fig. 6.11 Haar wavelet function

to the high-frequency band wavelet, has a narrow width and a strong fluctuation. However, the low-frequency band wavelet has the complete opposite attribute. The Time Domain Information of different frequency bands in the signal can be obtained by using the wavelet functions of different levels. In Eq. 6.63, k is the position parameter. The variation of position parameter causes the wavelet to shift along the time axis, forming the wavelet function of different 'position'. The local spectrum of a signal $x(t)$ in different periods can be achieved by transforming a signal with the wavelet function in a different 'position'. Haar wavelet is orthogonal when there is a translation of t. For a one-dimensional Haar wavelet, it can be regarded as the differential operational calculation, which means obtaining the difference of the part that is not equal to the average of the observed result. Obviously, the Haar wavelet is not a continuously differentiable function.

It can be seen from Fig. 6.11 that wavelets with different 'levels' have different duration. Therefore, the analysis length (window width) on the time axis can be adjusted automatically by applying different 'levels' of wavelet to transform a signal. The wavelets with large scaling parameters are used to analyze high-frequency bands and have narrow analysis length. Meanwhile, the wavelets with small scaling parameters are used to analyze low-frequency bands and have broad analysis lengths. These characteristics are entirely suitable for the analysis of different frequency components in the signal, which means when we analyze the details of the signal, a higher analysis frequency, and a narrower analysis length is required; when we analyze the general view of the signal, a low analysis frequency and wide analysis length are required. Wavelet transform meets the requirement of the analysis length and the analysis frequency so that it is called the 'Mathematical Microscope' in the signal analysis field. It is also an effective tool to analyze non-stationary signals.

If a dyadic wavelet has the orthogonal property, which meets the following equation,

$$\int \psi_{j,k}(t)\psi_{l,m}(t)dt = \begin{cases} 1 & j = 1,\ k = m \\ 0 & others \end{cases} \tag{6.64}$$

it is called the dyadic orthogonal wavelet. The dyadic orthogonal wavelet plays an important role in the field of wavelet transform due to its obvious advantages.

Haar dyadic wavelet has the orthogonal property, which means it is an orthogonal wavelet.

(2) **Mexican Hat Wavelet**

Mexican Hat Wavelet is the second-order derivative of Gaussian function, which is:

$$\psi(t) = \frac{2}{\sqrt{3}}\pi^{-1/4}(1 - t^2)e^{-t^2/2} \tag{6.65}$$

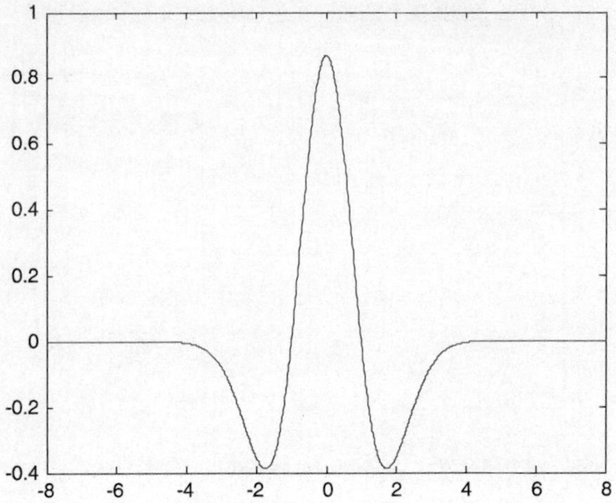

Fig. 6.12 The waveform of the Mexican hat wavelet

where the coefficient $2/\sqrt{3}\pi^{-1/4}$ is used to guarantee the normalization of $\psi(t)$, which means $\|\psi\|^2 = 1$. The wavelet uses the second-order derivative of the Gaussian smoothing function, whose waveform is similar to the paraboloid contour of a Mexican Hat and the wavelet is named for this property, as is shown in Fig. 6.12. It is also called Marr wavelet because of its wide applications in visual information processing and edge detection.

Based on the mth-order derivation of Gaussian function, a wavelet cluster can be gotten:

$$\psi_m = C_m(-1)^m \frac{\mathrm{d}^2}{\mathrm{d}t^m}\left(\mathrm{e}^{-t^2/2}\right) \tag{6.66}$$

where the constant in Eq. 6.66 is used to guarantee $\|\psi_m\|^2 = 1$. Although the Mexican hat wavelet is equivalent to the case of M = 2 in Formula (6.66), it cannot detect the different directions of the signal because it is isotropic.

The DOG wavelet, which is formed by the Difference of Gaussians (DOG) function, is a good approximation of the Mexican hat wavelet and is shown below:

$$\psi(t) = \mathrm{e}^{-t^2/2} - \frac{1}{2}\mathrm{e}^{-t^2/8} \tag{6.67}$$

(3) **Complex Gaussian Wavelet**

Complex Gaussian wavelet is made up of nth-order derivation of complex Gaussian function and is defined as follow:

$$\psi(t) = C_n \frac{\mathrm{d}^n}{\mathrm{d}x}\left(\mathrm{e}^{-jt}\mathrm{e}^{t^2}\right) \tag{6.68}$$

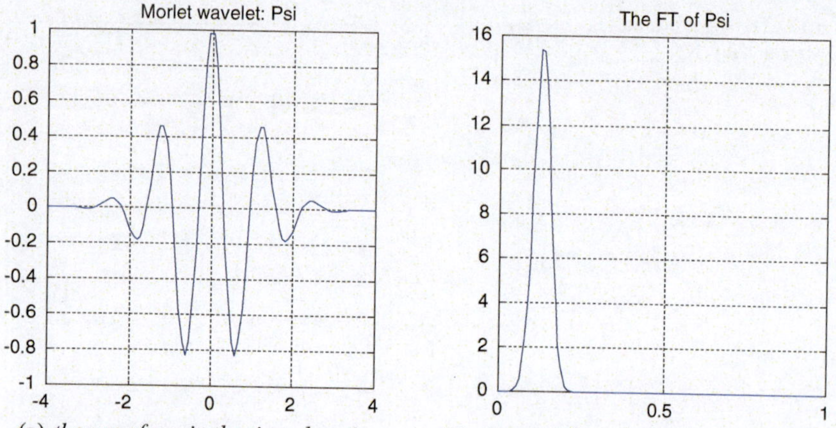

(a) *the waveform in the time-domain* (b) *the waveform in the frequency-domain*

Fig. 6.13 Morlet Wavelet

where *Cn* is used to make the energy of the wavelet function normalized.

(4) **Morlet Real Wavelet**

Morlet real wavelet is defined as:

$$\psi(t) = \pi^{-1/4}\cos(5t)e^{-t^2/2} \tag{6.69}$$

It is a single frequency sine signal with a Gaussian envelope. This wavelet is not compact support and t can be treated as $(-\infty, \infty)$ in theory. However, $\psi(t)$ and its Fourier Transform $\Psi(\Omega)$ has the good concentration property in both the time-domain and frequency-domain and they are shown in Fig. 6.13.

Morlet wavelet is neither orthogonal nor double orthogonal and can be used for the continuous wavelet transform. However, this wavelet is symmetric and has a wide application.

(5) **Morlet complex wavelet**

Morlet wavelet is the most commonly used complex wavelet, whose equation is shown in Eq. 6.70, and the waveform is shown in Eq. 6.20:

$$\psi_0(t) = (\pi f_B)^{1/2} e^{j2\pi f_C t} e^{-t^2/f_B} \tag{6.70}$$

The Fourier Transform of Eq. 6.70 is:

$$\psi_0(f) = e^{-\frac{(f-f_0)^2}{f_B}} \tag{6.71}$$

Normally, the conditions that $\omega_0 \geq 5$, $\omega_0 = 5$ are widely used. f_B is the bandwidth and f_C is the central frequency (Fig. 6.14).

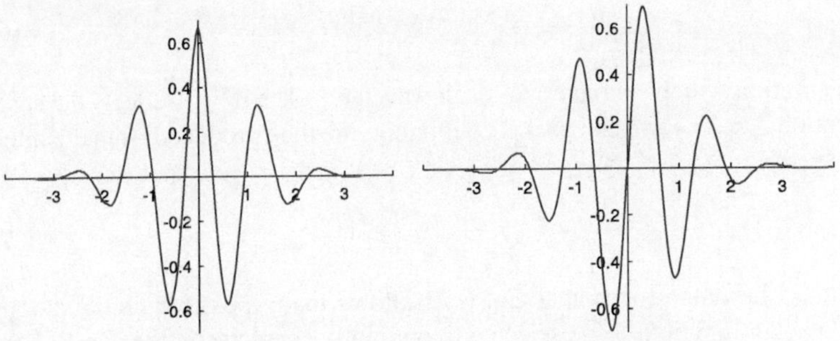

Fig. 6.14 The waveform of Morlet complex wavelet

(6) **Orthogonal Wavelet**

At present, there are four main kinds of orthogonal wavelet, Daubechies wavelet, symmetric wavelet, Coiflets wavelet, and Meyer wavelet. These orthonormal wavelets $\psi(t)$, which are different from the aforementioned "classical wavelets", cannot generally be given by a simple expression but are generated by a weighted combination of $\varphi(t)$, which is called a 'Scaling Function'. Scale function is another important concept of the wavelet transform. Based on the discussion in the next section, wavelet function $\psi(t)$, scaling function $\varphi(t)$ correlates with a low pass filter $H_0(z)$ and a high pass filter $H_1(z)$ at the same time, where $H_0(z)$ and $H_1(z)$ are combined to construct an analysis filter bank with two channels. These contents form the theoretical basis of the multi-resolution analysis of wavelet transform. Therefore, in the discussion of orthogonal wavelet, the scaling function $\varphi(t)$, the analysis filter bank with $H_0(z)$, $H_1(z)$ and the synthetic filter bank with $G_0(z)$, $G_1(z)$ are involved.

6.6.5 Wavelet Packets Decomposition

The DWT may lead to a loss of useful information at high frequency since successive details are no longer analyzed. We need to double the sampling rate for higher frequency analysis, which however involves more data and computation. Wavelet packet decomposition [33] is a generalization of wavelet decomposition at higher frequencies. In the wavelet packet decomposition, each detail coefficient vector is also decomposed into two parts using the same approach as DWT in approximate coefficients. This offers the analysis at the high frequencies.

Basically, in WPD, for fixed values of j and k, $W_{j,n,k}$ analyzes the fluctuations of the signal roughly around the position $2^j \cdot k$, at scale 2^j and oscillation parameter n. WPA operates by approximating a signal with scaled and translated wavelet packet functions,

$$W_{j,n,k} = 2^{-j/2} W_n(2^{-j/2}t - k) \tag{6.72}$$

The notation which characterizes each wavelet packet $W_{j,n,k}$ reflects the scale 2^j and location $2^j k$. The wavelet packet coefficients are then produced from the integral,

$$d_j^k = \int W_{j,n,k}(t)\, f(t)\, dt \tag{6.73}$$

It should be emphasized that Eq. 6.73 allows many possible combinations of wavelet packet functions to be selected to optimally characterize the signal. Several criteria have been developed to select the best basis for these purposes [33].

6.6.6 *Some Remarks on Wavelet Transform*

6.6.6.1 Regularity

The regularity condition requires that the mother wavelet has to be locally smooth and concentrated in both the time and frequency domains. To ascertain a good local property in the frequency domain, the $\left|W_\psi\ f(a, b)\right|$ is required to decay rapidly with the decrease of scale a. As a result of this requires the first m moments of $\psi(x)$ equal to zero. These are the vanishing moments. If the Fourier transform of the wavelet is M times differentiable the wavelet has M vanishing moments [26].

$$\int x^m \psi(x) dx = 0,\ m \in [0, M] \tag{6.74}$$

This means the wavelet is "blind" to the polynomial segment with the degree $\alpha \leq M - 1$. Physically speaking, the higher degree of polynomial means higher frequency (faster variations of the signal), as a result, the higher moment indicates that the wavelet is a high-pass filter.

6.6.6.2 Accuracy of Signal Approximation

It has been proved [26] that the accuracy of wavelet approximation of function (signal) $f(t)$ in the vicinity of t_0 has the following upper bound,

$$|\langle f, \psi_a\rangle - f(t_0)| \leq \frac{2C}{M!}\left(\frac{R}{a}\right)^M \tag{6.75}$$

where R is the support area, C is a constant related to f.

Two facts can be derived from this equation:

(1) The accuracy of wavelet approximation $|\langle f, \psi_a \rangle - f(t_0)|$ depends on the support area R of wavelet function ψ_t and scaling parameter a. If support area R is compact (limited and close bounds) support and the scaling parameter a approach infinity, $a \to \infty$, then $|\langle f, \psi_a \rangle - f(t_0)| \to 0$.
(2) In the case of the scaling parameter $a > R$, the accuracy of $|\langle f, \psi_a \rangle - f(t_0)|$ increases with the increase of vanish moment M.

As a result, from the function approximation point of view, we hope the wavelet ψ_t with smaller support R and higher vanish moment M. At the same time, with smaller R, we get a better time resolution. According to the Uncertainty Principle, the frequency resolution is decreased.

When choosing a particular wavelet, we thus face a trade-off between the number of vanishing moments and the support size. Generally, if f has few isolated singularities and is regular between singularities, we may like to choose a wavelet with higher vanishing moments to produce a large number of small wavelet coefficients; if the density of singularities increases, it might be better to decrease the size of its support at the cost of reducing the number of vanishing moments. The Daubiches wavelet has the shortest support $\propto 2M$ for all wavelet families with M vanishing moments [26].

6.6.6.3 Continuous or Discrete Analysis

When the signal is recorded in continuous time or on a very fine time grid, both analyses are possible. Which should be used? It depends; each one has its advantages. DWT ensures space-saving coding and is sufficient for exact reconstruction. CWT is often easier to interpret since its redundancy tends to reinforce the traits and makes all information more visible. This is especially true of very subtle information. Thus, the CWT gains in "readability" and ease of interpretation but loses in terms of saving space.

6.6.6.4 The Selection of the Continuous Wavelet Basis Function

The selection of the wavelet basis function can be considered in the following three aspects:

(1) The selection between complex wavelet and real wavelet: The complex wavelet analysis method can get not only amplitude information but also phase information. As a result, complex wavelet analysis is suitable for analyzing and calculating the normal characteristics of signals. Meanwhile, real wavelet analysis is best for peaks or discontinuities detection.
(2) The selection of effective support zone for continuous wavelets: The continuous wavelet basis functions are rapidly fading outside the effective support zone. The larger the effective support zone is, the better the frequency resolution is. The smaller the effective support area is, the better the time resolution is.

(3) The selection of the wavelet waveform: If the time–frequency analysis is needed, a smooth continuous wavelet should be used. The reason is the smoother the basis function is in the time domain, the better the localization characteristic is in the frequency domain. If signal detection is needed, a wavelet, which is similar to the signal waveform, should be utilized.

6.6.6.5 Relationship Between Scale and Frequency

As wavelet analysis is time-scale analysis, the scale is not directly related to frequency. Abry [34] provided a way to associated them. The idea is to associate with a given wavelet a purely periodic signal of frequency F_c. The frequency maximizing the *FFT* of the wavelet modulus is called center frequency F_c. The center frequency captures the main wavelet oscillations. When the wavelet is dilated by a factor a, this center frequency becomes F_c/a. We then can associate to the scale the frequency,

$$F_a = \frac{F_c}{a \cdot T_s} \tag{6.76}$$

F_a is the pseudo-frequency corresponding to the scale *a*, in *Hz*.

6.6.6.6 Choice of Scale

A small scale value permits us to perform a local analysis; a large scale value is used for a global analysis. Combining local and global is a useful feature of wavelet analysis. The wavelets having a compact support are used in the local analysis. This is the case for Haar and Daubechies wavelets. As wavelet analysis constraints to Heisenberg uncertainty principle, it is impossible to reduce arbitrarily both time and frequency localization. The resolution increases as the scale decreases. The greater the resolution, the smaller and finer is the details that can be accessed.

The real signal has limited bandwidth in practice and a wavelet at a certain scale is equivalent to a band-pass filter, which must overlap with the signal being analyzed in the frequency domain. In engineering, we approximate the frequency with the largest power in the wavelet power spectrum as the central frequency of the wavelet and choose the appropriate scale to keep the central frequency within the bandwidth of the signal being analyzed.

Figure 6.15 is a typical multiresolution wavelet analysis of processing monitoring signals. For a given decomposition series, the fluctuation amplitude of the signal decreases with the increase of the scale, which corresponds to the low-frequency analysis of the signal. Whereas, it corresponds to the high-frequency analysis when the scale is small. The sampling frequency of a signal is 6000 Hz and its frequency division at different scales is shown in Fig. 6.16.

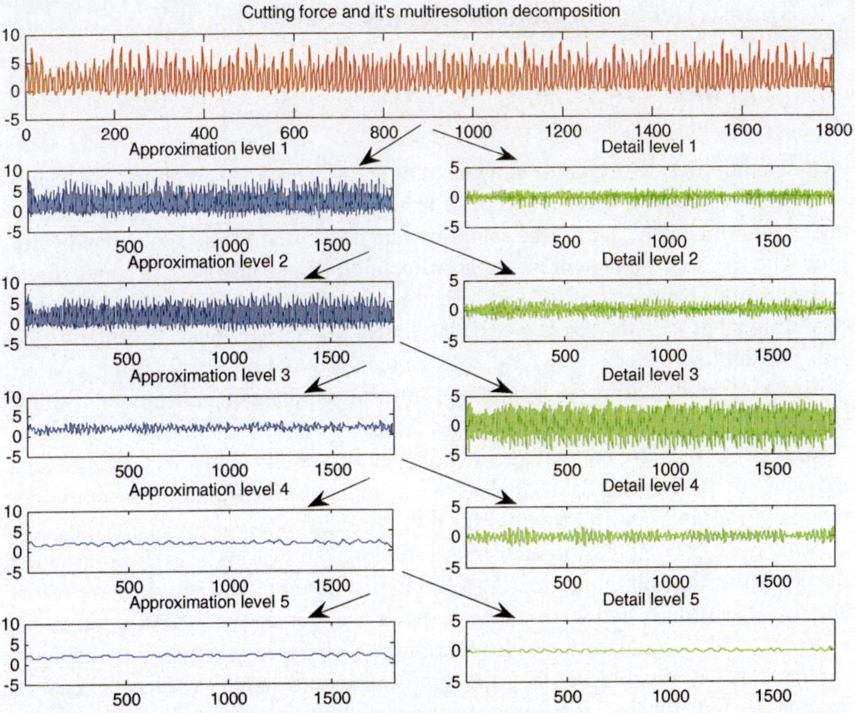

Fig. 6.15 The wavelet decomposition of a signal with different scales

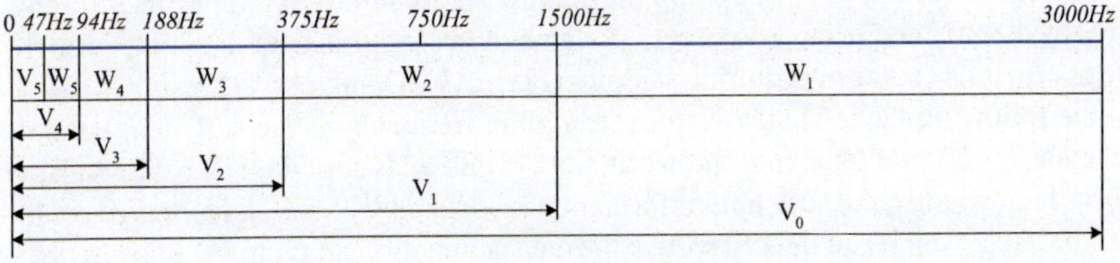

Fig. 6.16 The wavelet decomposition of a signal: frequency ranges of different scales

It should be noted that the corresponding frequencies of wavelet analysis are Pseudo-frequencies for different scales, not true frequency components and its resolution is quite different from that of Fourier analysis, such as power spectrum.

6.7 Sparse Decomposition of Signals

6.7.1 Compressive Sensing

Shannon sampling theory is the fundamental theory for data sampling. The theory states that the sampling rate should be at least 2 times the bandwidth to reconstruct a signal without error. That is, the sampling rate is decided by the signal bandwidth. While with the development of information technology and advances in applications, the spectrum of bandwidth has been widening rapidly, and as a result, the sampling rate and signal processing speed is to be increased largely. This leads to difficulties in dealing with ultra-wideband signals from the data acquisition, storage transmission, e.g., the AE signal, cutting force, and high-resolution image in real-time machining monitoring.

On the other hand, in the real application, people usually apply data compression approaches to the sampled data with less byte to represent the signal and discard large amounts of unimportant information for less storage space and faster transmission. This kind of data compression after high sampling rate wastes a lot of information induced by the high sampling rate. As a result, is it possible that people make use of some signal's property and set up a new framework under another transformed space, and represent the signal without information loss while largely below the Nyquist sampling rate? If it could be, then the approach would largely reduce the cost of data storage and transmission, simplify the signal processing stages and save the processing time.

Compressive sensing [23, 24] provides a good solution to these ideas: It samples the information instead of sampling the signal. The scheme overcomes these problems by directly acquiring a compressed signal representation without going through the intermediate stage of acquiring samples (Fig. 6.17). Compressive sensing is based on the following facts: (1) the Nyquist sampling frequency is the sufficient but not necessary condition for signal reconstruction; (2) Besides applying the bandwidth as a priori knowledge, the structural information associated with the signal/image could also be used, which can largely reduce the data acquisition amount; (3) It is possible to recover K-sparse signal in the original N-dimension by $K + 1$ measurement with overwhelming probability.

A signal is said sparse if most of its entries are equal or close to zero. The signal's sparsity can be measured by its l_0-norm. In the Hilbert space, the l_p norm for a given signal $y = [y_1, y_2, \ldots, y_n]^T$ is defined as

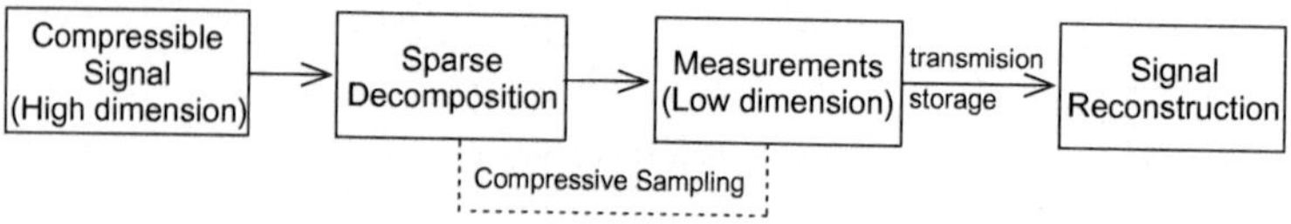

Fig. 6.17 The compressive sensing scheme

$$\|y\|_p = \left[\sum_{i=1}^{n} |y_i|^p\right]^{1/p} \quad (p \geq 1) \tag{6.77}$$

In particular, $\|y\|_1 = \sum_{i=1}^{n} |y_i|$, $\|y\|_2 = \sqrt{\sum_{i=1}^{n} |y_i|^2}$, $\|y\|_\infty = \max_{1 \leq i \leq n} \{|y_i|\}$.

The so-called l_0 norm, which is excluded from the above definition domain, is an extension of the l_p norm definition when $p \to 0$. Specifically, the sparsity of a discrete signal (vector) y is defined by,

$$\|y\|_0 = \sum_{i=1}^{n} |y_i|^0 = \text{number of nonzeros in } y \tag{6.78}$$

Consider a general linear measurement process that the measurements y is obtained from inner products between x and a collection of basis $\Phi \in R^{n \times k}$ $(n < k)$, the compressive sensing problem is stated as,

$$\min_x \|x\|_0 \text{ s.t. } \mathrm{y} = \Phi x, \tag{6.79}$$

Sufficient conditions have been raised that the signal can be recovered with high probability when Φ satisfies the Restricted Isometry Property (RIP) [23].

The three main aspects of compressive sensing are sparse representation, measurement coding, and recovery algorithms. Finding an optimum sparse decomposition space is the prerequisite for analysis and recovery. Candès and Tao [23] have shown that for a smooth signal its Fourier transform, wavelet transform, and total variation norm, the corresponding coefficients are sparse and the original signal can be recovered. There is a vast literature that deals with these problems both from theory to applications, i.e., the CS resources at Rice University [35] and the L1-MAGIC website maintained at Georgia Institute of Technology [36]. They contain a lot of papers, technical reports, and collections of MATLAB routines for solving the convex optimization programs central to compressive sampling. However, how to construct the orthogonal basis of specific signal and form the sparsest representation is still an open question to be further studied.

6.7.2 *Sparse Decomposition Over Pre-defined Dictionaries*

Sparse decomposition (also called sparse coding or sparse representation) means to represent a signal as a linear combination of a few atoms of a given (often overcomplete) dictionary [37]. Sparse decomposition was first accomplished by Olshausen and Field for estimating a whole basis of image patches [38], and later elaborated by Elad [39] in image processing.

Without loss of generality, in the Hilbert space, one can always find a set of functions $\{\varphi_i\}$ in $L_2(R)$, such that the signal y can be expressed as the linear combination of N elementary waveforms, signal atoms $\{\varphi_i\}$, such that,

$$y = \Phi x = \sum_{i=1}^{N} x_i \varphi_i \tag{6.80}$$

where x_i are the representation coefficients of y in the dictionary $\Phi = (\varphi_1, ..., \varphi_N)$. Clearly, x and y are equivalent representations of the same signal, with y in the time (or any other domain) and x in the Φ domain. The signal y is said to be K-sparse in Φ if it is a linear combination of only K basis vectors; that is, only K of the x_i coefficients in Eq. 6.80 is nonzero. In practice, x is considered compressible if it has just a few large coefficients and other coefficients are relatively small. Sparse decomposition techniques have been studied in-depth and applied successfully from signal classification to image recognition [26, 39, 40, 41].

Figure 6.18 gives a demonstration of sparse representation with over complete dictionary representation (sparse coding). The original measurement $y_{900\times1}$ has 867 nonzero coefficients, which is a dense vector. When it is represented in terms of an over complete dictionary $\Phi_{900\times1800}$, the coefficient vector $x_{1800\times1}$ is sparse with K = 100 in a higher dimension.

The two key points in the development of sparse decomposition are sparsity and incoherence [23, 26].

(a) **Sparsity**: The signal is compressible and has sparse representation in some domains, with most of its coefficients are zeros. Sparsity indicates that the information carried by the signal is much smaller than its bandwidth.
(b) **Incoherence**: Incoherence means that the basis must be significantly different. The coherence between the two orthonormal matrices ϕ_i and ϕ_j is defined as [23],

$$\mu(\phi_i, \phi_j) = \sqrt{n} \max_{1\le k,j\le n} |\langle \phi_i, \phi_j \rangle| \tag{6.81}$$

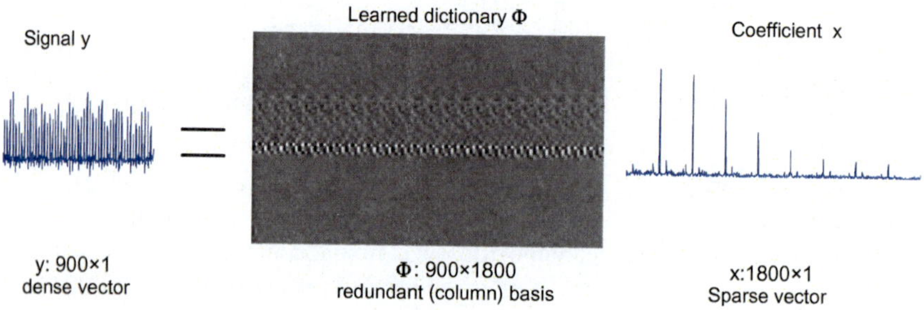

Fig. 6.18 Sparse representation of the signal vector

Incoherence refers to the bases with small coherence, which means the bases are less similar. As a result, a dictionary with incoherent bases has large varieties and "richer vocabulary" in the dictionary.

The research on sparse decomposition focus on two aspects: (1) How to construct a certain redundant dictionary that adapts to the real problems; (2) How to design the fast and effective decomposition approach. It is noted that sparse decomposition is different from "compressive sensing" in the previous section. The dictionaries used herein may not satisfy the signal recovery assumptions of compressive sensing. There exist several well-known orthonomal bases, such as Fourier Basis, Wavelet Basis, or Discrete Cosine Transform that are computationally efficient and span sparsely on certain types of signals. There are also over complete dictionaries that are built from these common bases. These include local cosine dictionaries, over complete wavelets, curvelets, and Gabor dictionaries [39]. These customized dictionaries are appealing because their corresponding representation can be found efficiently by leveraging their compatibility with fast transforms. Under certain circumstances, more elaborate dictionaries can be constructed by taking the union of different dictionaries.

In seeking for sparse representation, this formula (6.79) is NP-hard and is usually solved approximately, with either an iterative greedy algorithm such as Matching Pursuit (MP) [42], Orthogonal Matching Pursuit (OMP) [43], or global optimization algorithms, such as Basis Pursuit (BP) [44, 45] and FOCUSS [46].

6.7.3 Greedy Algorithms

(1) **Matching Pursuit (MP)**

Greedy algorithms are the most effective and widely studied algorithms for constructing sparse approximations of the signal. Matching Pursuit is a greedy algorithm that iteratively incorporates in the reconstructed signal the component from the measurement set that explains the largest portion of the residual from the previous iteration [42]. The stop criterion is reached when the residual reaches a magnitude below a set threshold. For a given over-complete dictionary $\Phi = \{\phi_k, k = 1, 2, ...K\}$, every atom is normalized as $\|\phi_k\|_2^2 = 1$. In the initialization, project the signal to the atom $\phi_{k_0} \in D$, then the residue R_f,

$$f = \langle f, \phi_{k_0} \rangle \phi_{k_0} + R_f \tag{6.82}$$

As R_f is orthogonal to ϕ_{k_0} then,

$$f = \left\| \langle f, \phi_{k_0} \rangle \right\|^2 + \left\| R_f \right\|^2 \tag{6.83}$$

To minimize the residue $\left\| R_f \right\|^2$, project R_f^m to the atom ϕ_{k_m}, and get,

$$R_f^m = \langle R_f^m, \phi_{k_m} \rangle \phi_{k_m} + R_f^{m+1} \tag{6.84}$$

Due to the orthogonal property of R_f^m and ϕ_{k_m},

$$\left\| R_f^m \right\|^2 = \left\| \langle R_f^m, \phi_{k_m} \rangle \right\|^2 + \left\| R_f^{m+1} \right\|^2 \tag{6.85}$$

By summation of the above two functions with m,

$$\| f \|^2 = \sum_{m=0}^{M-1} \left\| \langle R_f^m, \phi_{k_m} \rangle \right\|^2 + \left\| R_f^M \right\|^2 \tag{6.86}$$

It has been proven that the $\left\| R_f^m \right\|$ will exponentially converge to 0 when m approaches infinity [37].

(2) **Orthogonal Matching Pursuit (OMP)**

The MP algorithm works perfectly when the dictionary is a set of orthonormal bases. But in the case of over-complete bases for a redundant dictionary, the algorithm is sub-optimal as the projections are not orthogonal. In [43], the authors proposed an improved version, known as Orthogonal Matching Pursuit (OMP), which resolves the issue. OMP is based on MP but also applies an additional Gram-Schmidt orthogonalization to the residual against all measurement vectors selected in previous iterations [43]. By doing so the performance of the algorithm is improved and provides better reconstruction compared to plain MP although the complexity is increased.

With the selected atom ϕ_k, OPM algorithm applies the Gram-Schmidt procedure,

$$u_m = \phi_{k_m} - \sum_{p=0}^{m-1} \frac{\langle \phi_{k_m}, u_p \rangle}{\left\| u_p \right\|^1} u_p \tag{6.87}$$

Then the residue R_m is projected to u_m but not to ϕ_{k_m},

$$R_m = \frac{\langle u_m, R_m \rangle}{\| u_m \|^2} u_m + R_{m+1} \tag{6.88}$$

The signal f is represented with M-term as,

$$f = \sum_{m=0}^{M-1} \frac{\langle u_m, R_m \rangle}{\| u_m \|^2} u_m + R_M \tag{6.89}$$

It is noted that OMP never selects the same atom twice because the residual is orthogonal to the atoms that have already been chosen. Both MP and OMP have been shown to converge to a solution that fully explains the data and the noise; however, only OMP is guaranteed to converge to a sparse solution [43]. From the convergence

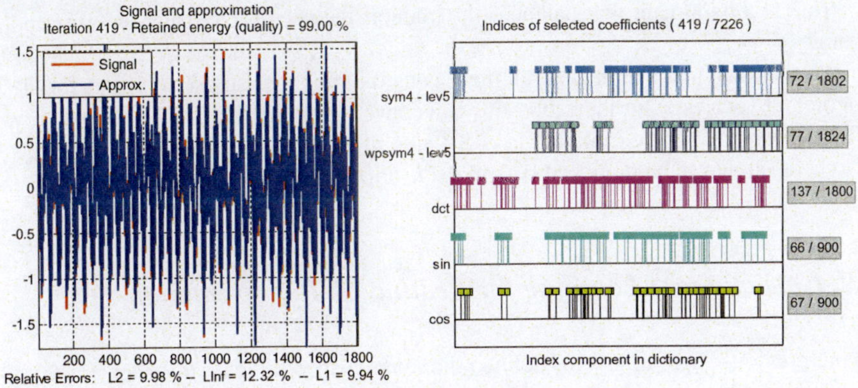

Fig. 6.19 The signal approximation with OMP in a pre-designed dictionary

speed, the OMP approach is faster than MP. As a result, the OMP representation will be sparser than the MP representation, but the precision is much lower than that of the minimum norm and needs strict constraints on the dictionary. Figure 6.19 shows an example of signal representation with the OMP algorithm. The coding is found to be very sparse with a pre-defined over compete dictionary which is a union of different othornormal bases such as sin wave and wavelet packets.

(3) **Basis Pursuit (BP)**

Because of the redundancy of the dictionary, the atoms are not linear independent. As a result, various linear combinations meet the requirement, even with the different number of atoms. Elad and Bruckstein [47] have proved that if there is a sparse representation of the signal s in the dictionary Φ, and

$$\|c_0\| \le \sqrt{2} - 0.5/\mu \tag{6.90}$$

Then the function (12) has a unique solution, and the solution of l_1 minimization is the same as that of l_0 minimization (11). As a result, we can solve this problem with l_1 minimization to replace the original l_0 minimization with the same solution (Basis Pursuit),

$$\arg\min_x \|x\|_1 \; subject\ to\ y = \Phi x \tag{6.91}$$

In the approximate sense, global optimization relaxes the sparseness constrain to an l_1-norm, converting the problem into a convex optimization problem and solved as a linear programming problem. An efficient approach to this is to solve it as a Lasso-type problem by Tibshirani [48],

$$\min_x \|y - \Phi x\|_2^2 + \lambda\|x\|_1 \,, \tag{6.92}$$

where λ is a parameter that balances the tradeoff between reconstruction error and sparsity.

It is a known fact that in general, the l_1 constraint induces sparse solutions for the coefficient vectors x. Furthermore, this is a convex problem that can be solved very efficiently using for example the LARS-Lasso algorithm [49]. This alternative has also been shown to be more stable than the l_0 approach.

6.7.4 Dictionary Learning for Redundant Representation

In the previous section, the pre-defined dictionaries also indicate that they may not fit well to the data outside their scope, and therefore a predefined analytic dictionary might not be adequate to describe arbitrary signals. Dictionary learning could be used to address this issue. This learning option starts by building a training database of signal instances, similar to those anticipated in the application, and constructing an empirically learned dictionary. Such a dictionary can then be used in the application as a fixed and redundant dictionary. Many dictionary learning algorithms have been developed for this purpose, such as the method of optimal directions (MOD) [50], and (K-SVD) [51], and online learning [52].

The MOD approach [50] is a two-stage iterative update procedure. It is close to the K-means outline with a sparse coding stage that uses either OMP [43] or FOCUSS [46] followed by an update of the dictionary. The MOD algorithm works as follows.

(1) Assuming that the sparse coding for each example is known and defining the errors as $e_i = y_i - \Phi x_i$, the overall representation error is,

$$\|E\|_F^2 = \|[e_1, e_2, ...e_N]\|_F^2 = \|Y - \Phi X\|_F^2 \tag{6.93}$$

(2) Assuming X is fixed, and seeking an update to Φ such that the above error is minimized by taking derivative of the above equation w.r.t.Φ,

$$(Y - \Phi X)X^T = 0 \tag{6.94}$$

(3) Update $\Phi^{(n+1)}$ as,

$$\Phi^{(n+1)} = YX^{(n)T}(X^{(n)}X^{(n)T})^{-1} \tag{6.95}$$

To date, the most prominent and successful dictionary training algorithm is the K-SVD [51]. The K-SVD algorithm is a generalization of the K-Means method that is used to design optimal codebooks for vector quantization. In [51] the authors have demonstrated that the K-SVD provided improved performance over the MOD, while also being computationally more tractable. An example of K-SVD dictionary learning implementation will be studied in the next chapter.

On the other side, with the redundant dictionary, the sparsity of the signal's representation is increased, and the signal could be recovered with higher probability. While in reality, as the number of atoms in the dictionary is usually huge, then the stability and complexity of optimization algorithms should further be studied.

References

1. Oppenheim AV, Schafer RW (2010) Discrete-time signal processing, 3rd edn. Prentice Hall
2. Bendat JS, Piersol AG (2010) Random data: analysis and measurement procedures, 4th edn. Wiley
3. Woyczynski W (2011) A first course in statistics for signal analysis, 2nd edn. Springer
4. Cohen L (1989) Time–frequency distribution—a review. Proc IEEE 77:941–981
5. Candy JV (2006) Model-based signal processing. Wiley
6. Oppenheim AV, Willsky AS, Nawab SN (1996) Signals and systems, 2nd edn. Prentice Hall
7. Duhamel P, Vetterli M (1990) Fast Fourier transforms: a tutorial review and a state of the art. Signal Process 19:259–299
8. Kammler DW (2008) A first course in Fourier analysis. Cambridge University Press
9. Allen RL, Mills DW (2004) Signal analysis time, frequency, scale, and structure. Wiley
10. Kamarthi SV, Pittner S (1997) Fast Fourier and wavelet transform for flank wear estimation–a comparison. Mech Syst Signal Process 11:791–809
11. Dowling MJ (1993) Application of non-stationary analysis to machinery monitoring, ICASSP-93. In: IEEE international conference on acoustics, speech, and signal processing, pp 59–62
12. Dimla SDE, Lister PM (2000) On-line metal cutting tool condition monitoring I: force and vibration analyses. Int J Mach Tools Manuf 40:739–768
13. Elbestawi MA, Papazafiriou TA, Du RX (1991) In-process monitoring of tool wear in milling using cutting force signature. Int J Mach Tools Manuf 31(1):55–73
14. Gong W, Obikawa T, Shirakashi T (1997) Monitoring of tool wear states in turning based on wavelet analysis. JSME Int J (Series C) 40:447–453
15. Kwak JS, Ha MK (2004) Detection of dressing time using the grinding force signal based on the discrete wavelet decomposition. Int J Adv Manuf Technol 23(1–2):87–92
16. Lee BY, Tarng YS (1999) Application of the discrete wavelet transform to the monitoring of tool failure in end milling using the spindle motor current. Int J Adv Manuf Technol 15:238–243
17. Li X, Guan XP (2004) Time-frequency-analysis-based minor cutting edge fracture detection during end milling. Mech Syst Signal Process 18:1485–1496
18. Tansel IN, Rodriguez O, Trujillo M, Paz E, Li W (1998) Micro-end-milling—I: wear and breakage. Int J Mach Tools Manuf 38:1419–1436
19. Yesilyurt I (2006) End mill breakage detection using mean frequency analysis of scalogram. Int J Mach Tools Manuf 46(3–4):450–458
20. Zhu KP, Hong GS, Wong YS, Wang WH (2008) Cutting force denoising in micro-milling tool condition monitoring. Int J Prod Res 46(16):4391–4408
21. Franks LE (1969) Signal theory. Prentice-Hall, USA
22. Moon TK, Stirling WC (2000) Mathematical methods and algorithms for signal processing. Prentice Hall, USA
23. Candès E, Romberg J, Tao T (2006) Robust uncertainty principles: exact signal reconstruction from highly incomplete frequency information. IEEE Trans Inform Theory 52(2):489–509
24. Donoho DL, Compressed sensing. IEEE Trans Inform Theory 52(4):1289–1306
25. Lim JS, Oppenheim AV (eds) (1988) Advanced topics in signal processing. Prentice-Hall
26. Mallat SG (2008) A wavelet tour of signal processing: the sparse way, 3rd edn. Academic Press
27. Mallat SG (1989) A theory of multiresolution signal decomposition: the wavelet representation. IEEE Trans Pattern Mach Intell 11(7):674–693

28. Wu Y, Du R (1996) Feature extraction and assessment using wavelet packets for monitoring of machining process. Mech Syst Signal Process 10(1):29–53
29. Mori K, Kasashima N, Fu JC, Muto K (1999) Prediction of small drill bit breakage by wavelet transforms and linear discriminant functions. Int J Mach Tools Manuf 39:1471–1484
30. Yoon MC, Chin DH (2005) Cutting force monitoring in the end milling operation for chatter detection. IMechE 2005 Proc IMechE, vol 219. Part B J Eng Manuf 455–465
31. Zhu KP, Wong YS, Hong GS (2009) Wavelet analysis of sensor signals for tool condition monitoring: a review and some new results. Int J Mach Tools Manuf 49(4):537–553
32. Vetterli M, Herley C (1992) Wavelets and filter banks: theory and design. IEEE Trans Signal Proc 40(9):2207–2232
33. Wickerhauser MV, Coifman RR (1992) Entropy based methods for best basis selection. IEEE Trans Inform Theory 38(2):719–746
34. Abry P (1997) Ondelettes et turbulence. Multirésolutions, algorithmes de décomposition, invaria nce d'échelles. Diderot Editeur, Paris
35. Compressive Sensing Resources (2014). http://dsp.rice.edu/cs
36. l1-Magic (2014). http://users.ece.gatech.edu/~justin/l1magic
37. Lewicki MS, Sejnowski TJ (2000) Learning overcomplete representations. Neural Comput 12(2):337–365
38. Olshausen BA, Field DJ (1996) Emergence of simple-cell receptive field properties by learning a sparse code for natural images. Nature 381(6583):607–609
39. Elad M (2010) Sparse and redundant representations: from theory to applications in signal and image processing. Springer
40. Bruckstein AM, Donoho DL, Elad M (2009) From sparse solutions of systems of equations to sparse modeling of signals and images. SIAM Rev 51(1):34–81
41. Cheng H, Liu Z, Yang L, Chen X (2013) Sparse representation and learning in visual recognition: theory and applications. Signal Process 93:1408–1425
42. Mallat S, Zhang Z (1993) Matching pursuits with time-frequency dictionaries. IEEE Trans Signal Process 41(12):3397–3415
43. Pati YC, Rezaifar R, Krishnaprasad PS (1993) Orthogonal matching pursuit: recursive function approximation with applications to wavelet decomposition. In: 27th Asilomar conference on signals, systems and computers, vol 1, pp 40–44
44. Chen S, Donoho D (1995) Atomic decomposition by basis pursuit. In: SPIE international conference on wavelets. San Diego
45. Chen SS, Donoho DL, Saunders MA (1998) Atomic decomposition by basis pursuit. SIAM J Sci Comput 20:33–61
46. Gorodnitsky IF, Rao BD (1997) Sparse signal reconstruction from limited data using FOCUSS: a re-weighted minimum norm algorithm. IEEE Trans Signal Process 600–616
47. Elad M, Bruckstein AM (2002) A generalized uncertainty principle and sparse representation in pairs of bases. IEEE Trans Inform Theory 48(9):2558–2567
48. Tibshirani R (1996) Regression shrinkage and selection via the Lasso. J Royal Stat Soc Series B 58(1):267–288
49. Efron B, Johnstone I, Hastie T, Tibshirani R (2004) Least angle regression. Ann Stat 32(2):407–499
50. Engan K, Aase SO, Husoy JH (2000) Multi-frame compression: theory and design. EURASIP Signal Process 80(10):2121–2140
51. Aharon M, Elad M, Bruckstein AM (2006) The K-SVD: an algorithm for designing of over-complete dictionaries for sparse representations. IEEE Trans Signal Process 54(11):4311–4322
52. Mairal J, Bach F, Ponce J, Sapiro G (2009) Online dictionary learning for sparse coding. In: Proceedings of the international conference on machine learning (ICML)

Chapter 7
Tool Condition Monitoring with Sparse Decomposition

7.1 Introduction

In modern CNC machining, an effective tool condition monitoring (TCM) system can improve productivity, ensure workpiece quality and enhance the manufacturing intelligence [1, 2]. Due to its importance, tool condition monitoring (TCM) has been extensively studied [2]. The earlier study on TCM is mainly carried out with time series analysis, such as [3] and [4]. With these methods, a threshold was set for binary state detection. However, the threshold value varies with cutting conditions and is difficult to determine. To improve the performance of monitoring applications, intelligent methods have been developed. Among them, neural networks (NNs) are most studied and successful in applications, due to their capability in learning and nonlinear mapping of features and tool states [5]. Yen et al. [6] applied a self-organization feature map (SOM) for tool wear monitoring in micro-milling. By proper selection of spectral features from acoustic emission, accurate tool wear classification can be obtained, and the approach was found robust to noise or working conditions. To deal with the nonlinearity and uncertainty, Wang et al. [7] presented a kernel method for feature selection and fusion, which incorporated kernel techniques to the factor analysis, and then monitored the tool conditions with support vector regression. Hidden Markov models (HMMs) have been employed for machinery condition monitoring lately due to their excellent local dynamical reasoning capability [8]. Geramifard et al. [9] developed a multimodal HMM approach, in which each HMM captured and emphasized a different tool wear regiment and three weighting schemes were introduced to combine the outputs from multiple modes into one for tool wear monitoring. Besides NNs, other machine learning methods such as fuzzy method [10], particle filter [11], clustering [12], manifold learning [13], and fusion of intelligent classifiers [14] have also been applied to condition monitoring.

In the above condition monitoring schemes, it operates with the signal preprocessing, feature extraction, and classifier design sequentially, and results in low processing speed and high computational complexity. At the same time, different stages of working independently may also lead to bad estimation or even contrary

K. Zhu, *Smart Machining Systems*, Springer Series in Advanced Manufacturing,
https://doi.org/10.1007/978-3-030-87878-8_7

results [2]. A unified framework with a simplified procedure is necessary. In this aspect, sparse decomposition provides a possible solution. In the past decade, sparse decomposition techniques have been developed in signal and image processing [15]. It is found that better classification rate could usually be achieved with sparse representation of image and signal. Sparse coding is reached by spanning a signal (or an image) in an over complete dictionary, with which the signal is represented in a compact linear form. There are two dictionary construction methods in sparse coding, namely the preset basis methods (i.e., Fourier basis, wavelets, and Gabor function [16]) and learning methods (i.e., k-SVD learning [17] and the online learning [18]). For these methods, choosing a standard dictionary of atoms is usually not effective enough to represent the signal, while a dictionary learned from training samples can be more accurate. So the second method has been more widely studied in applications [17–19].

In manufacturing process monitoring and fault diagnosis, sparse decomposition has recently been studied. Cai et al. [20] proposed a sparse decomposition for gearbox fault diagnosis. With this approach, vibration signals were decomposed into high-oscillatory and low-oscillatory components for faulty feature extraction. Liu et al. [21] developed a shift-invariant sparse coding approach, and based on the learned dictionary, the extracted features were found informative in diagnosing the bearing vibration conditions. In [22] a locality preserving sparse modeling approach was proposed to identify multiple operation conditions, and shown to be robust for process monitoring. The study [23] applied the sparse decomposition approach for noise reduction and cutting condition monitoring. With the learned Gabor basis, the force showed good noise robustness and sensitivity to the state variations. A shift-invariant dictionary learning was also developed for the bearing fault diagnosis [24]. The impulse signals at different locations with the same pattern were represented by only one atom through shifted operations, and the dictionary was then learned by taking all possible shifts of a few short atoms. Gao et al. [25] developed a sparse non-negative matrix factorization approach for monitoring defects in waveguide imaging. The model was adaptively learned to regularize the sparsity of the factorization and enhance the detecting accuracy. To meet the challenge of extract multiple components from coupled noisy observations, [26] presented an approach to simultaneously identify multiple features for gearbox fault diagnosis. It exploited the prior information and identified that multiple faults with different morphological waveforms could be sparsely represented over the union of redundant dictionaries. The rationale and applications of sparse decomposition were surveyed in [27] for machining process monitoring.

For the above dictionary learning-based approaches, the discrimination power of atoms and coding coefficients were generally not learned. How to constrain the learning, and make the dictionary more suitable for classification is a problem to be addressed? Aharon et al. improved the learning process by adding irrelevant restrictions so that each type of dictionary corresponded to its class while keeping as far as possible to the others [17]. As a result, the sparse encoding can achieve better signal reconstruction and classification results. Some researchers introduced discriminant constraint [28] and using discriminant K-SVD algorithm [29] to learn a global shared dictionary for image recognition. However, in the shared dictionary, the relationship between atoms and categories was ambiguous. In order to overcome this weakness, it needs to learn from training samples a dictionary corresponding to its respective category. This idea was studied in [30, 31] with the Fisher's criterion for dictionary learning and successfully applied to image classification.

Based on these considerations, this approach develops a simplified model of tool condition monitoring based on discriminative sparse decomposition. It reaches both feature extraction and state classification in the same optimization process. Discrimination criteria are adapted to learn the overcomplete dictionary, and constrain the sparse coefficient to be minimized within the class and maximized between classes. The learned class-specific sub dictionaries are then cascaded to form a large discrimination dictionary. A similarity criterion with discriminant distance is then implemented to match the coding to its closest class. Additionally, with no reconstruction operation, it would run fast for tool condition monitoring.

7.2 Sparse Coding for Denoising (Heavy Non-Gaussian Noise Separation)

7.2.1 *Introduction*

In mechanical micromachining, the comparatively small cutting force signal is contaminated by heavy noise, and as a result, it is necessary to denoise the force signal before further processing it. The traditional denoising methods, based on Gaussian noise assumption, are not effective in this situation because the noise is found to contain high non-Gaussian component. This section presents a new approach for cutting force denoising in mechanical micromachining condition monitoring. Based on the non-Gaussian noise assumption, in [32] the authors applied the independent component analysis (ICA) approach to deal with this problem, which is effective but the extracted independent components are prone to variations under different initializations. The other problem is that the approach in [32] needs to compute at least three force components to extract the independent components; it will be not applicable in case fewer channels are monitored. In this paper, a new method is developed for force denoising, which takes advantage of the non-Gaussian noise properties in the time–frequency domain.

Sparse representation (also named compressive sensing, sparse coding in information theory [33, 34]) has received a lot of attention in recent years, from signal classification to image recognition [17, 34–37]. Sparse representation originates from wavelet analysis while it has widened and deepened the scope of wavelet analysis. The idea of sparse representation is to represent a signal sparsely in terms of the linear combination of atoms in an overcomplete (redundant) dictionary [38]. The dictionary's redundancy offers a wide choice of atoms to represent the signal so that the smallest linear combination of atoms could be selected to approximate the signal. The affront mentioned ICA approaches, have been found to be closely related to the sparse representation [30]. In this section, based on both the signal and noise properties in the time–frequency domain, a sparse decomposition is developed for cutting force denoising in micro-milling TCM.

7.2.2 *Noise Properties in Micro-milling*

Figure 7.1 shows a typical sample of the cutting force and noise in micro-milling. As can be seen from the Fig. 7.1, the SNR is very low and the absolute value of the noise is nearly 1/3 that of the force. At the same time, the frequency bands of noise and force are overlapped quite a lot (This will be shown in later sections). An effective noise removal procedure is an essential first step towards accurate estimation of the tool wear state in micro-machining.

As can be seen from Fig. 7.1, the signal-to-noise ratio (SNR) is very low and the amplitude range of the noise is nearly 1/3 of that of the force. The SNR is calculated as,

$$SNR = 10 \times \log_{10} \frac{s_i^2}{n_i^2} = 10.1\,\text{dB} \tag{7.1}$$

According to Eq. 7.1, the SNR is generally around (9–15 dB) for micro-milling in this study. This value is rather low and denoising is a prerequisite for a reliable

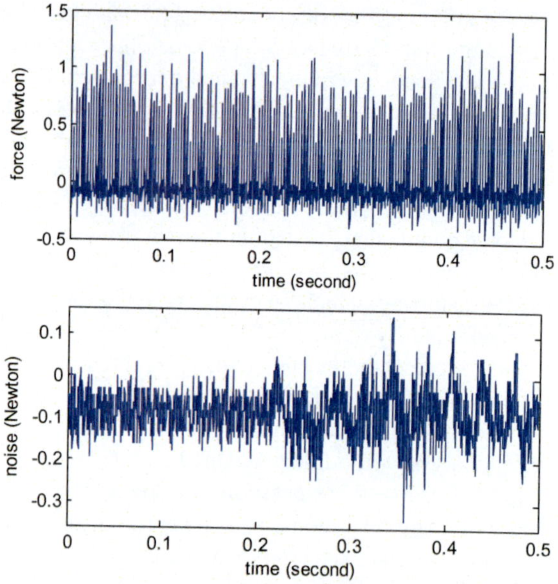

Fig. 7.1 Cutting force and noise in micro-milling. Both the force and noise are sampled at 6,000 Hz, and the 0.5 s accounts for 3000 sample points

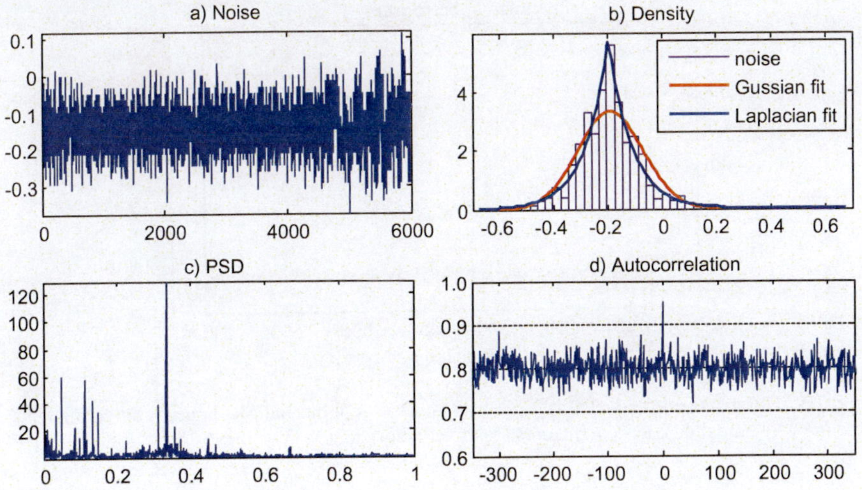

Fig. 7.2 Noise properties: **a** the noise **b** noise compared to Gaussian and Laplace distributions, **c** in the normalized frequency, **d** the autocorrelation coefficients

correlation of force-tool conditions. To find a suitable denoise approach, the noise statistical properties are identified in Fig. 7.2. The noise is collected during the recess of the tool hold during machining, which means the segment is in between two cutting processes and is pure noise.

The density estimation shows that the noise is super-Gaussian, with a longer tail than Gaussian distribution, and is well fitted with Laplace distribution as illustrated in Fig. 7.2b. The normalized frequency representation (power spectrum density, PSD) and autocorrelation show that the noise harmonics distribute widely in the frequency domain, and the autocorrelation is high with a value around 0.7. This means that the noise is neither Gaussian nor white. As a result, the traditional approaches, such as filtering and wavelet thresholding, are not effective under this case.

The noise is sampled with 6000 Hz, and the segment in Fig. 7.2a is just one second. Figure 7.2b shows that it is far from Gaussian. The interesting phenomenon in Fig. 7.2 is that only several amplitudes in the PSD are prominent and the others are very small. When the PSD is integrated along the time, it is the time–frequency representation (It is named spectrogram, and will be discussed in detail in later Section). Not like dense values in the time domain, the noise has sparse structure in the time–frequency domain as shown in Fig. 7.3. On the other side, the force signal also shows such a sparsity in the time–frequency domain, which will be discussed in the next section.

Based on the sparse property of noise and force in the time–frequency domain, a sparse representation approach is applied for noise separation in the micro-milling condition monitoring. The idea of this approach is based on the findings that both force and noise could be sparsely represented in the time–frequency domain, and that the dictionaries (time–frequency bases) can be learned differently for force and noise signals. The details are discussed in the following sections.

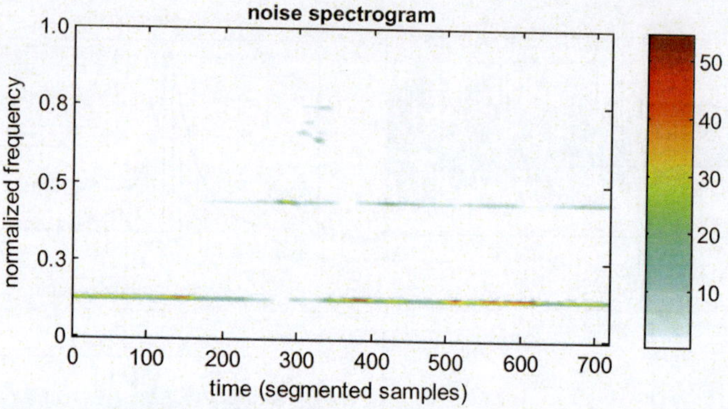

Fig. 7.3 The sparse property of noise spectrogram. Darker points indicate larger value coefficients, and the white area indicates coefficients are or very close to zero

7.2.3 Sparse Representation in the Time–Frequency Domain

For any signal $y(t)$, the Fourier Transform $\hat{y}(f)$ is obtained by inner product of $y(t)$ with a sinusoidal wave $e^{j2\pi ft}$,

$$\hat{y}(f) = \langle y(t), e^{j2\pi ft} \rangle = \int_{-\infty}^{\infty} y(t)e^{-j2\pi f}\,dt \tag{7.2}$$

The Fourier Transform transforms the signal $y(t)$ from the time domain into the frequency domain, and results in the amplitudes of the frequency f. To overcome the limitation of PSD's lack of local information, a time sliding window is applied to the Fourier transform. The resultant transform is denoted windowed or short-time Fourier Transform,

$$STFT_y(t, f) = \int_{t-T/2}^{t+T/2} y(\tau)g(\tau - t)e^{-2\pi jf\tau}d\tau \tag{7.3}$$

The corresponding energy density $|STFT_y(t, f)|^2$ is called a spectrogram, which are widely used for time–frequency analysis. The STFT coefficients represent the signal in the sparse time–frequency blocks at different positions. The STFT analysis is characterized by the window function, overlap and duration. With proper choosing of window size and overlapping interval, these time–frequency blocks $|STFT_y(t, f)|^2$ forms a redundant dictionary.

In this study, the noisy force is transformed to the frequency domain for sparse representation. In Fig. 7.4, a force component in the time domain is illustrated and its corresponding representation in the frequency domain by applying PSD. As can be

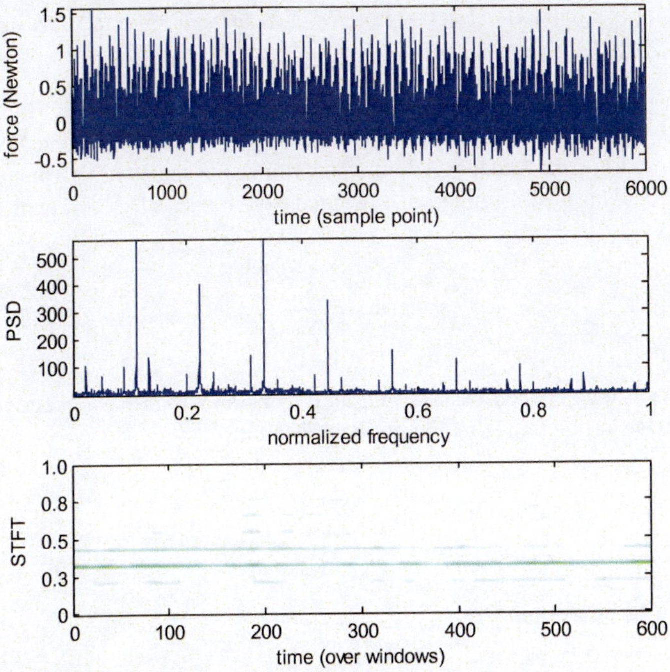

Fig. 7.4 The force and its sparse representation in PSD, and STFT domain

seen from the figure, the signal has non-zero coefficients for almost every time point in the time domain, but in the frequency domain, only some frequencies (harmonics) have large coefficient.

7.2.4 *Sparse Representation as a Convex Optimization Problem*

From the pattern classification point of view, the sparse representation is a reconstructive approach, as it is decomposed to contain sufficient information for reconstruction. While in this study, the aim is for noise separation and then the tool state estimation. The basis's discrimination power is concerned, besides the signal's reconstruction error from the commonly applied sparse decomposition approach. The basis is the combination of the pre-learned basis of the pure noise Ψ and the updating signal basis $\Phi^{(i)}$, $[\Phi^{(i)}; \Psi]$.

This work introduces a novel metric that includes both reconstruction and discrimination terms in the dictionary learning process, benefitting from the best of both discriminative and reconstructive worlds. It is illustrated to optimize objective function J,

$$\begin{aligned} J &= \max(Discrimination) + \min(Reconstruction) + \max(Sparsity) \\ &= \max(J_1) + \min(J_2) + \max(J_3) \end{aligned} \tag{7.4}$$

where the J_1 maximizes the discrimination of the basis in the STFT, and J_2 controls the reconstruction error, and the J_3 domain maximize the sparsity of estimated force.

A measure of basis discrimination is adapted from Fisher's Discriminant Analysis [39],

$$J(w) = \frac{S_B}{S_W} \tag{7.5}$$

where the S_B is the between-basis scatter and S_W is the within-basis scatter.

For a c-basis problem. Define within-basis scatter,

$$S_w = \sum_{i=1}^{c} S_i \tag{7.6}$$

and between-basis scatter,

$$S_B = \sum_{i=1}^{c} n_i (\mu_i - \mu_c)(\mu_i - \mu_c)^T \tag{7.7}$$

$$\text{where } S_i = \frac{1}{n_c} \sum_{i=1}^{n} (x - m_c)(x - m_c)^T \tag{7.8}$$

is the covariance matrix measure the variability of sample within each basis, $\mu_c = \frac{1}{n_c} \sum_{i=1}^{n} x_i$ is the mean of samples from basis c, and n is the number of data of each basis.

To find the maximization of $J(w)$, this question can be formulated equivalently as an optimization problem by minimizing the inverse of $J(w)$,

$$J_1 = \min_{w} \frac{1}{2} \frac{S_W}{S_B} \tag{7.9}$$

where the role of within and between scatter is switched.

The measure J_2 is the reconstruction error l2 norm that illustrates the robustness of the algorithm,

$$J_2 = \min \left\| Y_i - \varphi_i^T X_i \right\|_2^2 \tag{7.10}$$

The measure J_3 forces the sparsity of the representation the STFT,

$$J_3 = \min\left|Y_i - \varphi_i^T X_i\right|_1 \quad (7.11)$$

So finally the sparse representation problem is formalized as a convex optimization problem,

$$\min \frac{1}{2}\frac{S_B}{S_W} + \frac{1}{2}\lambda_1 \left\|Y_i - \varphi_i^T X_i\right\|_2^2 + \lambda_2 |X|_1 \quad (7.12)$$

In (7.12), λ_1 is the l_2-norm regularization parameter, and λ_2 is the l_l-norm regularization parameter.

How to learn the dictionary $\{\varphi_i\}$ is most important in this approach. The best dictionary is the one that leads to the sparsest representation. Hence it could be desirable to have a huge dictionary (i.e., $T >> N$), but it leads to a prohibitive computation time cost for calculating the x coefficients. Therefore, there is a trade-off between the complexity of our analysis (i.e., the size of the dictionary) and computation time. Pursuit algorithms can nearly reach optimal M-term approximations in incoherent dictionaries that include vectors that are sufficiently different [37].

In this work, an l_1 variation of K-SVD [17] is adapted to learn the dictionary by solving the optimization problem in Eq. (7.18). The optimization is carried out iteratively in three steps:

(1) Learn the dictionary Ψ from pure noise;
(2) Sparsely code the y given the current dictionary estimate;
(3) Update the dictionary atoms given the sparse representations.

The dictionary update is performed one atom at a time, optimizing the target function for each atom individually while keeping the remaining atoms fixed. The time-domain signal then can be recovered by inversion of STFT transformation. The detailed implementation of the algorithm is described in Table 7.1.

7.2.5 *Case Studies*

(1) **Results and discussions**

In this study, the STFT window is chosen as Gaussian. The window size is selected with 512 points, and with 64 points overlapping. The selection of optimal values for λ_1 and λ_2 is a delicate and difficult task. Their values are chosen with empirical studies. The regularization parameter λ_2 is directly related to the sparsity of noise in STFT: the higher the λ_2, the higher the sparsity, $\lambda_2 = 0.89$. The parameter λ_1 is chosen to control the reconstruction: the higher the value of λ_1, the better reconstruction, $\lambda_1 = 0.41$.

Figure 7.5 shows the force signals and their corresponding power spectrum at severe flank wear state. As can be seen from the top figure, the low-frequency components have relatively larger peaks and as a result, the desired frequency components

Table 7.1 The modified K-SVD algorithm

Initialization : Set the random normalized dictionary matrix

$$\Phi^{(0)} = abs(randm(m,k)),$$

$$X^{(0)} = abs(randm(k,n)),\ k < m,n .$$

Set $J = 1,\Phi^{(0)} = [\Phi^{(0)};\Psi]$

Repeat until convergence

Sparse Coding: Use OMP algorithm to compute the sparse representation $\{\varphi_i\}$ for each signal $\{y_i\}$

for $i = 1,2,\ldots,N$

$$\min \ \frac{1}{2}\frac{S_W}{S_B} + \frac{1}{2}\lambda_1 \left\| \varphi^T x_i - y_i \right\|_2^2 + \lambda_2 \left| x \right|_1$$

Dictionary Update: For $k = 1, 2, \ldots, K$

- Define the group of examples that use φ_k,

$$w_k = \{i \mid 1 \le i \le N, x_i(k) \ne 0\}$$

- Compute residual matrix

$$E_k = Y - \sum_{j \ne k} \varphi_j x^j$$

- Restrict E_k by choosing only the columns corresponding to those elements that initially used φ_k in their representation, and obtain E_k^R.
- Apply SVD decomposition $E_k^R = U \Delta V^T$.

Update: $\varphi_k = u_1, x_R^k = \Delta(1,1) \cdot v_1$

Set $J = J + 1$.

(the harmonics of rotation frequency *fr*) are largely affected. This phenomenon is even worse when the tool wear value is lighter. The denoised force and its corresponding power spectrum are shown in the middle figure. As can be seen from this figure, the minor lobes beyond the characteristic frequencies are largely compressed or eliminated. At the same time, the ratio between the first two characteristic frequencies *fr* and *2fr* is raised, which is an important indication of tool wear conditions.

To verify the results, the separated noise is shown in the bottom figure. The statistics of the noise are: mean = −0.2813, variance = 0.1862, skew = 0.0003, and kurtosis = 2.8930. This indicates the estimated noise is super-Gaussian and having heavy tails. It coincides with the reference noise distribution shown in Fig. 7.5, with heavy tail distribution fitted well by a Laplace distribution, which has skew = 0 and kurtosis = 3.0. Meanwhile, under the same mean and variance fit, the noise is far from Gaussian distribution.

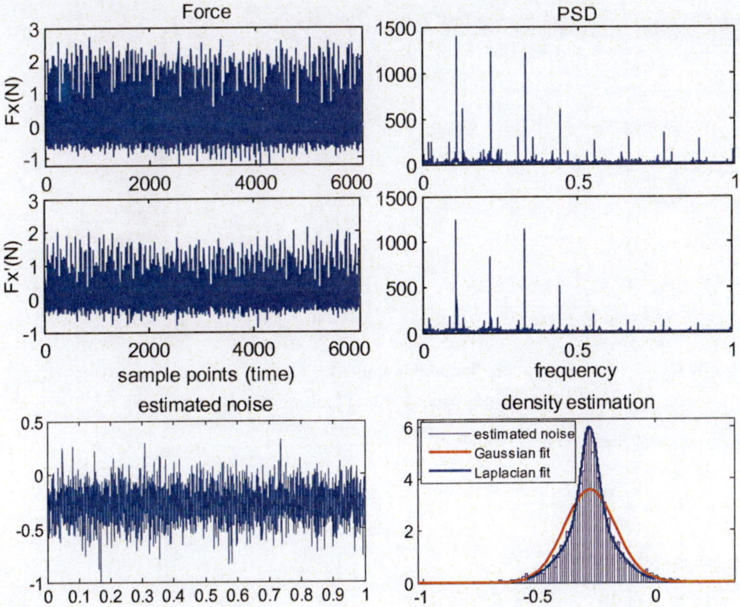

Fig. 7.5 Top: the sensor output and their corresponding power spectrum; Middle: after denoising; Bottom: the estimated noise density

The results show that the proposed approach could separate the heavy super-Gaussian noise well. This could be a good supplement to most current denoising algorithms based on the Gaussian noise assumption. The underlying assumption of this approach, the noise having sparse representation in the frequency domain, is a common phenomenon and adaptable to most machining force signals.

On the other side, it has been found that the learned dictionaries are quite dense and both of them have more than 95% nonzero elements, though most of them are small values (darker lines/points in Fig. 7.6a. The learned dense dictionary reflects the idea of sparse coding with the highly redundant dictionary, and its corresponding dense basis makes it possible for the discrimination between different signals with various elements. Figure 7.6b shows that the algorithm is fast and convergence quickly.

To demonstrate the denoising effect, Figs. 7.7 and 7.8 show the force spectrogram before and after denoising. This force is focused on the last few machining passes of Test 22 before the tool is severely worn. As it can be seen from Fig. 7.7, the force harmonics are completely immersed within the noise, and the difference cannot be discriminated from the figure when the tool flank wear dramatically increases. The sparse STFT denoised force spectrogram shows much better properties in Fig. 7.7. It is observed that after 80 s, $1Fr$, $2Fr$, and $3Fr$ ($Fr = 300$ Hz) of the reconstructed force increased dramatically, corresponding to the sharp increase of tool flank wear [23]. Comparing Figs. 7.8 to 7.7, it is clear that Fig. 7.8 shows better diagnostic

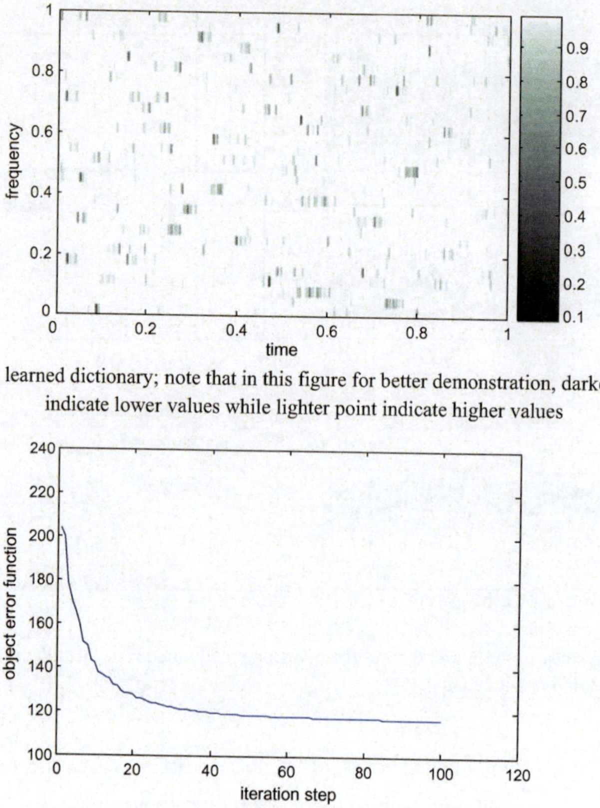

(a) The learned dictionary; note that in this figure for better demonstration, darker points indicate lower values while lighter point indicate higher values

(b) the convergence rate of the proposed dictionary learning

Fig. 7.6 Sparse representation and dictionary learning

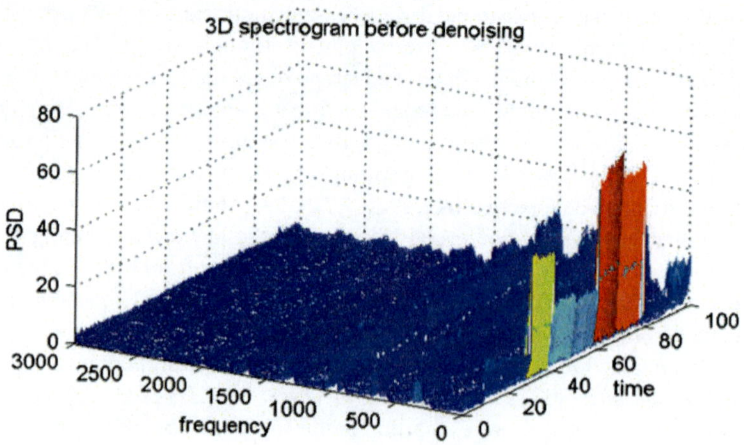

Fig. 7.7 Force spectrogram before denoising Fx

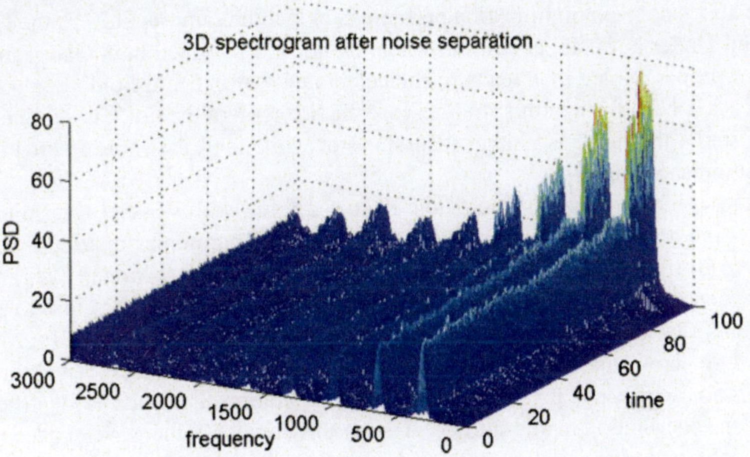

Fig. 7.8 Tool wear with sparse STFT denoised force

information since the harmonics stand out with the first three harmonics increasing dramatically according to the fast increase in flank wear.

On the other hand, besides denoising, the proposed algorithm can also be applied to tool condition estimation. The aim of this research is TCM in micromachining and denoising is only the first step. In fact, the proposed sparse decomposition algorithm can be furtherer slightly modified to learn various dictionaries from different tool conditions, and these dictionaries are constrained to be discriminant in the learning process. The learned basis can be applied to the tool state estimation when the force features are matched to the respective dictionaries. The detailed study of this approach is to be presented in [23].

(2) **Discussions**

(1) **Sparse decomposition and its relation to wavelet thresholding**

An important feature of compressed sensing is that it is robust to noise. For a signal contaminated with noise, the signal model is given by,

$$y = Ax + w \tag{7.13}$$

where w represents noise.

In the theory of sparse decomposition, the signal x in Eq. 7.13 may be estimated from noisy measurement y by solving the convex minimization problem, as follows.

minimize $\|x\|_1$,
subject to: $\|Ax - y\|_2$,
where ε is bound of the amount of noise in the data.

The above convex minimization problem is in fact the same as Eq. 7.2, the Lasso problem. Under a special condition, when the noise is assumed both Gaussian and white, a proper choice of ε leads to the celebrated wavelet thresholding approach [40]. It is a special form of this study in case the first term of the Eq. 7.12 diminishes. In this study, the noise is neither Gaussian nor white, and it is not effective with wavelet thresholding.

Figure 7.9 shows the residue of the denoised force with wavelet thresholding, which is the noise separated. From Fig. 7.9c the PSD harmonics distribute nearly evenly in all frequency bands; this explains why it cannot be separated even in the low-frequency noise environment. In Fig. 7.9d, the autocorrelation is close to zero; this means that the estimated noise is purely random, not as shown in Fig. 7.9d. Figure 7.9b shows that it is typically a Gaussian noise. Donoho's universal thresholding is not as effective in non-Gaussian noise separation. The noise is identified to be Laplace distribution in this study, and under this condition, the wavelet decomposition coefficients for noise are not evenly distributed among all scales as Gaussian noise. As a result, some of the noise coefficients are not small in certain scales, and the threshold is too low under this condition. This is similar to the findings in [41].

(2) **Sparse representation and ICA**

In the ICA approach, the extracted signals are measured by their Gaussianity (kurtosis). Super-Gaussian is defined as with normalized kurtosis $K > 3$ (see [42] for the definition of kurtosis). The intuitive idea is that the super-Gaussian density has heavy tails and a peek at the mean. The data distribute densely around the meanwhile most other areas have small values or close to zero. This property is closely related to the definition of sparseness. Figure 7.10 shows an example of probability density

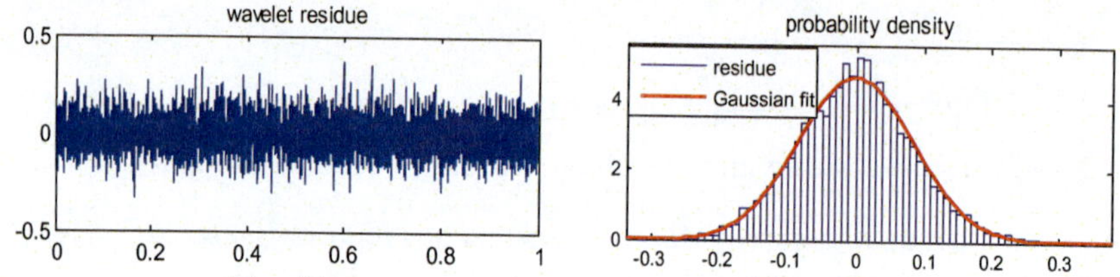

(a) residue

(b) noise distribution compared with Gaussian distribution

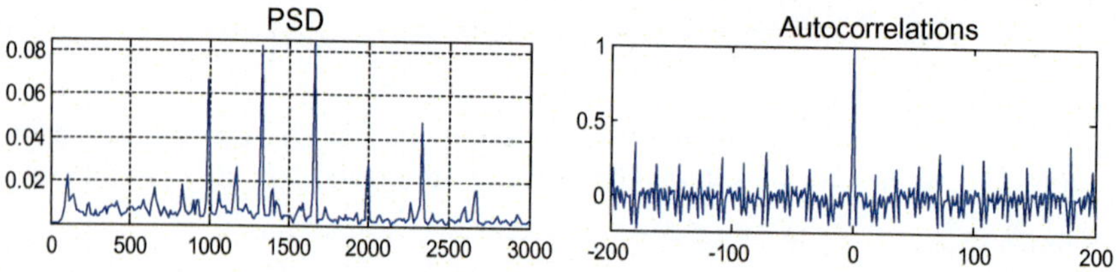

(c) the corresponding power spectrum density

(d) the autocorrelation coefficients

Fig. 7.9 Illustration of the wavelet residue

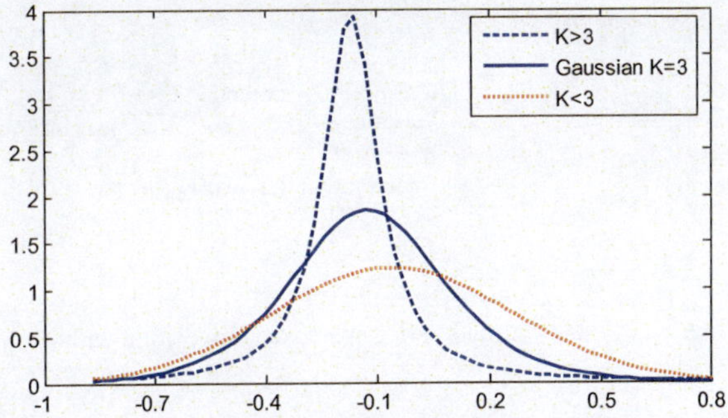

Fig. 7.10 Different distributions together with their kurtosis measure

functions with different values of kurtosis. Kurtosis can be used to check if a distribution is Gaussian. For symmetrical distribution, super-Gaussian has kurtosis higher than 3, while sub-Gaussian is lower than 3. In [30], Daubeches et al. have found that the solution to the ICA approach is the sparse solution.

In micro-milling there is machining noise that is relatively heavy, with non-Gaussian distribution, and has wide harmonics. This section has developed a denoising approach that imposes the constraints on the signal and force's representation in the time–frequency domain. The results show that the developed sparse coding approach separates the non-Gaussian noise and the noise is found to be consistent with the reference noise. The approach has potential in any non-Gaussian noise separation, provided it has sparse representation in the time–frequency domain. Further studies will be carried out on the approach to other precision machining types such as micro-drilling and micro-turning, and on how to simultaneously denoise signal and learn different dictionaries for tool condition estimation.

7.3 Sparse Representation for Tool State Estimation

The basic problem of tool condition monitoring (TCM) is to identify new samples under trained samples of different classes. In this sparse decomposition framework of TCM, the most important step is the sparse representation of training samples. This is achieved by learning a redundant dictionary from training data sets, in which the samples are represented as a sparse linear combination of dictionary atoms. It first extracts wavelet packet coefficients in multiple scales, and these packets are used as initial measurements. Fisher's discriminant criteria are then applied to the coding coefficient, to make sparse coding coefficients as small as possible within the same class and as large as possible between different classes. Finally, the dictionary coefficients are used to classify the test samples, which reaches fast state classification without signal reconstruction. Figure 7.11 illustrates such an idea, where state estimation is accomplished in the sparse coding scheme, and no feature extraction

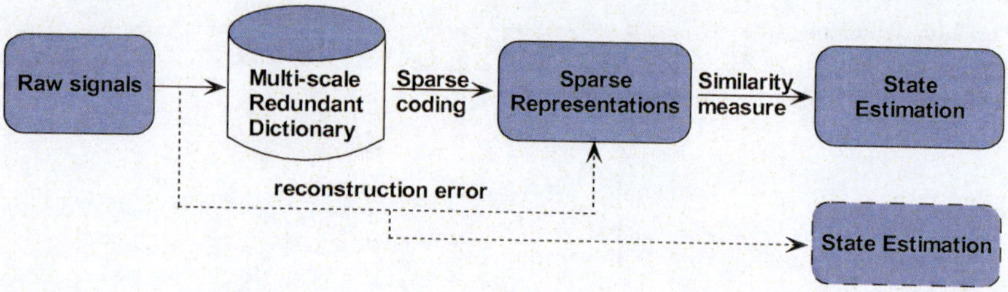

Fig. 7.11 State estimation in a sparse decomposition without reconstruction

is needed. The dashed line refers to the conventional sparse decomposition-based condition monitoring approach.

7.3.1 Sparse Coding of Wavelet Packet Decomposition Coefficients

In general, if the dictionary learning is carried out in the original data domain, it is difficult to make full use of local information. This local information is most important for condition monitoring and fault diagnosis, as it will identify where and when the faulty patterns happen. To this end, this study learns the dictionary from multiple-scale feature sets instead of original samples, to improve the monitoring capabilities of sparse representation. To learn the multi-scale features in the dictionary, this study applies the wavelet packet decomposition (WPD) [16] to the measurements. The WPD is a generalization of wavelet decomposition at higher frequencies, and efficient to process non-stationary machining signals. With the WPD, the signal $y(t)$ is approximated by the scaled and translated wavelet packet function $W_{m,b,n}$:

$$y(t) \approx \sum_{m} \sum_{b} \sum_{n} c_{m,b,n} W_{m,b,n}(t) \tag{7.14}$$

where the resolution level changes the scale 2^m and location $2^m n$. The wavelet packet coefficients are produced from this integral:

$$c_n^m = \int W_{m,b,n}(t)\, y(t)\, dt \tag{7.15}$$

It should be emphasized that Eq. 7.14 allows many possible combinations of wavelet packet functions to be selected to characterize the signal. In another sense, this is an over complete representation and the collection of basis is redundant. The sparse decomposition is adapted to make use of the benefits of these properties.

Sparse decomposition means to represent a signal as a linear combination of a few atoms of a given (often over complete) dictionary [17]. The code has to be a fair representation of the signal while it should also be sparse. Practically, a signal is treated as sparse if most of its entries are close to zero while the others are relatively large. A signal is k-sparse if it has only k nonzero elements. Without loss of generality, the signal y can be represented as the linear combination of T elementary waveforms, also called signal atoms,

$$y = \Phi x = \sum_{i=1}^{T} x_i \varphi_i \tag{7.16}$$

where x_i are called the coefficients of y in the dictionary $D = [\varphi_1, ..., \varphi_T]$.

The sparse representation problem is to find the sparsest coefficients under the dictionary [43],

$$\min_{x} \|x\|_0 \text{ s.t. } \mathrm{y} = \mathrm{D}x, \tag{7.17}$$

where $\|x\|_0$ is the l_0-norm of the coefficient vector $x \in R^k$, the number of non-zero elements. This problem is NP-hard, and hard even to approximate. Donoho [33] found that if the solution is sparse enough, the solution to the l_0-minimization of Eq. 7.17 can be unique and equal to the solution of the convex l_1-minimization problem. This is turned into a convex problem and various solutions were developed for Eq. 7.18, for example, the LARS-Lasso algorithm [44]. By substituting the l_0-norm in Eq. 7.17 with l_1-norm, it gets,

$$\min_{D,x} \|\mathrm{y} - \mathrm{D}x\|_2^2 + \lambda \|x\|_1 \tag{7.18}$$

where the l_1-norm is the sum of the magnitudes of elements of vector x.

With the WPD transform of measurement y, the Eq. 7.18 can equivalently be rewritten as,

$$\min_{D,x} \|\mathrm{Wy} \text{ - } \mathrm{D}x\|_2^2 + \lambda \|x\|_1 \tag{7.19}$$

where W denotes WPD operator. This formulation suggests that we can train the dictionary, not in the original measurement domain but the wavelet domain.

In the application, the dictionary is learned to have more columns than rows, in an over complete form to sparsify the coefficient x when the original data is dense. The dictionary D is viewed as a matrix whose columns are the atoms d_i, and in the application, the atoms are normalized to a unit l_2 norm, i.e., $\|d_i\|^2 = \sum_{n=1}^{N} |d_i[n]|^2 = 1$.

In tool condition monitoring, the basis of dictionary D allows recovering different states and coefficients allows assigning each measurement to its corresponding category. In the training, maximally overlapping segments are processed, and atoms

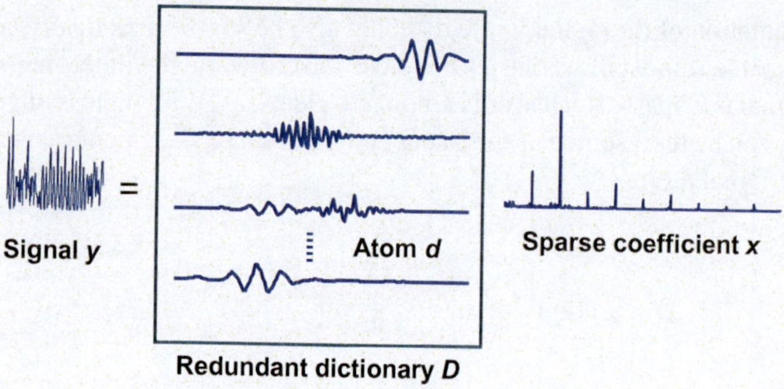

Fig. 7.12 Sparse representation of the signal

are then learned from different scales and sub-bands. This dictionary would reach shift-invariance, which is an important property for classification [15].

Figure 7.12 gives an example of sparse decomposition with an over complete dictionary representation. The signal y has 256 nonzero coefficients. In this over complete representation, the measurement y is composed of the sum of K columns of D, and the coefficient vector x is sparse with $K = 20$ while coded in the 1024×256 dimension dictionary D.

7.3.2 The Discriminant Dictionary Learning

The sparse coding above is a reconstructive approach, as the dictionary is learned to minimize the least reconstruction error. In condition monitoring, it is more concerned with the algorithm's classification performance instead of its reconstruction accuracy. The discriminant power of the coding is more important in this aspect. To make the dictionary discriminant to the sample features, this study adapted Fisher's discriminant analysis [45] to the coding coefficients x in the dictionary D. It achieves simultaneous sparse decompositions within each class, to extract its internal structure while keeping a global discrimination term among different classes. This proposed framework introduces a novel learning metric to include both re-construction and discrimination power in dictionary learning. With these considerations, the final objective function consists of three parts.

$$\begin{aligned} J &= \max(Discrim) + \max(Sparse) + \min(Recon) \\ &= \max(J_1) + \max(J_2) + \min(J_3) \end{aligned} \tag{7.20}$$

where the J_1 maximizes the class discrimination and J_2 maximizes the sparsely of the solution, and J_3 controls the reconstruction error.

Similar to the Sect. 7.2.4, this could be solved by introducing a discriminant function J_1,

$$J_1 = tr(S_W(X)) - tr(S_B(X)) + \eta \|X\|_2^2 \tag{7.21}$$

where η is the l_2-norm discrimination regularization parameter.

The other object functions are chosen to be the J_2 with the l_2 norm reconstruction error that illustrates the robustness of the algorithm, and J_3 with l_1 minimization that forces the representation coefficients sparse,

$$J_2 = \min|x|_1 \tag{7.22}$$

$$J_3 = \frac{1}{2}\min\left\|d^T x_i - W y_i\right\|_2^2 \tag{7.23}$$

By combining the Eqs. 7.21, 7.23 to the Eq. 7.20, it finally gets the optimization function,

$$J(D, X) = \arg\min \left\{ \begin{array}{c} (tr(S_W(X)) - tr(S_B(X)) + \eta\|X\|_2^2) \\ +\lambda_1\|X\|_1 + \frac{1}{2}\lambda_2\left\|d^T x_i - W y_i\right\|_2^2 \end{array} \right\} \tag{7.24}$$

$$s.t. \ \|d_i\|_2 = 1, \ d_i \geq 0.$$

In Eq. 7.24, λ_1 is the l_1-norm regularization parameter, and λ_2 is the l_2-norm regularization parameter. The cost function *J*(*D*, *X*) in Eq. 7.24 is not convex to (*D*, *X*), but it is convex if either *X* or *D* is fixed. So the problem can be solved by alternatively updating the *D* and *X*. The *J*(*D*, *X*) optimization criterion intends to learn the basis that good for tool state estimation when dictionaries under different states are learned.

In this study, the K-SVD approach [17] is adapted from the Orthogonal Matching Pursuit (OMP) [46] algorithm to learn the dictionary. In this learning process, the dictionary is updated one atom at a time, by optimizing the target function for each atom respectively while preserving the sparsity constraints in (7.24) and keeping the remaining atoms fixed. After getting all kinds of redundant sub-dictionary by the above method, they are combined into a large discrimination dictionary $D = [D_1, D_2 \ldots]$. Table 7.2 illustrates such a learning process.

Table 7.2 The multi-scale discriminant dictionary learning algorithm

Initialization:
Take the WPD decomposition of signal Y, and set the random normalized dictionary matrix:

$\Phi^{(0)} = abs(randm(m,k))\,, X^{(0)} = abs(randm(k,n))\,,\ k < m,n$, scale $j \in N$

for $j = 1 : J$

Sparse Coding:

Apply OMP algorithm to compute the sparse representation $\{\varphi_i\}$ for at scale j of WPD coefficients $\{Wy\}_j$ using (7.15).

Dictionary Update:

for k = 1, 2, ...,K

Compute residual matrix $E_k = d^T x_i - Wy_i$ according to (7.23).

Restrict E_k by choosing only the columns corresponding to those elements that initially used φ_k in their representation, and obtain E_k^R.

Apply SVD decomposition, and update the atoms and dictionary D_j.

end

$j = j + 1$

Build the cascade dictionary $D = [D_0, D_1, ...D_j]$.

end

7.3.3 *Fast Tool State Estimation Without Signal Reconstruction*

For most studies in condition monitoring, the signal reconstruction error is used for condition monitoring and fault diagnosis, such as in [20–27]. While practically the aim of TCM is to inference the tool state from signal features, with which the signal recovery and reconstruction is not necessary. So we can consider not reconstruct the original signal, but directly processing sampling values to complete the monitoring task, which will greatly reduce the computational cost, and improve online processing speed, and achieve fast state estimation.

Under this consideration, a similarity measure is important to differentiate these coding coefficients, which are usually under different scales. To overcome these problems and make use of the discriminant information, a variation as in Eq. 7.2 is adapted to measure the similarity by measuring the ratio of inter-class distance over intra-class ratio,

$$R = \frac{\text{Inter-class distance}}{\text{Intra-class distance}} = \frac{|S_b|}{|S_w|} \tag{7.25}$$

The ratio R has large values when samples are well centered around their mean within each class, while different classes are well separated. With the Eqs. 7.22 and 7.23, the above ratio results in the so-called Fisher's discriminant ratio (FDR), which is similar to the idea of Mahalanlobis distance [45],

$$FDR = \frac{|\mu_1 - \mu_2|^2}{s_1 + s_2} \tag{7.26}$$

For the multiclass case, an average form of *FDR* is used:

$$FDR(m, j, x) = \sum_{i=1}^{K} \frac{\left|\mu_i(x_j) - \mu_j^m\right|^2}{s_i(x_j) + s_j^m} \tag{7.27}$$

where the subscripts *i* refer to the mean and variance corresponding to the feature *x* under investigation for the classes *m*, and *j* is the scale of wavelet packet analysis. As the Eq. 7.27 tends to emphasizing separation between two classes, we modify this formula to emphasize the feature ranking more than two classes:

$$FDR(m, j, x) = \frac{1}{2^j} \sum_{j=1}^{C} \left(\frac{\sum_{i=1}^{K} \left|\mu_i^m(x_j) - \mu_j^m\right|^2}{\sum_{i=1}^{k} s_i^m(x_j)} \right) \tag{7.28}$$

In Eq. 7.28, $FDR(m, j, x)$ is averaged over all classes. The scaling factor $1/2^j$ is added to eliminate the dynamic variations of coefficients under different analysis scales, where coefficients are much larger in large scales and much less in small scales.

The value of the criterion $FDR(m, j, x)$ is computed for each scale of the coefficients. The possible classes are then ranked in order of descending values of $FDR(m, j, x)$. The test features are finally classified to the category with the lowest value (opposite to similarity),

$$Classify(x) = \underset{i \in \{1,2,\ldots C\}}{\arg\min} \{FDR(m, x, j)\} \tag{7.29}$$

The use of *FDR* is robust because it is independent of the underlying statistical distributions. When the feature distributions are constrained, for example, the distributions are Gaussian and covariance matrixes are equal, the *FDR* criteria converge to the K-L divergence [45].

7.3.4 Experimental Validation

In this study, the proposed approach is applied to the micro-milling tool monitoring. Micro-milling is one of the most important precision manufacturing processes for its capabilities in machining various metallic materials in complex geometries. It has

Table 7.3 The experimental conditions

Test	Tool diameter (mm)	Spindle speed (rpm)	Depth of cut (mm)	Radial depth of cut (mm)	Feed rate (mm/min)
1–2	0.5, 0.8	12,000	0.060	0.100	0.120
3–4	0.5, 0.8	12,000	0.080	0.100	0.120
5–6	0.5, 0.8	12,000	0.100	0.150	0.120
7–8	0.5, 0.8	18,000	0.100	0.150	0.150
9–10	0.5, 0.8	18,000	0.120	0.175	0.150
11–12	0.5, 0.8	18,000	0.100	0.175	0.150
13–14	0.5, 0.8	24,000	0.100	0.200	0.180
15–16	0.5, 0.8	24,000	0.120	0.200	0.180
17–18	0.5, 0.8	24,000	0.150	0.250	0.180
19–20	0.5, 0.8	30,000	0.120	0.250	0.150
21–22	0.5, 0.8	30,000	0.150	0.300	0.150
23–24	0.5, 0.8	30,000	0.120	0.300	0.150

wide applications in the manufacturing of high-performance aeronautical, optical, and medical devices. However, due to the nature of ultra-high rotation speed and intermittent cutting of the micro-scale tool, the tool conditions deteriorate quickly and limit the productivity greatly. Effective tool condition monitoring is most important to the quality and productivity in this aspect.

There were 24 milling tests conducted and studied (working conditions are listed in Table 7.3), which had different wear degrees under varied cutting conditions in a high-precision milling center. The micro-milling tools are two flutes Tungsten carbide helix end mills as shown in Fig. 7.13, with diameters 0.5 and 0.8 mm. The work materials are steel T4 and pure copper. The experiment parameters such as feed rate, depth of cut are mainly referred to as the recommended working conditions by the tool supplier.

The wear in the minor cutting edge is significant in micro-milling and more convenient to be monitored. In this study, a multi-category tool wear criterion is adapted to meet the high precision requirement of micro-milling. This criteria is adapted from [47], which was based on the cross-over points of the second-order derivative of Taylor's tool life curve. The specific values are determined in Table 7.4. The other failure types such as chipping and breakage are defined according to the optical observations. Figure 7.13a shows the experiment setup and the measurement setup. Figure 7.13b, c shows the bottom view of the micro-milling tool under different wear conditions.

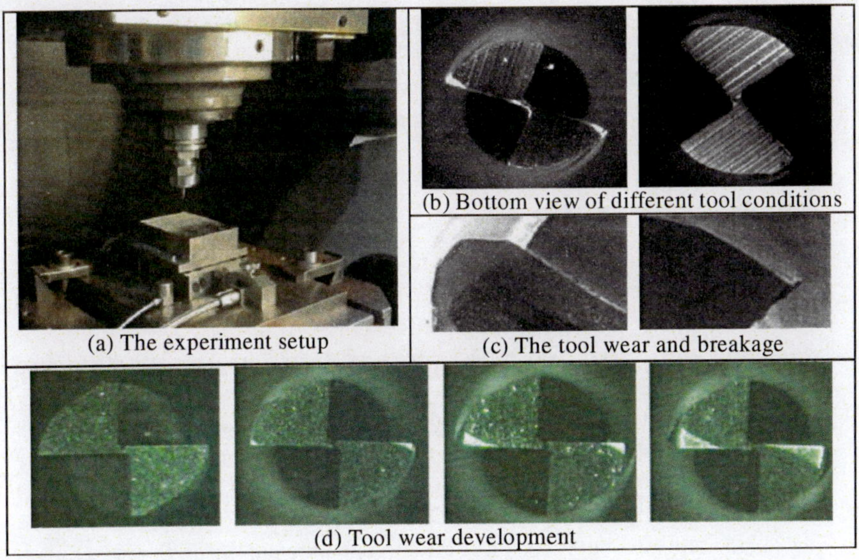

Fig. 7.13 The experiment setup and optical inspection

Table 7.4 Tool wear states

Tool diameter (μm)	Fresh state (μm)	Medium wear (μm)	Severe wear (μm)
500	0–20	20–50	>50
800	0–20	20–60	>60

7.3.5 *Results and Discussions*

(1) Dictionary Learning

To meet the precision requirement of high-speed milling, in this study the tool conditions are classified into five categories: slight wear, medium wear, severe wear, chipping, and breakage. Cutting forces contain reliable information on cutting conditions and the most effective for tool condition monitoring [48], and it is applied in this study. As the dynamic ranges change quite a lot under varied cutting conditions, a normalization step is first applied to the cutting force in the dictionary learning to eliminate this effect and generalize the results. This is achieved by re-scaling the force amplitude over the maxima of the segment and results in force amplitude between [0, 1]. Based on empirical analysis, the force signals are filtered with high pass 75 Hz to eliminate the outstanding low-frequency noises.

For the category i with n_i training samples, then they form the matrix $O = [O_{i1}, O_{i2}, ..., O_{in}]$, where each column represents a training sample o. In this study the forces are segmented into 3600 points as samples and then decomposed with 5-level WPD analysis into 62 packets, all the packets are learned from 100 training samples, and form a WPD dictionary with 62 × 100 × 5 basis. For these collections of basis, the algorithm in Table 7.2 is then applied to learn the most dictionary atoms. In order to simplify the procedure, all the categories of redundant dictionary D_i ($i = 1, 2, \ldots, K$) size is equal, i.e. the whole dictionary D is K times of sub dictionary. Based on the numerical optimization and empirical analysis, the optimum dictionary size is decided to be 32 × 20 × 5.

The algorithm steps are:

(1) Generate five categories of tool state training samples $O = [O_1, O_2, \ldots, O_5]$. The cutting force is sampled with a frequency of 24,000 Hz, the fundamental frequency is $fr = 133$ Hz, and segment length 3600.
(2) Take the 5-level wavelet packet decomposition, the training samples are formed $Y = [Y_1, Y_2, \ldots, Y_5]$.
(3) Train the five category tool states with their respective tool condition samples, and form the redundant dictionary $D_i = [1, 2, \ldots, 5]$, and integrate these cascade dictionaries into a discriminant dictionary $D = [D_1, D_2, \ldots, D_5]$.
(4) Input the test sample y, and take WPD, and forms the sample features Wy. Find the sparse representation vector of the force features in the dictionary D, with the K-SVD algorithm.
(5) With the sparse coding coefficients, find the distance (similarity values) to their respective categories according to (21), and assign the category to the smallest distance.

The discriminant dictionaries are learned with the modified K-SVD approaches and the results of a case study are shown in Fig. 7.14a. The optimization process converges within limited iteration steps, generally fewer than 300 steps (Fig. 7.14b until reaching the preset approximation error = 0.01. To show the algorithms' robustness and consistency in learning the samples, reconstruction error is also shown in Fig. 7.14b. It is found that the reconstruction is rather good though this reconstruction is not used for fast condition monitoring purposes in this study.

Some learned atoms are shown in Fig. 7.15. It is observed that very different atoms are found: from the single (multiple) localized atoms to constant-like atoms. In addition, such a dictionary naturally obtains some favor of shift-invariance, as similar patterns may appear in different locations in different atoms.

On the determination of regularization parameters: In dictionary learning, selecting optimal values for the l_1-norm regularization parameter λ_1 and the l_2-norm regularization parameter λ_2 is a delicate and difficult task. In our experiments, we set $\lambda_1 = 1 - N_0/N_f$. N_0 is the number of zeros, and N_f is the number of features. The sparsity regularization parameter λ_1 is directly related to the sparsity N_0 in the range $[1, N_f]$. The ratio is set to constrain the λ_1 in (0, 1). The higher the λ_1, the higher the sparsity. The sparsity K is chosen to be optimum at $N_0 = 180$, and correspondingly, the $\lambda_1 = 0.95$.

The parameter λ_2 is chosen to control the tradeoff between the discrimination and reconstruction of basis. With a higher value on λ_2, the algorithm favors the reconstruction approach, where reconstruction of DX to WY is concerned. With a lower value on λ_2, the algorithm prefers the discrimination approach, where discrimination

Fig. 7.14 The discriminant dictionary learning

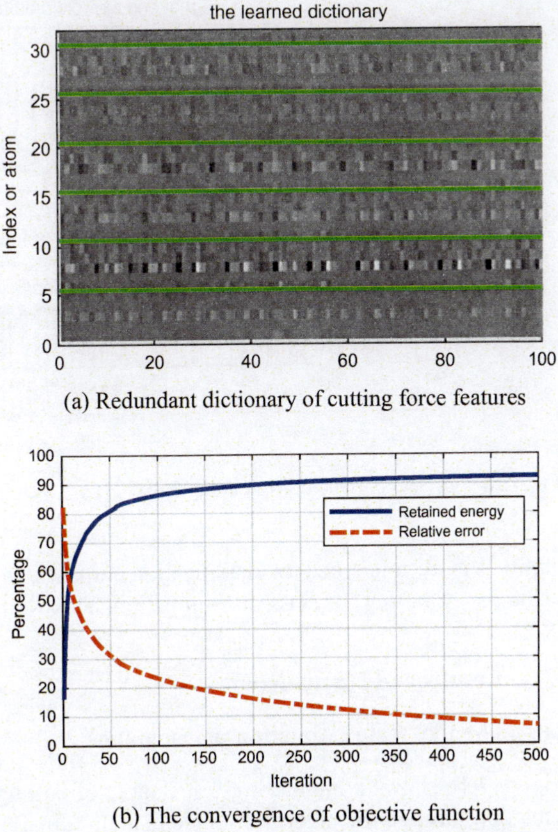

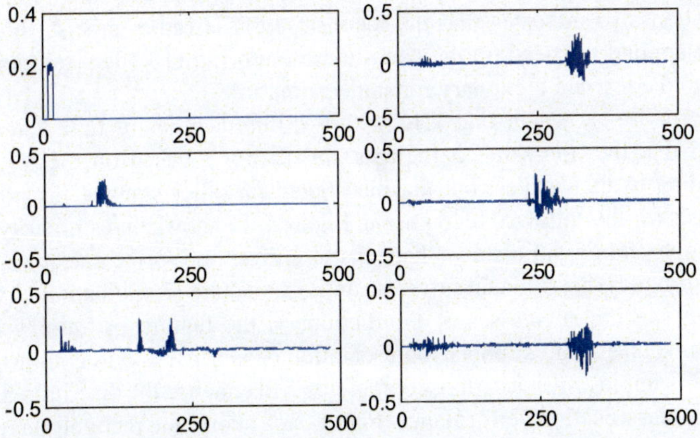

Fig. 7.15 The KSVD dictionary atoms after WPD: localized time and frequency properties

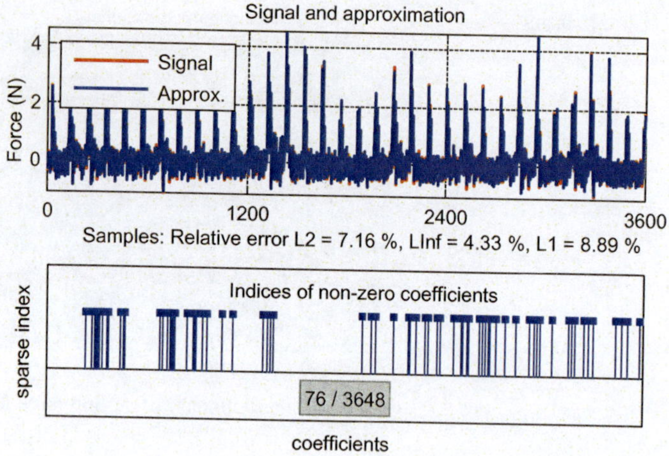

Fig. 7.16 Sparse representation of cutting force

ability is more concerned for classification purposes. In this study, we choose a lower value on λ_2 aiming for class discrimination. This value is chosen empirically $\lambda_2 = 0.16$.

(2) Results and Discussions

(a) The tool state estimation performance

Figure 7.16 shows a monitoring test sample of cutting force with its sparse representation through the algorithm implementation in Table 7.2. As can be seen from the figure, the coefficient is very sparse, and most nearly approximated to zero. The approximation also indicates that the reconstruction is rather precise though this step is not applied to speed up the condition monitoring task. The coefficients are matching to the learned dictionary for state estimation.

More tests are shown in Fig. 7.17 with 5 different types of tool state signals. It is found that the non-zero coefficients are mainly focused on the test sample corresponding to their states, but not distributed on other states. This provides a reliable basis for the followed recognition. Figure 7.17 shows that similarity of state 1, for example, minimum distance happens to correspond to the category of slight wear, and provides discriminant information for tool state identification.

For each specific subclasses of the dictionary, the feature of training samples from the same class with stronger representation capability, and in contrast, for the other class's samples, sparse coding coefficients will enhance the discrimination, and improve the classification performance. Figure 7.18 shows the recognition results of all 24 tests, respectively corresponds to type 1 (Test 1–24, slight wear), and type 2 (Test 1–24, medium wear), and type 3 (Test 1–24, Severe wear), and type 4 (Test 15–22, chipping), and type 5 (Test 21–24, breakage). The recognition results in the total recognition rate reached has 94%. The FDR measure is efficient to evaluate,

Fig. 7.17 State-sparse coefficients

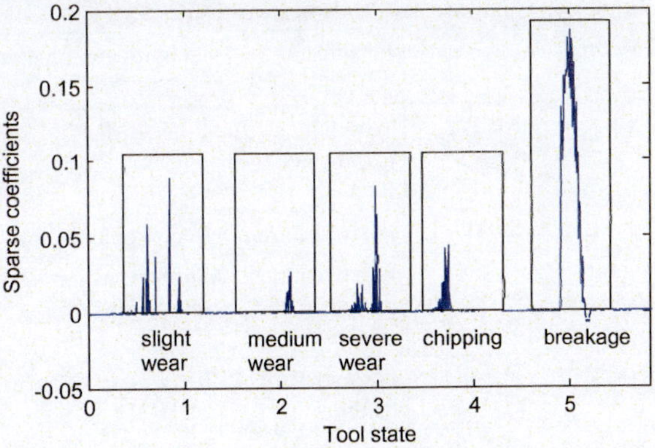

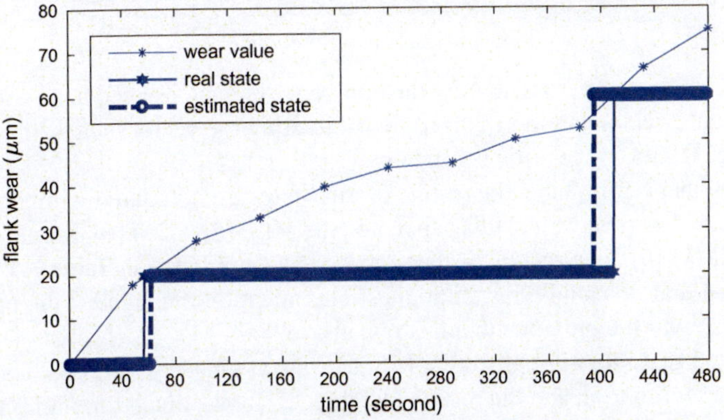

(a) The real wear value vs estimated state of Test 4

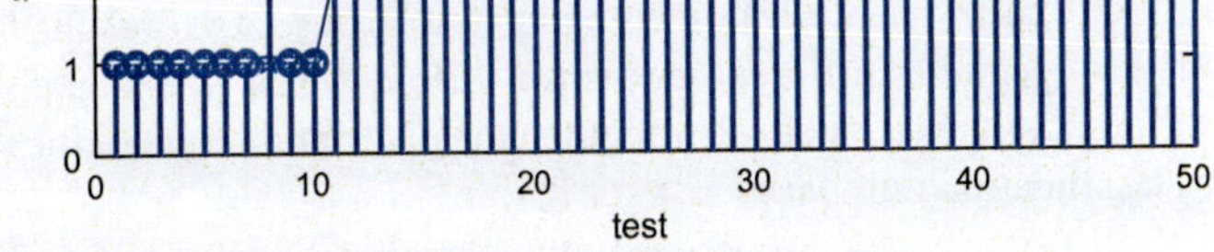

(b) Overall classification results

Fig. 7.18 The state recognition performance

Table 7.5 Comparison of classification results

Method		Features	Classification methods	Classification rate (%)	Refs
NN	MLP	Working parameters, force	5-layer MLP	75.5	[5]
	SOM	Force spectrum	Vector quantization	82.6	[6]
Kernel methods		Force features in time and frequency	Support vector regression	83.8	[7]
HMM		Force statistical features	Semi-nonparametric M^2HMM	87.2	[9]
Sparse decomposition		Force sparse coefficients	Fisher's discriminant ratio	94.0	This study

especially for sparse vectors, as only the non-zero dimensions need to be considered. If the attribute vectors are normalized by subtracting the vector means, the measure is equivalent to the correlation coefficient.

To show its performance, the results of this study are compared with those traditional approaches reviewed in this paper, i.e., the MLP [5], SOM [6], Kernel methods [7], and HMM [9]. The results of these approaches are listed in Table 7.5. It could be observed that this study has the highest classification rate against the other four approaches, with the only one that has rate above 90%.

It is noted that the original study of [5, 6] was on binary wear classification, i.e. sharp tool vs worn tool, and the results in Table 7.5 shows much lower classification results than originally shown in [5] and [6]. In the approach of [9], only the semi-nonparametric M^2HMM is implemented among the three methods, due to its best performance identified among them in [9].

(b) **On the discrimination ability**

As in this study, the signal has slow variations between different states, especially for different tool wear conditions. If mapped to the same dictionaries, the subtle difference may be lost and not good for classification. Under this condition, more constraints have been introduced in the optimization. The proposed algorithm learns discriminant dictionaries as well as incoherent properties. As a result, the signal features will be matched to different atoms for different tool state signals, and the inter-class differences are retained to strengthen the discriminant properties of signal features for classification purposes.

Figure 7.19 illustrates these properties with normalized features of different tool wear states. In the original space, the signals are mixed. When the features are learned and mapped to the learned basis 1 and 2, and the three states can discriminate; when they are mapped into the basis 2 and 3, more discriminant information will be found from the other states. More mapping of different basis combinations could find other

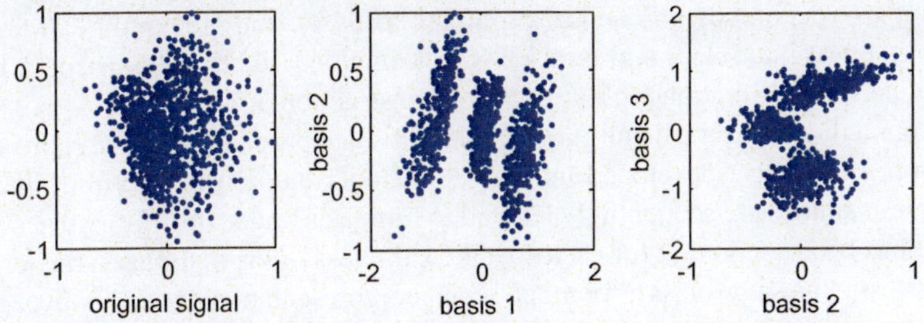

Fig. 7.19 Discriminant coefficients under different basis: minimize within-class scatter and maximize between class scatter

states. So in the higher dimension, the learned basis may represent state features and would be discriminant in tool state monitoring.

(c) On the multi-scale properties

In general, if the dictionary learning is carried out in the original data domain, it is difficult to make full use of local information. Figure 7.20 shows the learned atoms from raw signal without WPD. Compared to those of Fig. 7.15, the atoms' waveform variations cover the full-time scale and are rather stable in the segments, which describe much less local information and frequency variations. This local information is most important for condition monitoring, as it will identify where and when the changed patterns happen. This study improves the monitoring capabilities of sparse representation by learning the dictionary from multi-scale WPD features instead of original samples.

For multi-categories, most current approaches need one-to-one or one-against-all classifier, and the classification becomes very complex with the increase of categories, for example, the NN-based approaches. The sparse decomposition-based framework in this study has the advantage of a simple unified model for multi-category classification over traditional approaches. The algorithm makes use of both the benefits

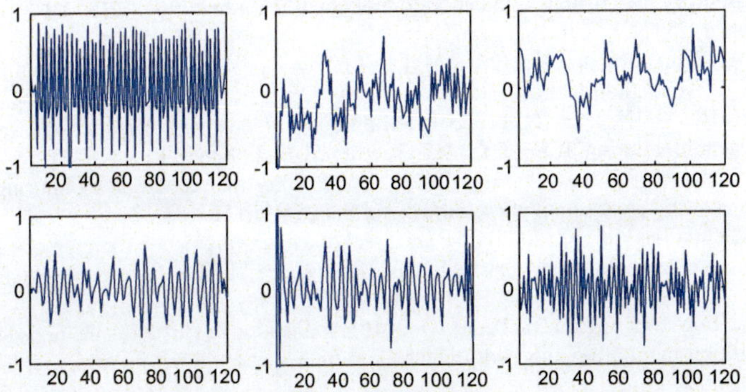

Fig. 7.20 The atoms learned from raw signal without WPD

of the sparse decomposition and discriminant analysis, and efficiently extracts the discriminant of inter-class and intra-class information. Additionally, it could reach fast online state estimation without signal reconstruction process.

Combined with discriminant power and multiscale property, a sparse representation approach has been developed and applied for tool condition monitoring. Different tool states samples are sequentially learned to form a cascaded discriminative dictionary. It has been found that for each specific sub-class of the dictionary, the learned features from the same class have much stronger representation capability over other class' samples. The sparse coding coefficients are verified to enhance the discrimination and improve the classification performance. The experimental validations have shown that the state recognition rate of 94% can be achieved in total with the learned dictionary. Additionally, the state recognition process is much simplified to reach fast estimation without signal reconstruction. Future works will be extended to the fusion of heterogeneous source signals such as vibration and images to increase the representation capability and comprehension of the dictionary.

References

1. Dornfeld D, Lee DE (2007) Precision manufacturing. Springer
2. Teti R, Jemielniak K, O'Donnell G, Dornfeld D (2010) Advanced monitoring of machining operations. CIRP Annals-Manuf Tech 59(2):717–739
3. Altintas Y (1987) In-process detection of tool breakages using time series monitoring of cutting forces. Int J Mach Tool Manuf 28(2):157–172
4. Kumar SA, Ravindra HV, Srinivasa YG (1997) In-process tool wear monitoring through time series modeling and pattern recognition. Int J Prod Res 35(3):739–751
5. Dimla D, Lister P, Leighton N (1997) Automatic tool state identification in a metal turning operation using MLP neural networks and multivariate process parameters. Int J Mach Tool Manuf 38(4):343–352
6. Yen C-L, Lu M-C, Chen J-L (2013) Applying the self-organization feature map (SOM) algorithm to AE-based tool wear monitoring in micro-cutting. Mech Syst Signal Process 34(1–2):353–366
7. Wang J, Xie J, Zhao R, Mao K, Zhang L (2016) A new probabilistic kernel factor analysis for multisensory data fusion: application to tool condition monitoring. IEEE T Instrum Meas 65(11):2527–2537
8. Soualhi A, Clerc G, Razik H et al (2016) Hidden Markov models for the prediction of impending faults. IEEE Trans Ind Electron 63(5):1781–1790
9. Geramifard O, Xu JX, Zhou JH et al (2014) Multimodal hidden markov model-based approach for tool wear monitoring. IEEE Trans Ind Electron 61(6):2900–2911
10. Ren Q, Balazinski M, Baron L et al (2014) Type-2 fuzzy tool condition monitoring system based on acoustic emission in micromilling. Inf Sci 255(1):121–134
11. Pang CK, Zhou JH, Yan HC (2015) PDF and breakdown time prediction for unobservable wear using enhanced particle filters in precognitive maintenance. IEEE T Instrum Meas 64(3):649–659
12. Torabi AJ, Er MJ, Li X, Lim BS, Peen GO (2016) Application of clustering methods for online tool condition monitoring and fault diagnosis in high-speed milling processes. IEEE Syst J 10(2):721–732
13. Li W, Zhang S, Rakheja S (2016) Feature denoising and nearest-farthest distance preserving projection for machine fault diagnosis. IEEE Trans Ind Inform 12(2):393–404

14. Kannatey-Asibu E, Yum J, Kim TH (2017) Monitoring tool wear using classifier fusion. Mech Syst Signal Process 85(2):651–661
15. Bruckstein AM, Donoho DL, Elad M (2009) From sparse solutions of systems of equations to sparse modeling of signals and images. SIAM Rev 51(1):34–81
16. Mallat SG (2009) A wavelet tour of signal processing: sparse way, 3rd edn. Academic Press, New York
17. Aharon M, Elad M, Bruckstein A (2006) The K-SVD: an algorithm for designing overcomplete dictionaries for sparse representation. IEEE Trans Signal Process 54(11):4311–4322
18. Mairal J, Bach F, Ponce J (2009) Online dictionary learning for sparse coding. In: Proceedings of the 26th ICML, Canada, pp 689–696
19. Tosic I, Frossard P (2011) Dictionary learning. IEEE Signal Process Mag 28(2):27–38
20. Cai G, Chen X, He Z (2013) Sparsity-enabled signal decomposition using tunable Q-factor wavelet transform for fault feature extraction of gearbox. Mech Syst Signal Process 41(1):34–53
21. Liu H, Liu C, Huang Y (2011) Adaptive feature extraction using sparse coding for machinery fault diagnosis. Mech Syst Signal Process 25(2):558–574
22. Peng X, Tang Y, Du W, Qian F (2017) Multimode process monitoring and fault detection: a sparse modeling and dictionary learning method. IEEE Trans Ind Electron 64(6):4866–4875
23. Zhu K, Vogel-Heuser B (2014) Sparse representation and its applications in micro-milling condition monitoring: noise separation and tool condition monitoring. Int J Adv Manuf Technol 70(1–4):185–199
24. Yang B, Liu R, Chen X (2017) Fault diagnosis for a wind turbine generator bearing via sparse representation and shift-invariant K-SVD. IEEE Trans Ind Inform 13(3):1321–1331
25. Gao B, Woo WL, Tian G, Zhang H (2016) Unsupervised diagnostic and monitoring of defects using waveguide imaging with adaptive sparse representation. IEEE Trans Industr Inf 12(1):405–416
26. Du Z, Chen X, Zhang H, Yan R (2015) Sparse feature identification based on union of redundant dictionary for wind turbine gearbox fault diagnosis. IEEE Trans Ind Electron 62(10):6594–6605
27. Zhu K, Lin X, Li K, Jiang L (2015) Compressive sensing and sparse decomposition in precision machining process monitoring: from theory to applications. Mechatronics 31(10):3–15
28. Mairal J, Bach F, Ponce J (2007) Discriminative learned dictionaries for local image analysis. In: IEEE CVPR, Anchorage, pp 1–8
29. Zhang Q, Li B (2010) Discriminative K-SVD for dictionary learning in face recognition. In: IEEE CVPR, San Francisco, pp 2691–2698
30. Yang M, Zhang L, Feng X, Zhang D (2011) Fisher discrimination dictionary learning for sparse representation. In: IEEE ICCV, pp 543–550
31. Yang M, Zhang L, Feng XC, Zhang D (2014) Sparse representation based fisher discrimination dictionary learning for image classification. Int J Comput Vision 109(3):209–232
32. Zhu KP, Hong GS, Wong YS, Wang WH (2008) Cutting force denoising in micro-milling tool condition monitoring. Int J Prod Res 46(16):4391–4408
33. Donoho DL (2006) Compressed sensing. IEEE Trans Inf Theory 52(4):1289–1306
34. Plumbley MD, Blumensath T, Daudet L, Gribonval R, Davies ME (2010) Sparse representations in audio and music: from coding to source separation. Proc IEEE 98(6):995–1005
35. Zibulevsky M, Pearlmutter BA (2001) Blind source separation by sparse decomposition in a signal dictionary. Neural Comput 13(4):863–882
36. Mairal J, Bach F, Ponce J, Sapiro G, Zisserman A (2008) Supervised dictionary learning. Neural Inf Process Syst (NIPS) 21:1033–1040
37. Huang K, Aviyente S (2007) Sparse representation for signal classification. The Neural Inf Process Systems (NIPS) 19:609–617
38. Mallat SG (2008) A wavelet tour of signal processing: the sparse way, 3rd edn. Academic Press
39. Candès E, Romberg J, Tao T (2006) Robust uncertainty principles: exact signal reconstruction from highly incomplete frequency information. IEEE Trans Inf Theory 52(2):489–509
40. Cohen A, Dahmen W, DeVore R (2009) Compressed sensing and best k-term approximation. J Am Math Soc 22(1):211–231
41. Theodoridis S, Koutroumbas K (2003) Pattern recognition. Academic Press

42. Elad M, Aharon M (2006) Image denoising via sparse and redundant representations over learned dictionaries. IEEE Trans Image Process 15(12):3736–3745
43. Lewicki MS, Sejnowski TJ (2000) Learning overcomplete representations. Neural Comput 12(2):337–365
44. Efron B, Johnstone I, Hastie T, Tibshirani R (2002) Least angle regression. Ann Stat 32(2):407–499
45. Duda RO, Hart PE, Stork DS (2001) Pattern classification, 2nd edn. Wiley
46. Pati YC, Rezaiifar R, Krishnaprasad PS (1993) Orthogonal matching pursuit: recursive function approximation with applications to wavelet decomposition. In: The 27th annual Asilomar conference on signals, systems, and computers, pp 40–44
47. Zhu KP, Mei T, Ye DS (2015) Online condition monitoring in micro-milling: a force waveform shape analysis approach. IEEE Trans Ind Electron 62(6):3806–3813
48. Jemielniak K, Arrazola PJ (2007) Application of AE and cutting force signals in tool condition monitoring in micro-milling. CIRP J Manuf Sci Tec 1(2):97–102

Chapter 8
Machine Vision Based Smart Machining System Monitoring

8.1 Machine Vision System for Machining Process Monitoring

8.1.1 Introduction

Machine vision based monitoring system refers to the monitoring system that captures the target attributes (pixel distribution, brightness, colour, etc.) using visual devices, and transmit and process the digital image to carry out a variety of detection and control operations for equipment action. The advantage of machine vision is that with a proper setup it can reaches high precision non-destructive monitoring in the machining process, and improves the flexibility and automation of production. Moreover, machine vision is easy to realize automation integration and software integration, which forms the basis of intelligent manufacturing [1].

The machine vision system consists of image sensors, image processing algorithms, and pattern recognition tools [2]. Machine vision systems are widely used in the manufacturing industry for non-destructive testing of industrially manufactured products. For example, in the manufacturing process, the machine vision system is used metal surface defect [3–5], metal additive manufacturing monitoring [6, 7], manufacturing process assembly and control [8–10], etc. Recently, methods based on machine vision systems have also been extensively studied for tool condition monitoring, machining process monitoring, and control [11–23].

A typical machine vision system consists of various functional components as shown in Fig. 8.1. According to the image acquisition process, from bottom to top, it has a pyramid structure. The suitability and quality of the lowest level are critical to the entire visual inspection task chain. The amount and quality of the information contained in the collected image data depend entirely on the image acquisition, as the information not obtained in this image acquisition step is difficult to recover or even impossible to recover in the subsequent image processing. In the machine vision inspection system, the light emitted by the light source interacts with the measured

K. Zhu, *Smart Machining Systems*, Springer Series in Advanced Manufacturing,
https://doi.org/10.1007/978-3-030-87878-8_8

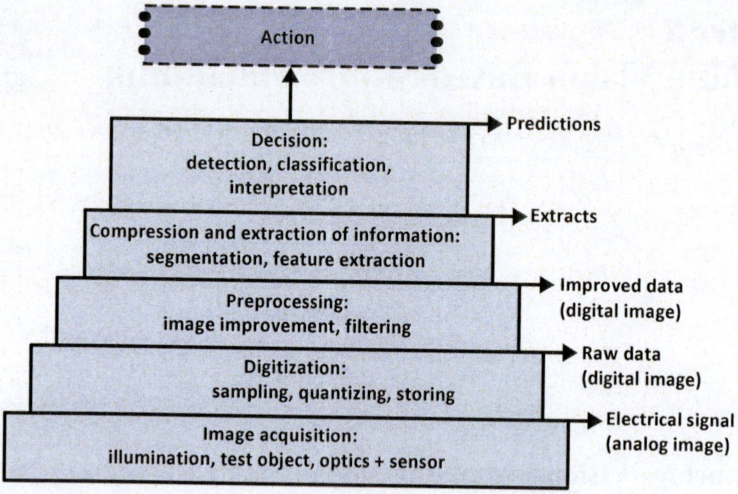

Fig. 8.1 The machine vision monitoring system for manufacturing process [2]

object, passes through the imaging optics, and finally reaches the sensor where it is converted into an electrical signal. This electrical signal (usually a voltage or current) is discretized and has a limited amplitude. It can be saved as a digital image and processed by a computer. In addition to information related to visual inspection, raw data usually contains interference and irrelevant components, such as noise, background, and so on. The subsequent image restoration process attempts to retain relevant information and filter out irrelevant signal components. The improved image data is obtained, and meaningful area segmentation or the processing is performed to extract the features (parameters) related to the monitoring task. Finally, a decision can be made based on this compressed information. Depending on the visual inspection task, these decisions can be detection (for example, defects), classification (for example, different objects), or interpretation (for example, process parameter inference). Then, follow-up actions can be taken according to the decision. Control and adjustment also can be performed according to the decision such as discarding or retaining the test object or selectively changing the forming system parameters.

The key technologies of vision detection include imaging, automatic image acquisition, image preprocessing, location and segmentation, recognition, and detection. This chapter mainly focuses on the digital image processing techniques, by assuming the image acquired from a commercial camera.

8.1.2 The State-of-the-Art

There are two main types of application of machine vision methods in machining system monitoring [11–23]: one is the direct monitoring method based on the tool surface image, the other is the indirect monitoring method based on the workpiece surface.

(1) **Visual monitoring method based on tool surface image**

The surface of the tool is a direct reflection of the wear state of the tool, so the direct monitoring method based on the image of the tool surface has the advantages that other monitoring methods can't compare. The direct monitoring system based on the image of the tool surface generally consists of lens, CCD, camera, light source, image capture card, computer, and so on. Since the chip is blocked on the surface of the tool during the cutting of the workpiece, the visual monitoring method based on the image of the tool surface generally cannot be carried out during the cutting process, so it is only an intermittent quasi-on-line monitoring method. However, because this method directly detects the geometry of the wear part of the tool, it has the advantages of intuitive, accurate, and reliable detection results, and therefore has a wider application prospect.

Wang et al. [19] proposed an intermittent measurement method for milling flank wear based on continuous image analysis. The continuous images were taken during the rotation of the spindle. In order to minimize the blurring of the images, the researchers used high-speed cameras and low-speed rotation (20 rpm). The tool wear state monitoring method based on the image series is studied. Unlike shooting static tool images, the tool can be monitored without stopping the machine. This method shows its application potential in industrial monitoring. Prasad and Ramamoorthy [20] studied the method of using stereo vision to measure the rake face of the tool. The three-dimensional dimensions of the rake face were measured. Multiple sets of three-dimensional images are used to realize the visualization of tool wear geometry. A method of the artificial neural network is proposed to predict the wear status of the tool. The widely used multi-layer perceptron is used, and the neural network of the back propagation algorithm is used for training. A relatively small amount of data can be used to estimate the wear of the tool. The cutting speed, feed rate, and tool engagement are used as input parameters, and the flank width and rake face depth are used as output parameters. Devillez et al. [21] proposed the method of measuring the rake face using white light interferometry for tool inspection, using an optical profiler to obtain three-dimensional images of the rake face under various cutting conditions, and measuring the geometric parameters of the rake face (including length, width, depth, and volume), to evaluate the state of tool wear.

Wang et al. [22] studied the three-dimensional measurement of the rake face using the phase-shift method and carried out a three-dimensional reconstruction measurement of the crescent area of the tool through fringe analysis. The specific method is to project four fringes with different phase shifts onto the rake face to obtain four gray-scale images. Through calculation and phase demodulation, the entire size of the crater is obtained, including crater depth (KT), crater width (KB), crater center distance (KM), and crater front distance (KF). Using this method to evaluate blades with various crater surfaces, Fig. 8.2 shows the three-dimensional image and size of the rake face obtained by the phase shift method. This structured

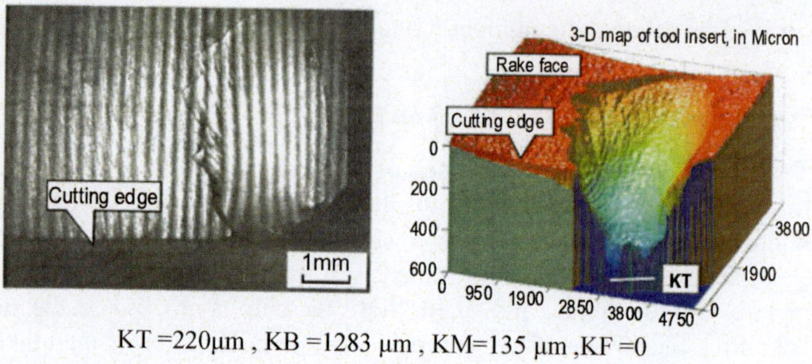

Fig. 8.2 Three-dimensional profile of rake surface obtained by phase-shifting method [22]

light-based method is robust to ambient light and has certain practical value for actual factory monitoring.

(2) **Visual monitoring method based on workpiece surface texture image**

The surface texture of the workpiece refers to the definition of parameters such as the surface morphology or geometric characteristics of the processed workpiece. It includes some features that appear in the surface profile, such as roughness, waveform, and defects. The surface texture of the workpiece is a negative image of the state of the cutting edge of the tool. When the blade is sharp, the texture of the workpiece is clear and has good continuity; when the blade is blunt, the texture of the workpiece is disordered, discontinuous, and has broken marks. Different processing methods and tools have different texture characteristics.

Since the surface of the workpiece is a negative image of the surface shape of the tool, it is directly affected by the shape of the cutting edge of the tool. Therefore, observing the texture of the surface of the workpiece to be processed can also determine the cutting edge state of the tool. The visual monitoring method based on the image of the workpiece surface is an indirect monitoring method that analyzes the surface texture of the processed workpiece through computer vision, thereby judging the state of the tool. Kassim et al. [23] proposed a fast Hough transform for connectivity, which can easily detect all linear parts of the binary edge image of the processed surface texture, and found that the features extracted from image segmentation are closely related to tool wear (Fig. 8.3), and uses a multilayer perceptron neural network to estimate the flank wear of various machining processes. Using this method based on the improved Hough transform can effectively analyze the quality of the machined surface, which can be used to monitor tool wear.

The computer vision based tool wear monitoring is widely studied in recent years. With the rapid development of optical three-dimensional measurement technology, especially structured light technology, wear state measurement based on three-dimensional depth images will receive more attention and become a research hotspot in the future. It is foreseeable that the development of computer technology and the proposal and improvement of various image analysis algorithms, which will

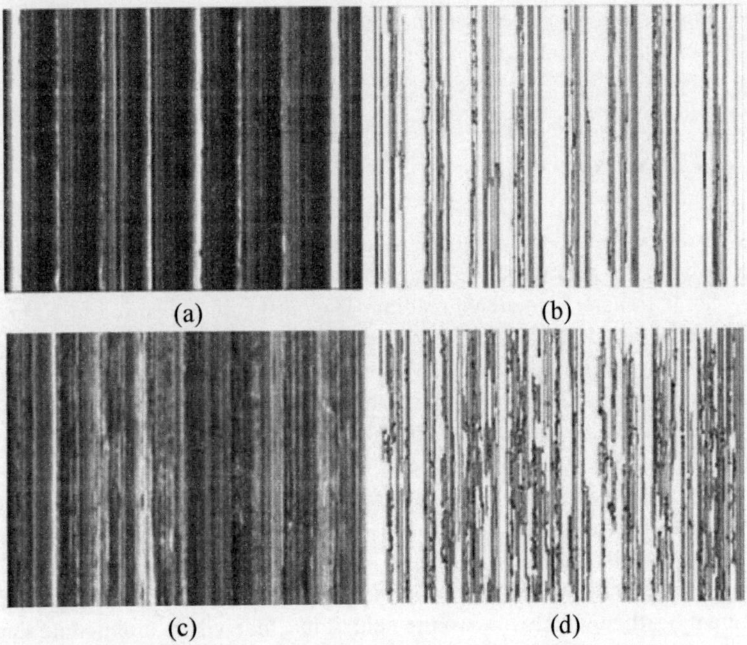

Fig. 8.3 Workpiece surface texture and image segmentation [23]. **a** The surface stripe processed by sharp tool. **b** The COHT segmentation results of **a**. **c** The surface stripe processed by the worn tool, and **d** The COHT segmentation results of **c**

surely drive the development and wide application of tool state visual monitoring and even the entire tool monitoring technology.

8.2 Digital Image Acquisition and Representation

8.2.1 Image Acquisition of the Monitored Objects

The most critical component of a machine vision-based monitoring system is an industrial camera. Its most essential function is to convert an optical signal into an orderly electrical signal and then process the obtained data according to different applications. Compared with traditional civilian cameras (cameras), industrial cameras have high image stability, high transmission capacity, and high anti-interference ability, etc. Most of them are based on Charge Coupled Device (CCD) or Complementary Metal Oxide Semiconductor (CMOS) chips. Choosing the right camera is an important link in the design of machine vision systems. The choice of the camera not only directly determines the resolution and image quality of the collected images, but also directly related to the operating mode of the entire system [24].

8.2.2 CCD Sensor

CCD is currently the most commonly used image sensor for machine vision. It is a typical solid-state imaging device that integrates photoelectric conversion, charge storage, charge transfer, and signal reading. The outstanding feature of CCD is that charge is used as a signal, unlike other devices that use current or voltage as a signal. This type of imaging device forms charge packets through photoelectric conversion, then transfers and amplifies the output image signal under the action of driving pulses. A typical CCD camera consists of an optical lens, timing, and synchronization signal generator, vertical driver, and analog/digital signal processing circuit. As a functional device, CCD has the advantages of no burns, no hysteresis, low voltage operation, and low power consumption.

CCD sensors [24] currently dominate in industrial image processing. The first batch of CCD sensors was developed in the early 1970s and they quickly replaced the camera tube, then widely used in the field of machine vision. Solid-state sensors are usually composed of photosensitive photodiodes arranged in a matrix type or line sensors arranged in a row. Since digital cameras usually do not have a mechanical aperture, the photodiode is constantly exposed to light. To obtain a usable image, the exposure time must be defined both at the start and the end. Therefore, some mechanism is needed to limit the exposure time, that is, the time that incident light is useful for acquiring images. The exposure time starts from removing the charge in the photodiodes of all pixels. As shown in Fig. 8.4, a vertical shift register or transfer register is set between the gaps of the photodiodes. At the end of the exposure time,

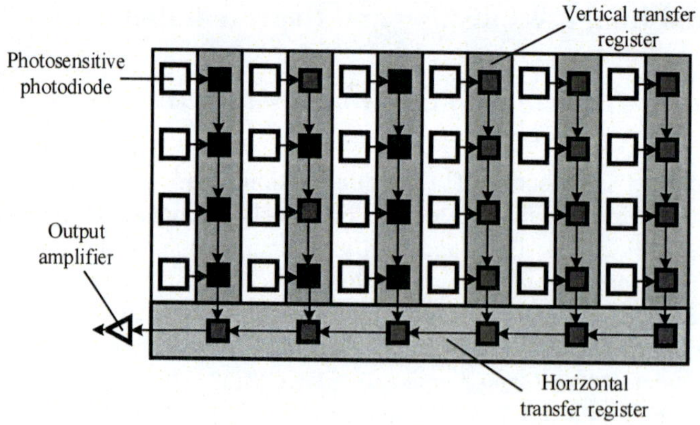

Fig. 8.4 The structure of CCD sensors

all charges representing image information move from the photodiodes of all pixels to the transfer register at the same time. Since the transfer register is usually protected by a thin metal foil from incident light, no more charge is generated. In the transfer register, the charge can be moved in a pixel-wise manner. The sensor named after this method is a charge-coupled device. Charges move from the vertical transfer register to the horizontal transfer register row by row and read them out pixel by pixel. In this way, the image information captured by the sensor is recognized as the charge sequence of each pixel.

8.2.3 CMOS Sensor

The development of CMOS image sensors first appeared in the early 1970s. In the early 1990s, CMOS image sensors developed rapidly with the development of very large-scale integrated circuit (VLSI) manufacturing technology [24]. CMOS image sensor integrates photosensitive element array, image signal amplifier, signal reading circuit, analog-to-digital conversion circuit, image signal processor, and controller on a chip, it also has the advantage of programming random access to local pixels. CMOS image sensors have been widely used in high-resolution and high-speed applications due to their good integration, low power consumption, high-speed transmission, and wide dynamic range.

The main difference between CMOS and CCD sensors is that CCD sensors (except for multi-tap sensors) have only one charge amplifier for the entire sensor, while CMOS sensors have their charge amplifier for each pixel. These amplifiers directly convert the charge generated during the exposure process into a voltage in the pixel. Because of this conversion, CMOS sensors are also called active pixel sensors. The brightness information of each pixel is read out as a voltage instead of the charge of the CCD sensor. The circuit for reading the voltage of each pixel is shown in Fig. 8.5. The selection transistors of the pixels in a row are switched from the vertical control to select the image row for readout. Each active selection transistor puts the voltage of its pixel on the corresponding column line. Then the horizontal controller switches column by column to the output amplifier of the sensor and finally reads out the pixel voltages in sequence. Each pixel has its charge amplifier, which is used to convert all collected charges into voltage. These charge amplifiers are different from each other in amplification and offset, and therefore the output voltage of each pixel having the same charge also changes frequently.

8.2.4 Representation of Digital Images

An image can be defined as a two-dimensional function $f(x, y)$, where x and y are space (plane) coordinates, and the amplitude of f at any pair of space coordinates (x, y) is called the intensity or grayscale of the image at that point [25, 26]. When x, y

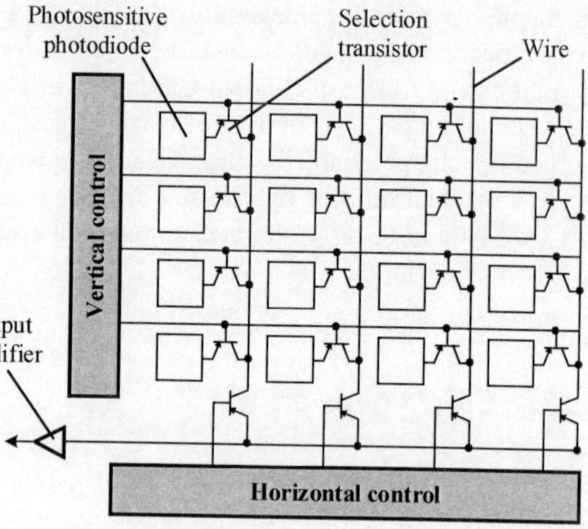

Fig. 8.5 The structure of CMOS sensor

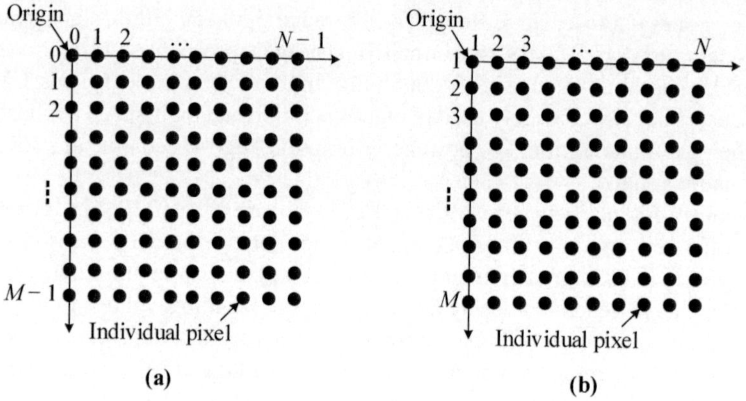

Fig. 8.6 **a** The pixel in the coordinate. **b** The pixel in the Matlab

and gray value f are finite discrete values, we call the image a digital image. Digital image processing refers to processing digital images with the help of computers. Note that digital images are composed of a limited number of elements, and each element has a specific position and amplitude. These elements are called picture elements, or pixels, the latter is widely used to denote digital image elements [26]. Converting an image to digital form requires the digitization of coordinates and gray levels. The digitization of coordinate values is called sampling, and the digitization of grayscale is called quantization. The result of sampling and quantization is a real matrix. Suppose that an image with M rows and N columns is obtained after sampling an image $f(x, y)$. Then the size of this image is $M \times N$, and the coordinates are discrete values. A digital image can be expressed as a matrix as follows:

$$f(x, y) = \begin{bmatrix} f(0,0) & f(0,1) & \cdots & f(0, N-1) \\ f(1,0) & f(1,1) & \cdots & f(1, N-1) \\ \vdots & \vdots & \vdots & \vdots \\ f(M-1,0) & f(M-1,1) & \cdots & f(M-1, N-1) \end{bmatrix} \tag{8.1}$$

The two sides of the formula quantitatively express a digital image in an equivalent way (Fig. 8.6). On the right is a matrix of real numbers, and each element in this matrix is called a pixel. Note that the coordinate convention used to represent the array in the widely used Matlab image processing toolbox is different from the previous paragraph, and the coordinate origin is (1, 1).

8.2.5 Digital Image Processing

Digital image processing refers to the process of converting image signals into digital signals and using computers to process them to extract useful information for application purposes [25]. Briefly, an image can be considered as a two-dimensional function $f(x, y)$. The (x, y) represents the position, and the function value represents the gray value or intensity of the image at that position. When x, y, and f are all discrete values, we call the image a digital image, that is to say, the gray value is composed of a limited number, and each gray value has its specific position and amplitude value. Digital image processing means that we use computers to process these digital images. Two important application areas of image processing:

(1) Improving image information for human understanding,
(2) Processing images for the convenience of storage, transmission, and presentation, to achieve the purpose of facilitating automatic identification.

Typically, image processing techniques are applied to extract three levels of computer vision information, namely low-level, intermediate, and high-level processing. Low-level processing involves some basic operations, such as image noise reduction, contrast enhancement, image sharpening, etc. The input and output of low-level processing are images. Intermediate processing involves a wide range, such as segmenting an image (separating different areas or objects in the image) and then classifying different objects. Intermediate processing uses images as input, but the output is derived from the different features extracted from these images, such as the contour information of the image or the identification of each object. And high-level processing involves "understanding" the content on the image to form some cognitive functions.

In this book, the scope of digital image processing is defined as: the input and output of the processing are both images, the processing of extracting features in the image, and the recognition of each target in the image. Common methods of digital image processing:

(1) **Image transformation**: Due to the large image array, processing directly in the spatial domain involves a large amount of calculation. Therefore, various image transformation methods are often used to convert spatial domain processing into transform domain processing, such as Fourier transform, Walsh transforms, discrete cosine transform, and other indirect processing techniques, which can not only reduce the amount of calculation but also obtain more effective Processing (such as Fourier transform can be digitally filtered in the frequency domain). At present, the newly researched wavelet transform has good localization characteristics in both the time domain and the frequency domain. It also has a wide range of effective applications in image processing.
(2) **Image coding compression**: Image coding compression technology can reduce the amount of data describing the image (i.e., the number of bits) in order to save image transmission, processing time and reduce the memory capacity occupied. Compression can be achieved without distortion or under allowable distortion conditions. Encoding is the most important method in compression technology, and it is the earliest and relatively mature technology in image processing technology.
(3) **Image enhancement and restoration**: The purpose of image enhancement and restoration is to improve image quality, such as removing noise and improving image clarity. Image enhancement does not consider the reasons for image degradation but highlights the part of the interest in the image. If the high-frequency component of the image is enhanced, the outline of the object in the image can be clear and the details are obvious; if the low-frequency component is enhanced, the effect of noise in the image can be reduced. Image restoration requires a certain understanding of the reasons for image degradation. Generally speaking, a "degradation model" should be established according to the degradation process, and then some filtering method should be used to restore or reconstruct the original image.
(4) **Image segmentation**: Image segmentation is one of the key technologies in digital image processing. Image segmentation is the extraction of meaningful features such as edges and regions in an image, which is the basis for further image recognition, analysis, and understanding. Although many methods of edge extraction and region segmentation have been studied, there is no effective method that is generally applicable to various images. Therefore, the research on image segmentation is still deepening, and it is one of the hotspots in image processing.
(5) **Image description (descriptor)**: Image description is a necessary prerequisite for image recognition and understanding. As the simplest binary image, its geometric characteristics can be used to describe the characteristics of the object. The general image description method uses two-dimensional shape description, which has two methods: boundary description and area description. For special texture images, two-dimensional texture feature descriptions can be used. With the in-depth development of image processing research, the study of three-dimensional object description has begun, and methods such as volume

description, surface description, and generalized cylindrical description have been proposed.

(6) **Image classification (recognition)**: Image classification (recognition) belongs to the category of pattern recognition, and its main content is image segmentation and feature extraction after certain preprocessing (enhancement, restoration, compression), to make judgment classification. Image classification often uses classic pattern recognition methods, including statistical pattern classification and syntactic (structural) pattern classification. In recent years, the newly developed fuzzy pattern recognition and artificial neural network pattern classification have also received more and more attention in image recognition.

8.3 Machine Vision System for Micro Milling Tool Condition Monitoring

8.3.1 The Micro Milling Tool Condition Monitoring

Tool condition monitoring (TCM) is a key issue in micromachining for part quality control because the excessive tool wear and abnormal tool conditions will significantly decrease the size accuracy of the part and shorten the tool durability as well. In view of this, a novel configuration of machine vision system for online tool condition monitoring is presented to improve the part quality and extend the micro tool life. The vision system is committed to automating on-machine vision inspection for monitoring progressive wear. This inspection system uses a tele-centric lens with light source and a camera to minimize the errors in imaging. The control system drives a three-dimensional motion platform carrying the imaging device to probe and grab the in-focus image at the predetermined time interval of machining. In addition to the flank wear, three new wear variables are explored to enhance the robustness in the prediction of tool wear state. Effective image processing algorithms are developed to reduce downtime. The effectiveness of the prototype system and the developed algorithms for tool wear extraction are verified by cutting experiments using two-flutter micro-milling tools, and the experimental results show that this novel on-machine vision inspection system is convenient and effective to measure the amount of progressive wear and reflect the trend of tool life.

Micro-milling operations are extensively applied in producing miniature components with three-dimensional (3D) features especially for high precision parts of metallic alloys [27]. The TCM is very necessary to avoid the tool's premature failure or extremely unpredictable tool life in a computer numerical control (CNC) machine [28] because the part quality is principally dependent on the cutting tool wear condition. The stable form of tool wear is flank wear which is empirically depicted as three stages including initial wear, steady-state and severe wear. An accurate and reliable prediction of the start point in the severe wear stage is always important in concern. For example, Tansel et al. [29] propose to reduce the cutting feed rate according to the detected micro tool wear to decrease the surface damage and increase the micro tool life. On the other hand, if a tool breaks during the machining, surface damage is inevitable, which leads to wasted workpiece and less productivity. In micromachining, the micro tools may be broken in a few seconds under abnormal cutting conditions or excessively worn state. The development of subsystem for tool wear inspection is an urgent demand and it is the main topic here.

The tiny shaft of a micro-milling tool is driven at extremely high rotational speeds with cutting materials in micro-scale. These features lead to critical issues in micro-milling [30], such as size effects, relatively large vibrations, single-toothed cutting phenomena, and micro-burrs, which in turn cause uncertainty in measurement for feedback control. Theoretically, the micro tool cutting-edge profile is an important factor that directly influences the final quality of a machined surface. The problem is that measuring the sharp edges of cutting tools is very challenging work because it involves a small radius that requires high lateral resolution and high angles. It is also important to be able to measure a diverse range of heights at a nanometre scale. Furthermore, the uncertainty in measurement will cause serious noise in micro-tool wear monitoring [31] which is very necessary to avoid excessive tool wear and maintain part tolerances [32]. The problem is that the existing methods or studies in the literature are mainly artificial inspecting or monitoring the indirect tool-state signals, which are seriously influenced by the noise in micro-milling. Most of the investigators attempt to extract the indirect signals, such as forces and AE (acoustic emission), and then construct the relationship to the tool life, which is very difficult in practice considering the mentioned noise and contact disturbance.

Though various direct and indirect tool wear monitoring techniques have been developed, there is still no consensus on the best choice for micro-milling tool wear monitoring especially considering the lack of automation and robustness. Comparatively, Signal-to-Noise in micromachining will decrease in an indirect method because of the size effects and relatively large vibrations [33], and the machine vision as a direct method has many attractive advantages [18]. Especially, the vision method can avoid contact disturbance from measuring setups to the tiny tools and can directly measure the progressive tool wear for analysis. In a machine vision system, digital image processing algorithms and light source arrangement are very important factors for high precision tool wear inspection. At the same time, noise in imaging, defocussing, camera vibration, incline from shooting surface, speckles in the image due to micro metal particles, and stray lights reflected by object surfaces should also be considered with countermeasures. Szydłowski et al. [34] design an image fusion system based on wavelet transform to resist defocussing in imaging of micro milling tools.

Considering the features above, convenient and effective tool condition monitoring systems especially for micro milling tools are still need to perfect or develop. Moreover, even the variables such as flank wear and crater wear, suggested in ANSI/ASME B94.55M-1985 standard for the extents of conventional tool wear indication are insufficient [35, 36]. To date, whether such variables are suitable for micromachining cases is still not ascertained in the literature, and there is no such standard for micromachining at all. Accordingly, multivariate is a good idea to enhance the robustness of tool life indication [36], such as the work by Dutta et al.

[37] that takes advantage of flank wear of tools and machined surface textures in analysis.

Because of the issues mentioned above, this paper proposes a machine vision system, especially for micro-milling tools. The time intervals are identified or added in CNC program codes for the progressive wear inspection. To overcome the difficulties in on-line microimaging of tool cutters, the vision inspection module runs automatically by program control of motors carrying the imaging device. The novel algorithms are also designed considering the characteristics of the images to extract the wear robustly. In experimental verifications, the developed prototype can run automatically in conjunction within a micro-milling center using two-flutter micro-milling tools. Moreover, several variables indicating the wear state are explored to enhance its robustness.

8.3.2 *Tool Wear Inspection System*

(1) **Experimental setup**

The total system is illustrated in Fig. 8.7, and the vision inspection module is mainly arranged below the original CNC structures. The kernel framework of a precision milling center remains unchanged, as schematically depicted inside the dashed rectangle in Fig. 8.7. Specifically, the tool images are captured in process with an Olympus CCD (charge-coupled device) with telecentric microlens after cleaning,

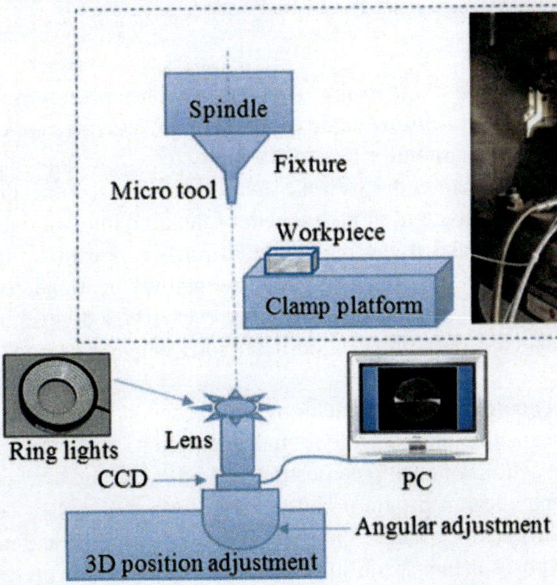

Fig. 8.7 Experimental setup

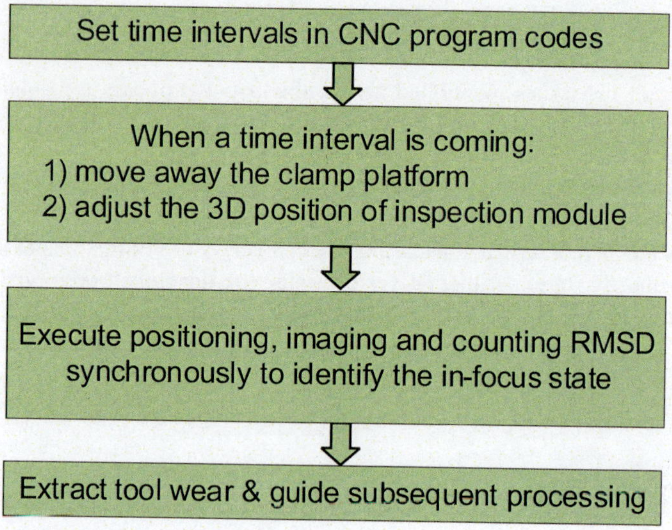

Fig. 8.8 Procedures of tool wear monitoring

which has more than 200 times enlargement. These raw images are then processed with the developed image processing system written in C++ program. The angular adjustment is a tripod ball-head stand commonly used in videography; the 3D position adjustment is assembled with three linear motors and the z-direction (up-down) is controlled with a higher resolution of 0.65 μm per pulse by C++ program codes. The working flow of the system is illustrated in Fig. 8.8.

(2) **Procedures**

As presented in the working flow chart in Fig. 8.8, the vision inspection module cooperates with the existing micro-milling center by inserting CNC program codes to realize an automatic tool wear monitoring. At the first step, time intervals are identified or added in CNC program codes to wake up tool wear inspections during processing. For insurance, one minute is enough length of the time interval for verification in the prototype system and it can be shortened further by estimating the necessary motor pulses in position adjustment. In experimental verification later, the sustained machining time is 100 s, which means that after every 100 s of machining, the tool wear will be inspected once in 60 s until obvious damage or condition variation.

When a time interval is coming during machining in the second step, drive the clamping platform together with the workpiece to make enough room for tool wear inspection. In practice, the actions of position adjustment can be achieved by controlling the number of motor pulses in program codes. The number of pulses corresponding to the lens in-focus position can be calibrated before machining and minor adjustment to the in-focus position in inspection is carried out by algorithm presented later. For example, a linear reciprocating motion can be accomplished by driving

Fig. 8.9 Representative images along with tool life history

several positive or negative pulses within the linear motor driving codes. The rest steps are mainly based on image processing, which will be presented later.

(3) Tool wear characterization

The resolution of the image is 1024 × 768 with 8 bits per pixel as shown in Fig. 8.9 without pre-processing. The diameter along the major axis of the two-flutter micro-milling tool (C-CHES 2008–0180) is 800 μm by tool manufacturer and the diameter along the minor axis is 660 μm by measurement. In this manner of light source arrangement, the worn regions at the end of the tool are illuminated perpendicularly to avoid diffraction distortion at the edges. At the same time, the minor cutting edges will directly reflect the incident light and cause brightness because of the shiny cutter edges from friction. Due to this surface characteristic, the intensity of the reflected light from the contacting surface is much higher than that from the unworn tool surface or the background. The two white regions in the image can be used as good indicators of tool life, and algorithms will be developed to establish the relationship between the variables of the white regions and tool life history.

The amount of tool wear will be extracted quickly with the designed algorithm depicted in the next section to guide subsequent machining. As verification, several images are also shown in Fig. 8.9 for a visual representation of this relationship. As seen in Fig. 8.10, both the symmetric white regions corresponding to minor cutter

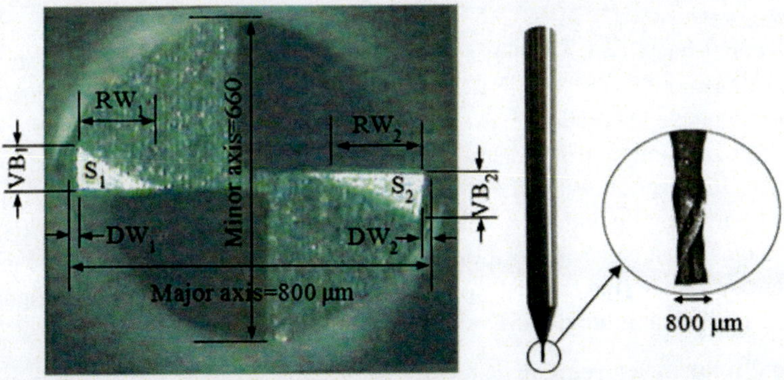

Fig. 8.10 Variables defined for micro-milling tool wear indication

edges are increasing with the extension of time. Therefore, variables for characterization of the incremental worn areas manifested with white pixels will be identified to explore the potential in revealing tool life.

Castejón et al. [36] are aware of the importance of finding out the best description of the wear regions in the image, and then nine geometric descriptors are introduced. However, their descriptors are calculated from basic variables in inspection, and the measuring accuracy of each variable is not fully considered. In the situation here, there are four variables, area of wear land (S_1 and S_2), flank wear (VB_1 and VB_2), radial wear (RW_1 and RW_2), and diameter wear (DW_1 and DW_2) which are available to indicate the extent of wear as shown in Fig. 8.10. In practice, the flank wear is widely accepted for tool life indication in the literature, and the area of wear land can be simply extracted without considering the rotation angle at imaging time. In addition, both the radial wear and diameter wear are useful indicators for tool life prediction [38], which are all verified in experiments later.

8.3.3 *Tool Wear Inspection Method*

(1) **Image acquisition**

A novel mechanism is developed to identify and grab the in-focus image by real-time counting root mean square deviation (RMSD) in the general Region of Interesting (ROI). In the developed program, the three threads of motor driving, image grabbing, and RMSD counting are executed synchronously. The in-focus state is corresponding to the maximum of RMSD.

$$RMSD = \sqrt{\frac{1}{n}\sum_{i=1}^{n}(f_i - \overline{f})^2} \tag{8.2}$$

where f denotes grey value of digital image; $\overline{f}$ describes the averaging operation; n stands for the number of pixels in ROI. As shown in Fig. 8.11, the ROI is identified as the pixels inside the rectangle.

The in-focus state can be identified and kept as several pules opposite to the initial state in advance. With the tool wear increasing, the in-focus state can be adjusted by moving up and down, meanwhile real-time counting RMSD of ROI in digital images. That is to say, position adjustment, grabbing an image, and counting RMSD are executed synchronously to save time, and the in-focus image is recognized according to the maximum RMSD without suspending the motors.

(2) **Tool wear measurement**

The developed algorithm includes steps listed in Fig. 8.12 for an overview and the main idea of each step is explained. There are mainly two motivations in the total algorithm. The first is to correct the angle by rotation image so that the flank wear

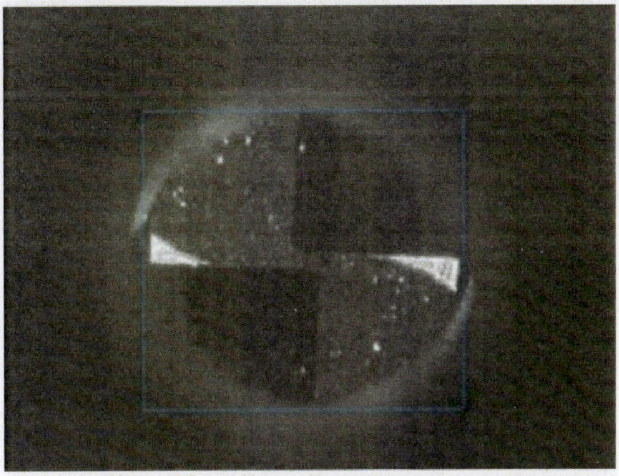

Fig. 8.11 Image in real-time acquisition state

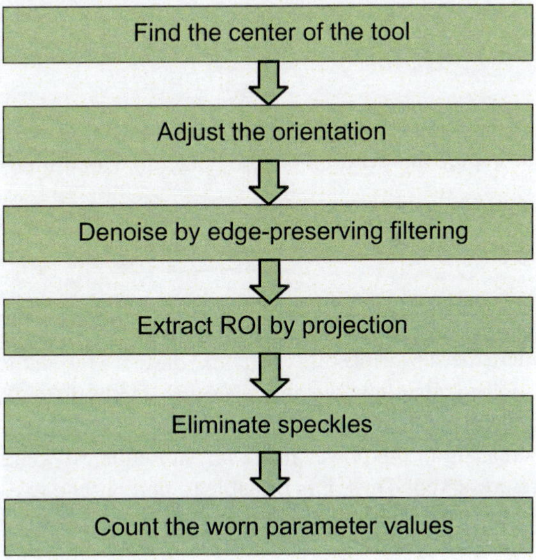

Fig. 8.12 Schematic of the image processing algorithm

can be identified conveniently (flank wear is corrected to be vertical). The next is to suppress noises and detect wear variables accurately. Several subroutines are developed especially considering the characteristics of the tool wear images. In addition, the processing speed is also very important because the effective tool life is even less than twenty minutes.

(3) **Identification of the tool center**

The tool in the grabbed image is different angles and positions. To extract the wear variables conveniently, adjusting the angle of the image is a good choice so that the wear can be counted in the horizontal or vertical direction. Thus, the exponential

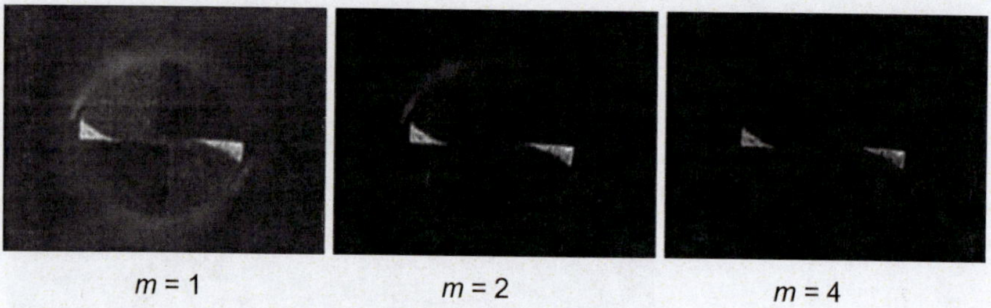

Fig. 8.13 Effects of exponential transform

transform of an image is introduced for image enhancement to find the center of the tool in the image. As seen in Fig. 8.13, the worn regions are distinguished from other regions with larger grey values. As a result, this distinguishes can be further enhanced by the exponential transform:

$$g_i = f_i^m \tag{8.3}$$

where f denotes the grey value of a given pixel i, m stands for the power index. After this exponent operation pixel by pixel, the total image will be re-scaled for showing by:

$$f_i = \frac{255}{g_{\max} - g_{\min}}(g_i - g_{\min}) \tag{8.4}$$

where the subscripts (max and min) describe the statistical operations upon all pixels.

In order to show the effects of this step by formulas 8.2 and 8.3, the results from different power indexes are shown in Fig. 8.13. The white regions are highlighted with increasing in the power index m. Subsequently, only the two largest white regions are preserved. Thus, it is possible to find out the two centers by counting the center of grey gravity in four quadrants. The center of grey gravity can be calculated by:

$$\begin{cases} x_c = \dfrac{\sum_{i=1}^{w}\sum_{j=1}^{h} x_i g_{ij}}{\sum_{i=1}^{w}\sum_{j=1}^{h} g_{ij}} \\ y_c = \dfrac{\sum_{i=1}^{w}\sum_{j=1}^{h} y_i g_{ij}}{\sum_{i=1}^{w}\sum_{j=1}^{h} g_{ij}} \\ g_{ij} = \begin{cases} 0, & f_{ij} < T \\ 1, & f_{ij} \geq T \end{cases} \end{cases} \tag{8.5}$$

where w and h stand for the width and height of the image respectively, the threshold value T can be 200 in practice and it is not critical in magnitude after exponential transform as seen in Fig. 8.13.

(4) **Adjusting the orientation**

The center of grey gravity operations can be carried out within the total image to obtain the total center and then carried out within four quadrants partitioned by total center respectively. In this way, the two center positions are obtained and used to ascertain the angle A:

$$A = \arctan \frac{y_{c1} - y_{c2}}{x_{c1} - x_{c2}} \tag{8.6}$$

As mentioned, the single-toothed cutting is difficult to avoid and through trying in experiments, the extents of wear in the two flutters are still not the same. Therefore, the two white-worn regions are not of the same size, and the angle corrections by rotating the images can only support accuracy within 5°. However, these results can be achieved quickly by the algorithm.

(5) **Denoising by edge-preserving filtering**

By now, there is no resistance to noise considered in the scheme of exponential transform and angle correction. Considering the speckles in the image due to micro metal particles and noise in imaging, an edge-preserving filter is necessary to suppress the noise in the image. There are several famous filters available for edge-preserving smoothing, including infinite symmetric exponential filter (ISEF) [39], bilateral filter [40], guided filter [41], and other intelligent filters [42]. Comparatively, the ISEF filter is more suitable for the application considering that the others are not linear translation-invariant (linear output) [42]. Moreover, the exponential function in ISEF is super than the widely used Gaussian function bilateral filters in approximation the steep boundaries in the image here [39]. Moreover, updating filter coefficients pixel by pixel results in a low arithmetic speed in bilateral filtering. While the guided filter is also slowed greatly in finding local direction before filtering, and intelligent filters may lead to an unsteady output or variables which should be prepared carefully. The improved two-dimensional ISEF filter is given by:

$$\begin{cases} f(r) = \dfrac{-\ln(b)}{2} b^r \\ r = \sqrt{x^2 + y^2} \\ c(x, y) = \dfrac{f(r)}{\sum f(r)}, \ \mathrm{r} \le \mathrm{R} \end{cases} \tag{8.7}$$

where x and y are coordinates with origin locating at the center of filter window,b is the smoothing coefficient which is assigned 0.95 in practice, the filter coefficients $c(x, y)$ are normalized within truncation radius R which is assigned 15 in discretization for acceleration. Inevitably, smoothing will corrode the edges with a clear boundary. The designed scheme below is powerful to remedy this insufficient.

$$g_i = Max\{f_{ISEF}, f_i\} \tag{8.8}$$

where f_{ISEF} stand for the result from ISEF filtering of the pixel i, f_i is the original grey value of the pixel i.

(6) **Extracting the ROI by projection**

In order to fast calibrate the scale from pixel to micron and correct the angle accurately in ROI extraction, the projection algorithm is designed. The presented projection operation is to count the lightness in each row or col following re-scaling for exhibition. The two steps in projection in x and y directions can be expressed as:

$$\begin{cases} Px(j) = \sum_{i=1}^{w} f(i, j), \ j \in [1, h] \\ Py(j) = \sum_{j=1}^{h} f(i, j), \ i \in [1, w] \end{cases} \tag{8.9a}$$

$$\begin{cases} PX(j) = \dfrac{w}{Px_{\max} - Px_{\min}}(Px(j) - Px_{\min}), \ j \in [1, h] \\ PY(j) = \dfrac{h}{Py_{\max} - Py_{\min}}(Py(j) - Py_{\min}), \ i \in [1, w] \end{cases} \tag{8.9b}$$

where $f(i, j)$ denote the grey value at the pixel (i, j), subscripts of max and min stand for the highest and the lowest values, w and h are the width and height of the image respectively.

As seen in Fig. 8.14, image (a) is an original image in which the major axis has a more or less 4° deviation to horizontality in the result of the second step. After ISEF filtering, the projected results in (b) and (c) are from rows (projection in x direction) and cols (projection in y direction) respectively, and scaled within the same size of the image for the better exhibition. The image (d) is a combination of (a), (b), and (c) to reveal the relative positions of significant points in projection, as marked in small circles.

In this way, the ROI is enclosed by significant points as the four squares formed by dashed lines in Fig. 8.14a. The cutter edges are wearing out but the cutter backs are not changed. Therefore, the high-precision angle for correction can be obtained by rotating a given degree of angle and probing the lowest of the center significant point in Fig. 8.14c. The scale between pixel and micron can be extracted from the distance between off-center significant points in Fig. 8.14b. This scale is calibrated as $660/478 \approx 1.38$ μm/pixel and experiments show that its fluctuation among images is negligible.

The result of angle correction is shown in Fig. 8.15a together with the scale values.

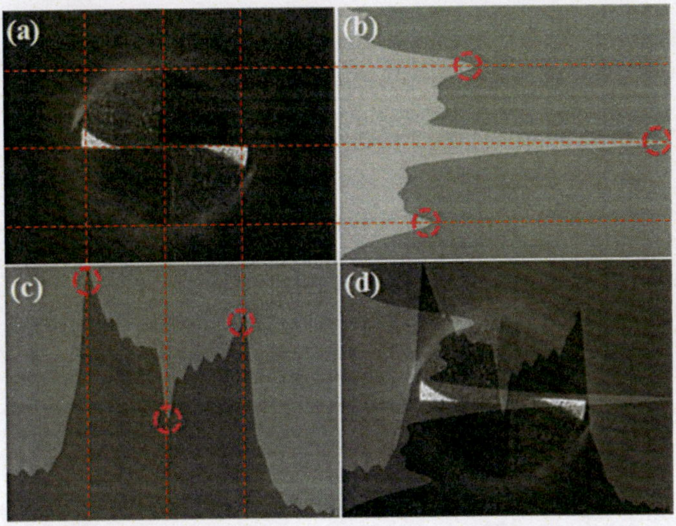

Fig. 8.14 Projection for ROI

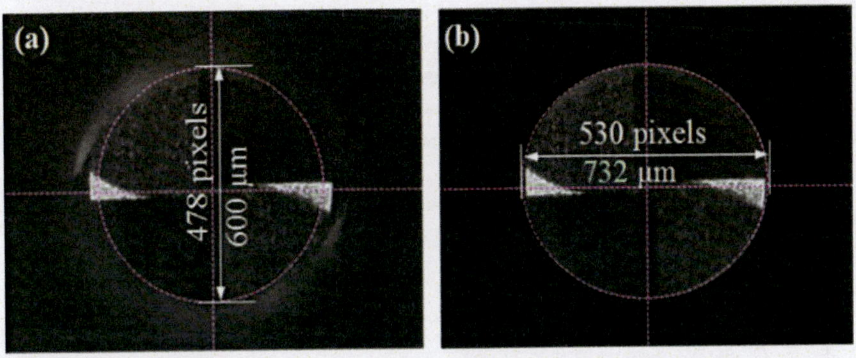

Fig. 8.15 Extracted ROI

The circle is drawn according to the extracted diameter of the minor axis. Then, by edge detection at flanks along the major axis (horizontal), the length of the major axis can be obtained according to the scale as shown in Fig. 8.15b. In this way, the diameter wear ($DW_1 + DW_2$) can also be calculated ($800 - 732 = 68$ μm).

(7) **Eliminating speckles**

After angle correction and edge-preserving filtering, a threshold value of 200 is possible to extract the wear regions. As a comparison, results from the mean filter and the designed edge-preserving filter both using the same filter window size of 15 are exhibited in Fig. 8.16. The regions from image binarization have been enlarged and located at the side of their original positions in the cutter images. The boundary

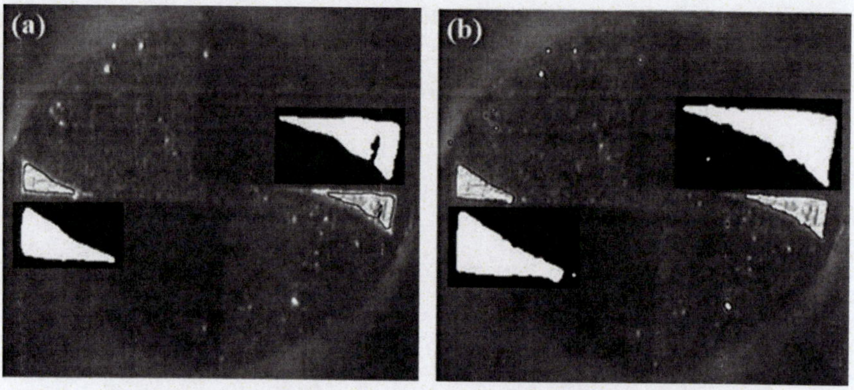

Fig. 8.16 Results of image binarization after mean filtering and the proposed method

lines of their white areas have been integrated with positional correspondence in the original cutter images. The worn areas manifested by white pixels are extracted insufficiently by mean filtering and binarization in Fig. 8.16a.

However, the presented edge-preserving filtering can give a very accurate result in Fig. 8.16b, which shows that the Eq. 8.7 is powerful to correct the distortions due to smoothing. Subsequently, the white speckles can be eliminated by comparing the size of the regions. The two largest white areas indicate worn regions and others can be eliminated. After the operations mentioned above, these variables can be obtained directly by counting the number of pixels in dimensions.

8.3.4 Experimental Verification

(1) **The experiment**

A total of 12 experiments were conducted to validate the designed tool wear monitoring system at different working conditions (Table 8.1). The machine tool was the five-axis MAKINO V55 vertical milling center driven by a 22 kw spindle drive motor. The micro tools used in this study were 800 μm diameter micro milling tools with the helix and shank taper angles 30 and 16°, respectively. They are Tungsten Carbide tools, with Titanium-Aluminum-Nitride coatings. The workpiece materials used were pure copper and steel T4. Tool wear image was originally captured and measured using the Olympus Toolmakers microscope (213 times enlargement). More experimental setup can also be found in [43].

(2) **Experimental results and discussion**

A group of results is shown in Fig. 8.17. The selected variables show steady trends with the time increment and can reveal the extent of wear in varying degrees. On the other hand, there is still no consensus in the literature with which is the best variable

Table 8.1 The experimental working conditions

Test	Spindle speed N (rpm)	Axial depth of cut a_p (mm)	Radial engagement a_e (mm)	Feed rate V_f (mm/min)
1	12,000	0.060	0.100	0.120
2	12,000	0.080	0.100	0.120
3	12,000	0.100	0.150	0.120
4	18,000	0.100	0.150	0.150
5	18,000	0.120	0.175	0.150
6	18,000	0.100	0.175	0.150
7	24,000	0.100	0.200	0.180
8	24,000	0.120	0.200	0.180
9	24,000	0.150	0.250	0.180
10	30,000	0.120	0.250	0.150
11	30,000	0.150	0.300	0.150
12	30,000	0.120	0.300	0.150

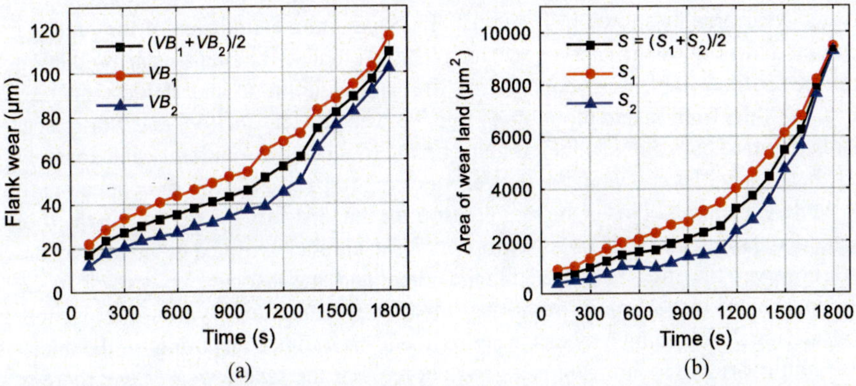

Fig. 8.17 Comparison of the extracted flank wear with area of wear land, **a** Flank wear, **b** Area of wear land

for micro tool wear indication. Many investigators attempted different descriptors [35, 36, 44, 45], and accurate measurement is undoubtedly a prerequisite. There is an agreement on the necessity of avoiding single-toothed cutting phenomena in micromachining [46, 47], however, it is very difficult to avoid inaction and the differences of worn amplitude with two flutters (VB1 and VB2) are obvious as shown in Fig. 8.17. The two flutters are the same in imaging because of symmetry, which may lead to wrong correspondence between wear value and flutter. Therefore, the results are all extracted discriminatively by comparing the wear size at each step. A reference point supports this rule too.

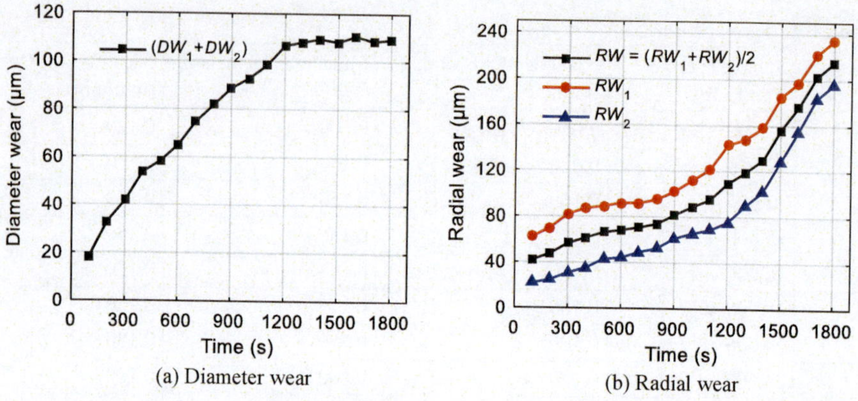

Fig. 8.18 Results of diameter wear and radial wear

Comparatively, the trends from flank wear Fig. 8.17a and areas of wear land Fig. 8.17b agree with the theoretical prediction and the results of the flank area wear predicted in [48]. It is observed that the more severe tool wear corresponding to greater wear area as shown in Fig. 8.17. The quantitative statistics of the areas of wear lands in Fig. 8.17b also show steady trends that will benefit wear prediction even better than flank wear considering the value amplitude and stable tendency. Because of the friction and ploughing effect between the cutting tool and workpiece, there are micro burs due to adhered chips on the tool ends, as randomly and sparsely distributed speckles shown in the figure, which are imaged as noises or connecting to the white wear lands. This is the main reason for the unsteady fluctuation on the tool wear curves in Fig. 8.18. In order to improve it, an air cock is aimed to blow away the burs, however, the effect is limited. On the other hand, it is also necessary to clean the tool end before inspection in traditional off-line measurement by tool microscope. Considering the cleaning may cause new damages and time consuming for the micro tools, multivariable monitoring is a good choice. For the severe wear stage, there is a jump of wear rate at about 1400 s, which is considered to be corresponding to the start of the severe wear stage. The fluctuations from the two flutters result from the single-toothed cutting phenomena and noise from burrs on the cutter [49] as shown in the Fig. 8.18.

There is no report concerning the diameter wear as shown in Fig. 8.18a, and only the total wear is given due to accuracy of measurement (the amounts of wear at the two flutters are different and thus it is difficult to ascertain them respectively before identifying the center position). The fluctuation after 1200 s is about 1 pixel. However, the curve also gives light to revealing the tool life from the early trend, especially the point of inflection at about 1300 s close to the time of wear rate jump in Fig. 8.18a. By artificial measurement carefully, it is also found that the diameter wear is indeed to be steady when the time is close to the tool's life. The Radial wear is also supplied in Fig. 8.18b though they are sensitive to the sharp corner of the delta-shaped white regions. Experiments show that the curvilinear trends reflect the progressively worn extent. At the same time, there is also no report concerning the difference between the two flutters and the experimental results indicate that the wear from the two flutters tends to be equal at the end of the steady-state. Tool run-out is believed to be one of the most important reasons for this asymmetric cutting [50]. Tool run-out plays an important role in micro-milling because the ratio between tool run-out and feed per tooth is very high, which seriously influences the instantaneous undeformed chip thickness. In some cases, tool run-out is so high that just one flute cuts the material, generating an asymmetric cutting [50]. Castro et al. [51] adopted a laser interferometer to measure the spindle rotation errors of machine tools with accuracy high enough. However, the special sphere affixed at the end of a wobble device which is clamped in the spindle will influence the normal movement due to associated mass, and the influence will be more obvious for micro tools. In our experiments, the cutter's coating thickness is reducing with machining and it is gradually becoming blunt, thus the further wear rate is relatively slowing. Furthermore, the differences of worn amplitude from the two flutters can also be explained as the differences of flutter cutters' contact area with the workpieces specially appearing in micromachining, as revealed by Li et al. [52].

Overall, the trend of wear area is steadier, and is possible to evaluate the life of micro tools, such as estimating the average tool life or worn stages roughly. It has integrated more factors to indicate the tool state, and the formats of flank wear, radial wear, and diameter wear are partly synthesized in it. At the same time, the flank wear, radial wear, and diameter wear are also meaningful to monitor and identify the abnormal conditions by identifying the jump of worn rate. Considering the small scale in micromachining, there are still a lot of unclear factors that have been simply considered as noises in the literature. Such noises may cause sudden abnormality such as tool breakage and tipping which are detrimental to the workpieces and even damage the part. Both steady trend prediction and abnormality identification are necessary. The further step is to improve machining operations, then to extend the tool life after monitoring the wear by the inspection system. The method proposed by Tansel et al. [29] is a good strategy, and it reduces the cutting feed rate when the abnormality is identified or the evaluated worn stage is coming. The method does save the tool's life in experiments as verified by Tansel. However, a quantitative assessment of efficiency and cost is difficult to carry out at present, and more instances in actual machining are necessary to collect statistical data in the future.

(3) **Accuracy analysis**

In experiments, there are mainly three factors that influence the repeatability of measurement. The first one is the stationary motion of the motor in imaging. Lowering the motor's speed during focusing is possible to improve the quality of the image to some extent. At the same time, the incline between the lens and the tool end should be avoided by system tests running before processing. The second factor is the denoising algorithm that is also concerned in many machine vision-based TCMs. In contrast,

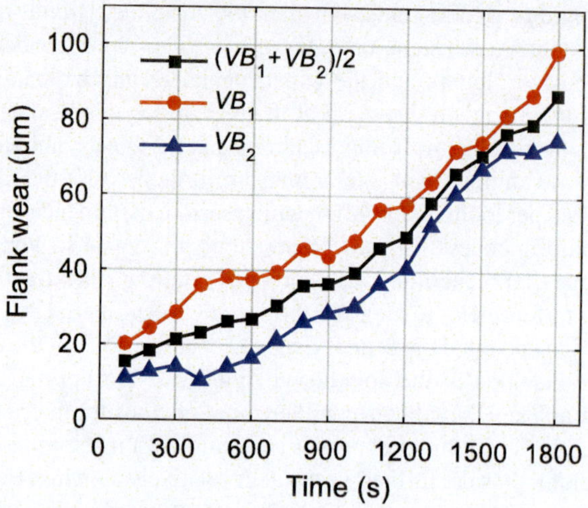

Fig. 8.19 Flank wear extracted by mean filtering

the proposed edge-preserving filtering supports a more accurate wear extraction than mean filtering, as seen in Fig. 8.17. The flank wear extracted by mean filtering is also shown in Fig. 8.18. Comparing with Figs. 8.17a and 8.19, the overall amplitude from mean filtering is reduced, as illustrated in Fig. 8.17. At the same time, the range of the curve fluctuates more greatly due to false boundary recognition as shown in Fig. 8.17a. In addition to filtering, the influence from threshold value is insignificant to disturb the curves' total trends.

The last main factor concerning the repeatability of measurement is the micro burrs adhered to the boundary of the cutters. This is part of the reason for the curve trend fluctuation in Figs. 8.17 and 8.18.

8.3.5 Conclusions

Premature failure is a major problem in micromachining. In order to successfully predict the micro tool life, the paper develops an automated machine vision system for tool condition monitoring. The image processing algorithms are developed according to the characteristics of micro-milling to extract the progressive tool wear. In addition to the flank wear, new variables are proposed to reflect the tool wear state. Experiments have shown that area of wear land is suitable to forecast the tool wear stages and other variables are possible to identify the abnormality in cutting. The strategy in the utilization of the inspection system is to reduce the cutting feed rate when the abnormality is identified or the evaluated worn stage is coming. There is no sudden failure that happened with the sample tools and the proposed algorithms support the developed system in these verification experiments. In the future, the more appropriate feed rate is possible to be studied in depth with the detected tool condition and further explore the potentiality of micro tools.

References

1. Davies ER (2012) Computer and machine vision: theory, algorithms, practicalities, 5th edn. Academic Press
2. Beyerer J, León FP, Frese C (2015) Machine vision: automated visual inspection: theory, practice and applications. Springer
3. Li Q, Ren S (2012) A real-time visual inspection system for discrete surface defects of rail heads. IEEE Trans Instrum Meas
4. Luo Q, Fang X, Liu L, Yang C, Sun Y (2020) Automated visual defect detection for flat steel surface: a survey. IEEE Trans Instrum Meas 69(3):626–644
5. Ren R, Hung T, Tan K (2018) A generic deep-learning-based approach for automated surface inspection. IEEE Trans Cybern 48(3):929–940
6. Scime L, Beuth J (2018) Anomaly detection and classification in a laser powder bed additive manufacturing process using a trained computer vision algorithm. Addit Manufact
7. Aminzadeh M, Kurfess TR (2019) Online quality inspection using Bayesian classification in powder-bed additive manufacturing from high-resolution visual camera images. J Intell Manuf 30(6):2505–2523
8. Chauhan V, Surgenor B (2017) Fault detection and classification in automated assembly machines using machine vision. Int J Adv Manufact Technol
9. Huang SH, Pan YC (2015) Automated visual inspection in the semiconductor industry: a survey. Comput Ind 66:1–10
10. Cheng Y, Jafari MA (2008) Vision-based online process control in manufacturing applications. IEEE Trans Automat Sci Eng 5(1):140–153
11. Castejón M, Alegre E, Barreiro J, Hernandez LK (2007) On-line tool wear monitoring using geometric descriptors from digital images. Int J Mach Tools Manuf 47(12–13):1847–1853
12. Lins RG, De Araujo PR, Corazzim M (2020) In-process machine vision monitoring of tool wear for cyber-physical production systems. Robot Comput-Integrat Manufact 61:101859
13. Zhang C, Zhang J (2013) On-line tool wear measurement for ball-end milling cutter based on machine vision. Comput Ind 64(6):708–718
14. Zhu K, Yu X (2017) The monitoring of micro milling tool wear conditions by wear area estimation. Mech Syst Signal Process 93:80–91
15. García-Ordás MT, Alegre-Gutiérrez E, Alaiz-Rodríguez R et al (2018) Tool wear monitoring using an online, automatic and low cost system based on local texture. Mech Syst Signal Process 112:98–112
16. Dutta S, Pal SK, Sen R (2016) On-machine tool prediction of flank wear from machined surface images using texture analyses and support vector regression. Precis Eng 43:34–42
17. Fernández-Robles L, Azzopardi G, Alegre E, Petkov N (2017) Machine-vision-based identification of broken inserts in edge profile milling heads. Robot Comput Integr Manuf 44:276–283
18. Dutta S, Pal SK, Mukhopadhyay S, Sen R (2013) Application of digital image processing in tool condition monitoring: a review. CIRP J Manuf Sci Technol 6(3):212–232
19. Wang W, Wong YS, Hong GS (2005) Flank wear measurement by successive image analysis. Comput Ind 56:816–830
20. Prasad KN, Ramamoorthy B (2001) Tool wear evaluation by stereo vision and prediction by artificial neural network. J Mater Process Technol 112:43–52.

21. Devillez A, Lesko S, Mozer W (2004) Cutting tool crater wear measurement with white light interferometry. Wear 256:56–65
22. Wang WH, Wong YS, Hong GS (2006) 3D measurement of crater wear by phase shifting method. Wear 261:164–171
23. Kassim AA, Mian Z, Mannan MA (2004) Connectivity oriented fast Hough transform for tool wear monitoring. Pattern Recognit 37:1925–1933
24. Tönshoff HK, Inasaki I (eds) (2001) Sensors in manufacturing. Wiley-VCH
25. Demant C, Streicher-Abel B, Garnica C (2013) Industrial image processing, 2nd edn. Springer
26. Gonzalez RC, Woods RE (2018) Digital image processing, 4th ed. New York
27. Chae J, Park SS, Freiheit T (2006) Investigation of micro-cutting operations. Int J Mach Tools Manuf 46:313–332
28. Patra K, Jha AK, Szalay T, Ranjan J, Monostori L (2017) Artificial neural network based tool condition monitoring in micro mechanical peck drilling using thrust force signals. Precis Eng 48:279–291
29. Tansel I, Nedbouyan A, Trujillo M, Tansel B (1998) Micro-end-milling—II. Extending tool life with a smart workpiece holder (SWH). Int J Mach Tools Manufact 38:1437–1448
30. Aramcharoen A, Mativenga PT (2009) Size effect and tool geometry in micromilling of tool steel. Precis Eng 33:402–407
31. Tansel IN, Arkan TT, Bao WY, Mahendrakar N, Shisler B, McCool D (2000) Tool wear estimation in micro-machining. Part I: tool usage–cutting force relationship. Int J Mach Tools Manufact 40:599–608
32. Lee HC, Chang YC, Huang YS (2014) A reliable wireless sensor system for monitoring mechanical wear-out of parts. IEEE Trans Instrum Meas 63:2488–2497
33. Tansel I, Trujillo M, Nedbouyan A, Velez C, Bao WY (1998) Micro-end-milling—III. Wear estimation and tool breakage detection using acoustic emission signals. Int J Mach Tools Manufact 38:1449–1466
34. Szydłowski M, Powałkaa B, Matuszakb M, Kochmańskia P (2016) Machine vision micro-milling tool wear inspection by image reconstruction and light reflectance. Precis Eng 44:236–244
35. Astakhov VP (2004) The assessment of cutting tool wear. Int J Mach Tools Manuf 44:637–647
36. Castejón M, Alegre E, Barreiro J, Hernández LK (2007) On-line tool wear monitoring using geometric descriptors from digital images. Int J Mach Tools Manuf 47:1847–1853
37. Dutta S, Kanwat A, Pal SK, Sen R (2013) Correlation study of tool flank wear with machined surface texture in end milling. Measurement 46:4249–4260
38. Jackson MJ, Robinson GM, Hyde J, Rhodes R (2006) Neural image processing of the wear of cutting tools coated with thin films. J Mater Eng Perform 15(2):223–229
39. Shen J, Castan S (1992) An optimal linear operator for step edge detection. CVGIP: Graph Models Image Proces 54:112–133
40. Tomasi C, Manduchi R (1998) Bilateral filtering for gray and color images. In: IEEE international conference on computer vision (ICCV), Bombay, India, Jan 7
41. He KM, Sun J, Tang XO (2010) Guided image filtering. In: European conference on computer vision (ECCV). Berlin, Germany, Sept 5–11
42. Jain P, Tyagi V (2016) A survey of edge-preserving image denoising methods. Inf Syst Front 18:159–170
43. Zhu KP, Mei T, Ye DS (2015) Online condition monitoring in micromilling: a force waveform shape analysis approach. IEEE Trans Industr Electron 62:3806–3813
44. Barreiroa J, Castejóna M, Alegrea E, Hernándezb LK (2008) Use of descriptors based on moments from digital images for tool wear monitoring. Int J Mach Tools Manuf 48:1005–1013
45. Liu TI (1990) A computer vision approach for drill wear measurements. J Mater Shaping Technol 8:11–16
46. Malekian M, Park SS, Jun MBG (2009) Tool wear monitoring of micro-milling operations. J Mater Process Technol 209:4903–4914
47. Fleischer J, Deuchert M, Ruhs C, Kühlewein C, Halvadjiysky G, Schmidt C (2008) Design and manufacturing of micro milling tools. Microsyst Technol 14:1771–1775

48. Su JC, Huang CK, Tarng YS (2006) An automated flank wear measurement of microdrills using machine vision. J Mater Process Technol 180(1–3):328–335
49. Lee KH, Dornfeld DA (2005) Micro-burr formation and minimization through process control. Precis Eng 29:246–252
50. Attanasio A (2017) Tool run-out measurement in micro milling. Micromachines 8(7):221
51. Castro HFF (2008) A method for evaluating spindle rotation errors of machine tools using a laser interferometer. Measurement 41(5):526–537
52. Li KX, Zhu KP, Mei T (2016) A generic instantaneous undeformed chip thickness model for the cutting force modeling in micromilling. Int J Mach Tools Manuf 105:23–31

Chapter 9
Tool Wear Monitoring with Hidden Markov Models

9.1 Introduction

In micro-machining, with the miniaturization of the cutting tool (<1 mm in diameter), and high speed (>10,000 rpm) used, the tool wears quickly. Tool wear is defined as the change of shape of the tool from its original shape during cutting, resulting from the gradual loss of tool material. Tool wear is a limiting factor in high-speed machining [1]. It is critical to monitor the tool wear in micro-machining due to the high precision requirement. Compared to conventional machining, different difficulties are encountered in the identification of tool conditions in micro-machining. Firstly, chip flow characteristics and vibration are not easily observable in micro-machining, due to the minimum chip thickness effect [2–4], and the very small vibration caused by variations of the cutting force [5, 6]. It is almost impossible to carry out direct approach with such a high spindle speed. Secondly, problems such as the scaling down of the size of the components and tools used, associated micro-chip formation, and tool wear have yet to be fully understood; Thirdly and especially for signal conditioning and processing, the noise component in the signal for monitoring micro-machining is usually very high [7–9] and difficult to separate [7, 9]. According to ISO 8688, the threshold for determining the tool life is maximum flank wear 0.3 mm in conventional machining [10]. This has to be redefined in micro-machining because the total cutting edge is less than 0.3 mm generally. According to Tansel et al. [7, 8] any sense of changes of the cutting edge is worn. From this point of view, it is a matter of different degrees of wear and hence, instead of a single indicator, multi-category identification and classification of tool wear for monitoring progressive wear of the tool are more appropriate.

This chapter presents the development of a general framework for multi-category tool flank wear estimation in micro-milling based on cutting force sensing. The cutting force has been found to correlate well with tool conditions in machining and most effective as sensor signal for tool wear monitoring [11–13]. It has been observed that the force signal is very low and highly non-stationary (periodically involves impulse). The signal-to-noise ratio is relatively very low [9]. Effective

K. Zhu, *Smart Machining Systems*, Springer Series in Advanced Manufacturing,
https://doi.org/10.1007/978-3-030-87878-8_9

signal processing for this highly non-stationary and noisy signal to provide a robust approach for tool condition monitoring (TCM) is very important in micro machining [9]. The approach is based on hidden Markov modeling (HMM) of tool conditions and involves the appropriate selection of HMM structures for tool wear classification. HMMs estimation results from cutting experiments performed on copper and steel are presented and their effectiveness and generalization potential are discussed.

Tool condition monitoring is widely studied in conventional machining. Byrne et al. [12] and Liang et al. [14] present good reviews of these methods. Generally, indirect measurement methods such as force, acoustic emission, and vibration are used to capture machining signals, and intelligent approaches such as neural networks and clustering are then applied to estimate the tool wear states. Of the classification methods, neural networks (NN) are the most widely studied, such as MLP [15, 16], adaptive resonance network (ART2) [17, 18], SVM [19] and self-organizing map (SOM) [20]. Hidden Markov models (HMMs) have recently been employed for TCM lately due to their excellent representation of temporary dynamics of signals and reasoning property in speech recognition [21–23]. The application of the hidden Markov model in TCM was first reported by Heck et al. [24] for tool wear detection and prediction in drilling. It was further studied for drilling [25, 26], turning [27, 28], and milling tool wear monitoring [29–34].

Compared to conventional machining, there has been little work in micro-machining TCM. There are different types of difficulties encountered in micro-machining TCM compared with conventional machining. Except for some work reported in micro-drilling tool condition monitoring [35, 36], most studies on micro-milling are those reported by Tansel and his collaborators [7, 8, 37–41], with neural network (NN) based approaches. For the application of NN approaches, however, the monitoring systems strongly depend on the network structure that is hard to generalize, and the amount of training data not always available. An inherent problem of NN is that the weights are sensitive to new inputs and the entire network has to be retrained if new features are added or state augmented.

In this chapter, a modeling framework based on hidden Markov models (HMMs) is implemented to classify force features of the different tool states. Previous applications of HMMs for drilling [34–36] and turning [27, 28] are not suitable for our problems. For HMMs in milling, the important issues for modeling HMMs, such as the selection of the number of Gaussian mixture components and HMM states, were not discussed in detail in [29, 30, 32–34]. The HMM implemented in this study is based on the concept of multi-rate modeling, but with several modifications to adapt to the conditions in micro-milling monitoring [42], which is discussed in the later sections accordingly.

9.2 HMM Based Methods

9.2.1 Hidden Markov Models

Markov process is a stochastic process that is used to model the evolution of random states as a function of time. For the first-order Markov process for a system with S distinct states, the current state of the model is dependent only on the previous state,

$$P\left(q_t = S_j | q_{t-1} = S_i, q_{t-2} = S_k, ...\right) = P\left(q_t = S_j | q_{t-1} = S_i\right) = a_{ij} \tag{9.1}$$

where q_t is the actual state at time t and a_{ij} is the transition probability between state i and j. The transition constraints to,

$$\sum_{j=1}^{N} a_{ij} = 1 \tag{9.2}$$

The hidden Markov Model (HMM) is a double-layered stochastic process, with an underlying finite-state hidden Markov process that associates with the observation process. For our application, the force features $Y = \{y_1, y_2, ..., y_T\}$ are the observation process, and the tool state sequence $S = \{s_1, s_2, ..., s_T\}$ is the hidden process. The probability of the observations Y is conditionally dependent on state S,

$$P(Y) = \sum_{S} P(Y, S) = P(y_1|s_1)P(s_1)\prod_{t=2}^{T} P(y_t|s_t)P(s_t|s_{t-1}) \tag{9.3}$$

In order to characterize an HMM completely, the following elements are needed. An HMM is identified with the parameter set $\lambda = (\pi, A, B)$ where:

- π is the initial state distribution vector, e.g. π is the probability of state i at the same arbitrary time $t = 0$.
- A is the state transition matrix, where $A = [a_{ij}]$, with a_{ij} being the probability of transiting to state j given current state i.
- B is the output distribution matrix, where $B = [b_{jk}]$, with b_{jk} being the probability of observing feature k given the current state j.

The estimation and selection of parameter set $\lambda = (\pi, A, B)$ are the essence of HMM modeling, which are discussed in the following sections.

In HMM based TCM, it is assumed that the sequence of features corresponding to each tool state is generated by a Markov model as shown in Fig. 9.1. A Markov model is a finite state machine that changes state once every unit time and each time t that a state j is entered, a speech vector o_t is generated from the probability density $b_j(o_t)$. Furthermore, the transition from state i to state j is also probabilistic and is governed by the discrete probability a_{ij}. Figure 9.1 shows an example of this process

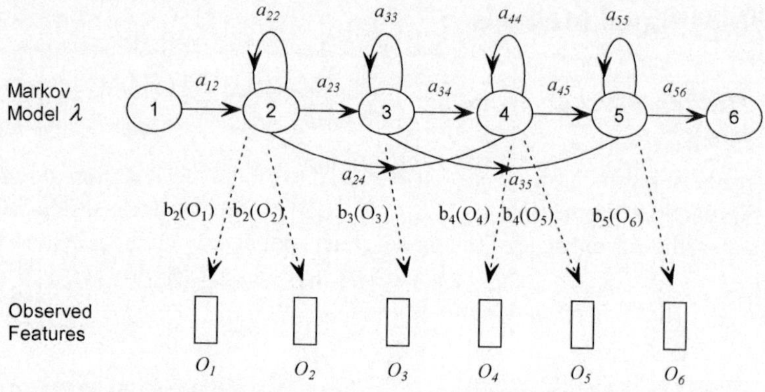

Fig. 9.1 The HMM generation model

where the six state model moves through thc state sequence $Q = 1; 2; 2; 3; 4; 4; 5; 6$ in order to generate the sequence o_1 to o_6.

9.2.2 Three Problems of Hidden Markov Models

Given the definition of HMMs above, three problems of interest must be addressed before they can be applied to real-world applications. They are listed below (See Appendix 1 for more details):

(1) **The Evaluation Problem**: Given a model λ and a sequence of observations $O = (O_1, O_2, ..., O_T)$, what is the probability $P(O|\lambda)$ namely, the probability of the model that generates the observation?

(2) **The Decoding Problem**: Given a model λ and a sequence of observations $O = (o_1, o_2, ..., o_T)$, what is the most likely state sequence $S = (q_0, q_2, ..., q_T)$, in the model that produces the observations?

(3) **The Learning Problem**: Given a model Φ and a set of observations, how can we adjust the model parameter Φ to maximize the joint probability (likelihood) $\prod_o P(O|\lambda)$.

If we could solve the evaluation problem, we would have a way of evaluating how well a given HMM matches a given observation sequence. Therefore, we could use HMM to do pattern recognition, since the likelihood $P(O/\lambda)$ can be used to compute posterior probability $P(\lambda/O)$, and the HMM with the highest posterior probability can be determined as the desired pattern for the observation sequence. If we could solve the decoding problem, we could find the best matching state given an observation sequence, or in other words, uncover the hidden state sequence. If we could solve the learning problem, we would have the means to automatically estimate the model parameter λ from an ensemble of train data. These three problems are tightly linked under the same probability framework. The efficient implementation of these algorithms shares the same principle of dynamic programming.

9.3 Hidden Markov Models Based Tool Condition Monitoring

9.3.1 HMM Description of Tool Wear Process and Monitoring

Like that of conventional milling, tool wear increases progressively with machining time, given specific working conditions in micro-milling. Generally, the dynamic process of the wear rate $v(t)$ can be modeled as a differential equation:

$$v(t) = \frac{dw(t)}{dt} \tag{9.4}$$

where $w(t)$ is the tool wear value at machining time t.

Approximating this function with difference function, we get

$$\frac{w(t+\Delta t) - w(t)}{\Delta t} = v(t) \Rightarrow w(t+\Delta t) = w(t) + v(t)\Delta t \tag{9.5}$$

For convenience and to facilitate physical interpretation, we rescale Δt to unit time step 1 and keep $v(t)$ as constant v but including a noise term $\varepsilon(t)$, which obeys a certain distribution.

$$w(t+1) = w(t) + v + \varepsilon(t) \tag{9.6}$$

If $\varepsilon(t) \sim N(\mu, \sigma^2)$ obeys a Normal distribution with μ and σ^2, then

$$p(w(t+1)|w(t)) = \frac{1}{\sqrt{2\pi\sigma^2}} \exp[-\frac{1}{2\sigma^2}(w(t+1) - w(t) - v)^2] \tag{9.7}$$

This means that we can predict the tool wear value $w(t+1)$ at the time $t+1$ given the current value $w(t)$ and wear rate. This is a first-order Markov process which refers that the current state only depends on the immediately preceding state in time. More formally,

$$p(w(t+1)|w(t), w(t-1), ...w(1)) = p(w(t+1)|w(t)) \text{ (hidden Process)} \tag{9.8}$$

This process is hidden to us and hence classed as *hidden process*, which we try to uncover. If the wear rate depends on second-order differential equations under the current wear, we can likewise approximate it with a second-order Markov process description, and this can be similarly repeated for higher-order representation. The second and higher-order Markov Model can describe more dynamics of the wear process but result in more complex models and more parameters to estimate. Generally, the first-order Markov Model is used to model dynamic systems.

For online estimation, we cannot measure the current tool wear value directly but it can be estimated from the observation of the force features $y(t)$. This relationship is very complex and typically non-stationary and uncertain. It is a function of $w(t)$:

$$y(t) = f(w(t)) \text{ (observation process)} \tag{9.9}$$

which represents the relationship of a certain feature $y(t)$ and a tool wear value $w(t)$. $y(t)$ is generally extracted from measurable sensor signals, so we call it the observation process.

The above double-layer stochastic processes can be illustrated in Fig. 9.2. The top three figures are the observed force features corresponding to different tool states shown in the pictures below. The variations are stochastic, while the upper is non-stationary and the bottom is stationary and Markov. Their relationships are modeled with HMMs, which will be discussed in detail in later sections.

We can formally specify the TCM problem as trying to find the most probable state given the machining signal features.

TCM aims to find,

$$\text{TCM}: \underset{tool\,state\,i}{\arg\max}\; p(tool\;state|signal\;features) \tag{9.10}$$

This is a dynamic inference problem since we do not estimate state only with prior knowledge, but also adapt to the current features. In the framework of HMMs, this kind of dynamic reasoning problem can be solved.

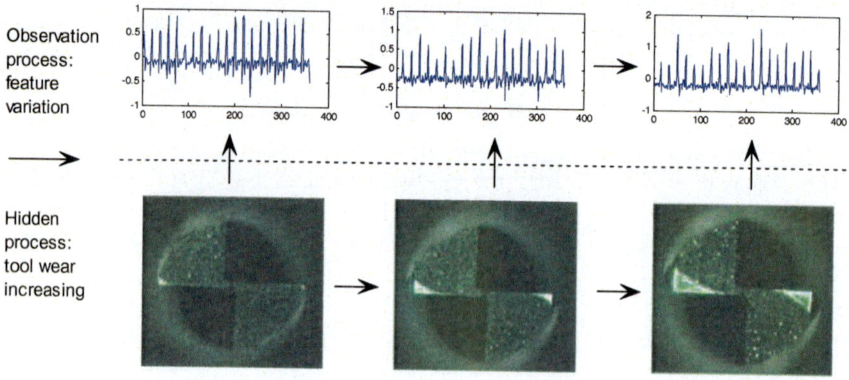

Fig. 9.2 Stochastic modeling of tool wear process

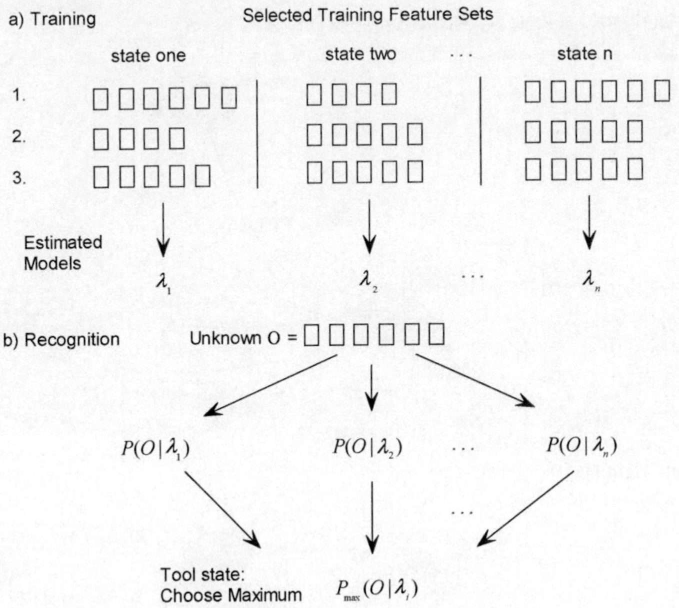

Fig. 9.3 Framework of Hidden Markov models for tool wear classification

9.3.2 *The Framework of HMMs for TCM*

The framework of HMMs for TCM is illustrated in Fig. 9.3. It first decomposes the signal into different wavelet packets, and selects the discriminant wavelet packets, and then uses these packets as features for HMMs training. Once the models are trained, we get the $\boldsymbol{\lambda} = (\boldsymbol{\pi}, \boldsymbol{A}, \boldsymbol{B})$ that represent different tool wear states. In the tool state recognition state, the extracted features from an unknown state are extracted and input to the HMMs to match the trained states, and then choose the most probable one.

9.3.3 *Hidden Markov Model Selection: Continuous Left–Right HMMs*

(1) **Left–Right HMMs**

The most widely used HMMs are the ergodic and the left–right HMMs. The ergodic HMM is a generic type of HMM and adapts to many kinds of problems. The ergodic HMM is fully connected with all states linked together, which infers that every state can be reached from every other state. The transition matrix is a full matrix as shown in Fig. 9.4.

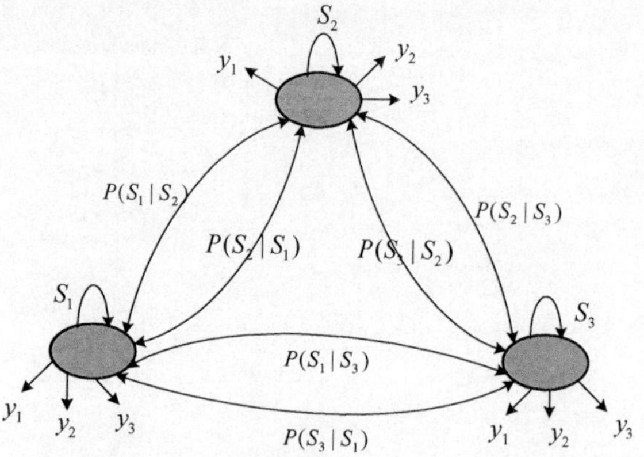

Fig. 9.4 Ergodic HMM tool state model

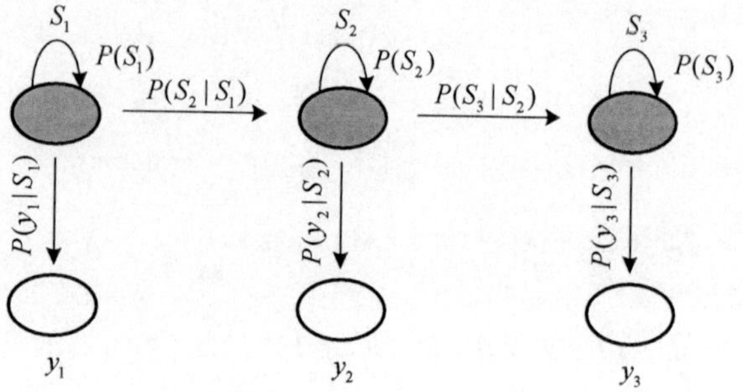

Fig. 9.5 Left–right HMM tool state model

However, ergodic HMMs are not feasible for TCM because the machining process is continuously changing and the state can't change back. Hence, there is the need to constrain the transition matrix. As shown in Fig. 9.5, the left–right HMM can transit to self and later state, but not to the former state. This transition matrix is thus constrained to an upper triangular matrix. The initial state of left–right HMM is $\pi_0 = [100]$ since the tool starts at the first state.

Not like that in ref. [31], the HMMs involve no jumping between states but progresses from one to another, i.e. the tool starts in the initial wear state, moves into the gradual wear state, which is relatively long in micro-milling, before advancing into the accelerated wear state.

$$\text{Transition matrix} \begin{bmatrix} a_{11}\, a_{12}\, a_{13} \\ a_{21}\, a_{22}\, a_{23} \\ a_{31}\, a_{32}\, a_{33} \end{bmatrix} a_{ij} \in [0, 1]$$

$$\text{Transition matrix:} \begin{bmatrix} a_{11}\, a_{12}\, 0 \\ 0\, a_{22}\, a_{23} \\ 0\, 0\, a_{33} \end{bmatrix}, \begin{array}{l} a_{ij} = P_{ij} = 0\; for\; |i - j| > 1\, \&\, i < j \\ \text{Initialization: } \pi_0 = [100] \end{array}.$$

(2) **Continuous HMMs and Gaussian Mixtures Modeling**

Besides classified with structures, the HMMs also classified discrete HMMs or continuous HMMs depending on whether the observation features are modeled as continuous or discrete. In the case of a discrete HMM, the feature space ***Y*** consists of a finite number of elements, $\boldsymbol{e_1}, \ldots, \boldsymbol{e_M}$, which are achieved through the Vector Quantization (VQ) [21, 22]. For most applications, the VQ of the features can degrade performance significantly [21] since it is an average over large regions. Moreover, the codebooks generated by the quantization process are constructed using training data from all classes. When a new class of shapes is added, we need to reconstruct the codebook and retrain all system modules. We choose continuous HMMs for our study.

In a continuous HMM, the feature (observation) space ***Y*** is considered to be infinite and continuous. Therefore, for each state ***j***, we aim to compute the probability $\boldsymbol{b_j(y)}$ for each vector ***y*** in the space. For a probability distribution mentioned above, however, it is not sufficiently flexible to accurately model the variation which occurs between different feature vectors which correspond to a state. This is particularly true if the models are used to characterize wear state from several features. Thus, Gaussian mixtures are typically used to model broad sources of variability. A ***K***-component Gaussian mixture is a linear combination of ***K*** Gaussian densities.

The d-dimensional Gaussian probability density function (pdf):

$$N_{(\mu,\Sigma)}(x) = \frac{1}{\sqrt{2\pi}^{d}\sqrt{\det(\Sigma)}} e^{-\frac{1}{2}(x-\mu)^T(x-\mu)} \tag{9.11}$$

with mean vector μ and covariance matrix Σ.

The most general representation of the PDF is a finite mixture of normal distributions with different means and variances for each state.

$$b_j(y) = \sum_{m=1}^{M} C_{jm} N(y, \mu_{jm}, \Sigma_{jm}) \quad 1 \le j \le N \tag{9.12}$$

where C_{jm} is mixture weights for the m-th mixture in state j and the weights are all positive and sum to one. N is Gaussian with mean μ_{jm} and covariance matrix Σ_{jm} for the m-th mixture component in state j.

By varying the number of Gaussians K, the weights c_k, and the parameters μ_k, Σ_k of each Gaussian density function, Gaussian mixtures can be used to describe any complex density function. Figure 9.6 is an example of the wavelet feature vector modeled as Gaussian Mixture. Figure 9.6a is the original one-dimensional signal and approximated with three weighted Gaussian components in Fig. 9.6b:

$$f(x) = 0.3 \times N_1(\mu_1, \Sigma_1) + 0.2 \times N_2(\mu_2, \Sigma_2) + 0.5 \times N_3(\mu_3, \Sigma_3)$$

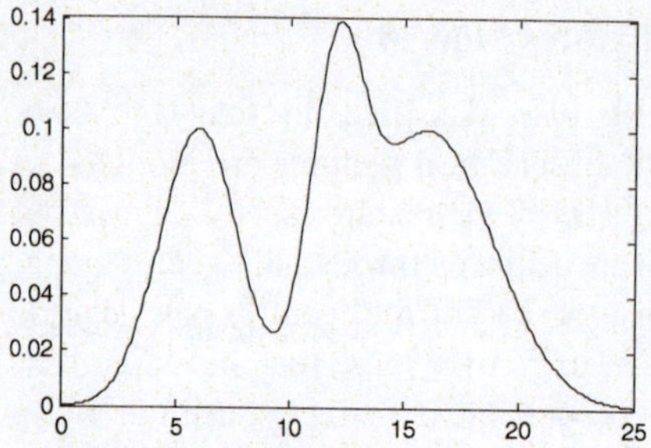

(a) original signal

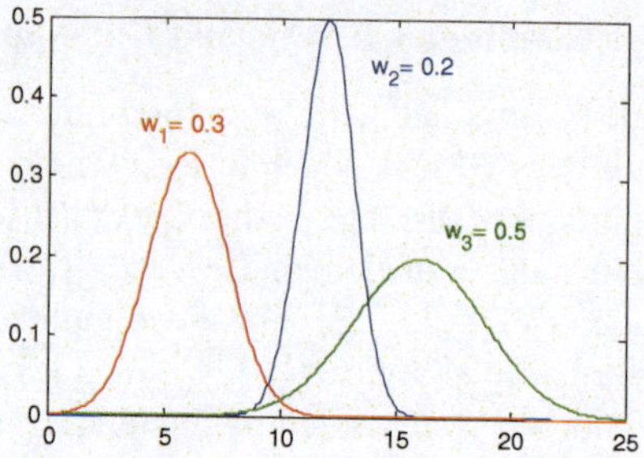

(b) approximated with 3 weighted Gaussians

Fig. 9.6 Gaussians mixtures

9.3.4 *Selection of the Number of Gaussian Mixture Components*

To model the wavelet feature vectors as Gaussian mixtures, the number of mixture components has to be decided. Theoretically, a GMM can approximate any signal to a certain precision when provided with enough mixture components. Too many components may not be desirable, however, as they over fit the data and involve needless computation. There are many criteria for the selection of the components [43, 44], including the Akaike information criterion (AIC), Bayesian information criteria (BIC), and Cross Validation (CV). The Bayesian Information Criterion (BIC) provides a reliable approximation to the integrated likelihood and most popular among all the criteria [45]. We apply the BIC for the selection. The Bayesian information criterion imposes a penalty for including too many terms in a regression model.

The Gaussian mixture model is characterized by the number of components K, the component weights w, and the vector means μ and covariance Σ: $\theta = (w_1, w_2, ...w_K, \mu_1, \mu_2, ...\mu_K, \Sigma_1, \Sigma_2, ...\Sigma_K)$. The way for choosing a model is to select this one maximizing the likelihood,

$$\hat{K} = \arg\max p(y|K) \tag{9.13}$$

An asymptotic approximation of the integrated likelihood leads to minimizing the so-called BIC criterion,

$$BIC_K = -2L_K + v_K \ln n \tag{9.14}$$

where v_K is the number of free parameters with K components, n is the number of observations, $L_K = \log f(y|K, \hat{\theta})$ is the maximum log-likelihood for K, and $\hat{\theta}$ is the ML estimate of θ.

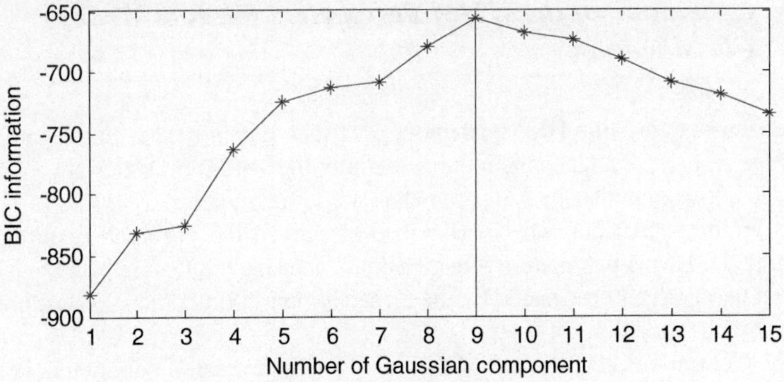

Fig. 9.7 Bayesian information criterion for mixture components

The BIC selects the best model for the data by balancing the complexity of the model with its fit to the data. The preferred model is the one with the highest value of the criterion. The larger the BIC value, the better the model.

The number of Gaussian components is identified to be 9 among the training tests for the 8-dimension feature vectors (The top 8 discriminant features extracted both from time and wavelet domains [9] (also refer to Sect. 9.3). For 9 components, the BIC information reaches the highest as illustrated in Fig. 9.7.

9.3.5 On the Number of Hidden States in Each HMM

The selection of the appropriate number of hidden states in the HMM is from a physical point of view. Generally, with one HMM modeled for one tool state, the states of HMMs lack the physical meaning as a single HMM for different tool wear states. The Tool wear state is characterized by the corresponding entire HMM in our models. The states in that HMM are the Markov transition state within the HMM, but not the tool wear states.

In the milling, the feature patterns are quite different even in the same tool wear state but among different cutting stages. So we train HMMs with the data based on each cutting pass, and constrain every HMM with three states as entry, progressing, and exit. In this sense, the HMMs can estimate both the tool wear state (i.e., which HMM match most) and the cutting state (i.e., which state in that HMM highest probable).

9.3.6 Estimation of the HMM Parameters for Tool Wear Classification

For the Gaussian mixture HMMs, then the HMMs are identified with the parameter set $\boldsymbol{\lambda} = (\boldsymbol{\pi}, \boldsymbol{A}, \boldsymbol{c}, \boldsymbol{\mu}, \boldsymbol{\Sigma})$. Once the number and topology of HMM states are chosen as discussed above, the distribution parameters are to be trained.

The Training problem can be described as: Given the observed features $\{y_i\}$, determine the HMM parameters λ that best characterize the features. These parameters can be iteratively estimated by the Expectation Maximization (EM) algorithm (Baum-Welch algorithm) [22].

In the Continuous HMM, the EM formulae find a new set of parameters in terms of the variables calculated by the Forward–Backward procedure [21]. The forward variable is defined as the probability of the partial observation sequence $\boldsymbol{Y1}, \boldsymbol{Y2}, \ldots, \boldsymbol{Yt}$ (until time t), and state Si at time t, given the model λ, as $\alpha_t(i)$. And the backward variable is defined as the probability of the partial observation sequence from $t + 1$ to the end, given state Si at time t and the model λ, as $\beta_t(i)$. Both $\alpha_t(i)$ and $\beta_t(i)$ are determined using the following forward–backward procedure:

Forward:

$$\alpha_1(i) = \pi_1 * b_i(Y_1) \tag{9.15}$$

$$\alpha_{t+1}(j) = \sum_i [\alpha_t(i) * a_{ij}] b_j(Y_{t+1}) \tag{9.16}$$

Backward:

$$\beta_T(i) = a_{iN} \tag{9.17}$$

$$\beta_{t-1}(i) = [\sum_j \beta_t(j) * a_{ji}] b_i(Y_{t-1}) \tag{9.18}$$

The probability of the observations sequence is:

$$P(Y|\lambda) = \sum_{i=1}^{N} \alpha_t(i)\beta_t(i) = \sum_{i=1}^{N} \alpha_T(i) \tag{9.19}$$

And the probability of being in state Si at time t, given the observation sequence $\boldsymbol{Y}$, and the model λ, as:

$$\gamma_t(i,k) = \frac{\alpha_t(i)\beta_t(i)}{\sum_{i=1}^{N} \alpha_t(i)\beta_t(i)} \cdot \frac{c_{jm} N(Y_t, \mu_{jm}, \Sigma_{jm})}{\sum_{m=1}^{M} c_{jm} N(Y_t, \mu_{jm}, \Sigma_{jm})} \tag{9.20}$$

The probability of being at state *Si* at time *t* and state *Sj* at time *t + 1* is:

$$\xi_t(i, j) = \frac{\alpha_t(i) a_{ij} b_j(Y_{t+1}) \beta_{t+1}(i)}{P(Y|\lambda)} \tag{9.21}$$

Each of the model parameters $\boldsymbol{\lambda} = (\boldsymbol{\pi}, \boldsymbol{A}, \boldsymbol{c}, \boldsymbol{\mu}, \boldsymbol{\Sigma})$ is maximized separately with EM, by setting its respective partial derivative equal to zero. The results are:

$$\pi_i = \gamma_0^i = [1\,0\,0] \tag{9.22}$$

$$\overline{a}_{ij} = \frac{\sum_{t=1}^{T} \gamma_t(i, j)}{\sum_{t=1}^{T} \sum_{j=1}^{N} \gamma_t(i, j)} \tag{9.23}$$

$$\overline{c}_{ij} = \frac{\sum_{t=1}^{T-1} \gamma_t(i, j)}{\sum_{t=1}^{T} \sum_{j=1}^{M} \gamma_t(i, j)} \tag{9.24}$$

$$\overline{\mu}_{ij} = \frac{\sum_{t=1}^{T-1} \gamma_t(i, j) Y_t}{\sum_{t=1}^{T} \gamma_t(i, j)} \tag{9.25}$$

$$\overline{\Sigma}_{ij} = \frac{\sum_{t=1}^{T-1} \gamma_t(i, j) \cdot [(Y_t - \mu_{ij})(Y_t - \mu_{ij})']}{\sum_{t=1}^{T} \gamma_t(i, j)} \tag{9.26}$$

It is noted that Eq. 9.23 is essentially the ratio between the expected number of transitions from state *i*. For the output probability re-estimation Eqs. 9.24–9.26, the numerator is the expected number of times the observation data emitted from state *j* with the observation symbol o_k, and the denominator is the expected number of times the observation data emitted from state *j*.

According to the EM algorithm, the Baum-Welch (Forward–backward) algorithm guarantees a monotonic likelihood improvement on each iteration, and eventually, the likelihood converges to a local maximum (Table 9.1).

Table 9.1 Expectation–maximization

1. Estimate an initial HMM model as $\boldsymbol{\lambda} = (\boldsymbol{\pi}, \boldsymbol{A}, \boldsymbol{c}, \boldsymbol{\mu}, \boldsymbol{\Sigma})$
2. Given λ and the observation sequence $\boldsymbol{Y}$, we calculate a new model $\boldsymbol{\lambda} = (\boldsymbol{\pi}, \boldsymbol{A}, \boldsymbol{c}, \boldsymbol{\mu}, \boldsymbol{\Sigma})$ such that $P(Y\vert\bar{\lambda}) > P(Y\vert\lambda)$
3. If the improvement $\frac{P(Y\vert\bar{\lambda}) - P(Y\vert\lambda)}{P(Y\vert\bar{\lambda})} < threshold,\ then\ stop$
Put $\bar{\lambda}$ instead of λ and go to step 1

9.3.7 Tool State Estimation with HMMs

The forward algorithm computes the probability that an HMM generates an observation sequence by summing up the probabilities of all possible paths, and it does not provide the best state sequence. It is desirable to find such a path. Since the state sequence is hidden in the HMM framework, the most widely used criterion is to find the state sequence that has the highest probability of being taken while generating the observation features. In other words, we are looking for the state sequence $S = (s_1, s_2, ...s_T)$ that minimizes $p(S, Y\vert\lambda)$. A formal technique based on dynamic programming, known as the Viterbi algorithm can be used to find the best path sequence for an HMM.

The Viterbi algorithm can be regarded as the dynamic programming algorithm applied to the HMM or as a modified forwarding algorithm. Instead of summing up probabilities from different paths coming to the same destination state, the Viterbi algorithm picks and remembers the best path. The best score accounts for the first t observations and ends in state S_i, which can be expressed as follows:

$$\delta_t(i) = \max P(q_1 q_2 ... q_t = i, Y_1 Y_2 ... Y_t\vert\lambda) \tag{9.27}$$

The complete procedure for finding the best state sequences is as follows (ψ is the variable that tracks the argument which maximized) (Table 9.2).

As shown in Fig. 9.8, this algorithm can be visualized as finding the best path

Table 9.2 Viterbi Algorithm

Initialization:
$\delta_1(i) = \pi_i b_i(Y_1) \quad 1 \le i \le N$
$\psi_1(i) = 0$
Recursion
$\delta_t(i) = \max\left[\delta_{t-1}(i) a_{ij}\right] b_j(Y_t) \quad 2 \le t \le T$
$\psi_t(i) = \arg\max\left[\delta_{t-1}(i) a_{ij}\right] \quad 1 \le i \le N$
Termination
$P^* = \max[\delta_T(i)] \quad 2 \le t \le T$
$q_T^* = \arg\max[\delta_T] \quad 1 \le i \le N$
Path backtracking
$q_T^* = \psi_{t+1}(q_{t+1}^*) \quad t = T-1, T-2, ...1$

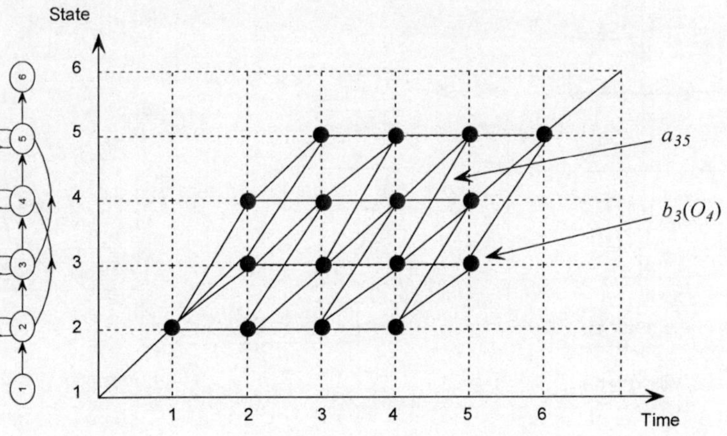

Fig. 9.8 The Viterbi algorithm for state recognition

through a matrix where the vertical dimension represents the states of the HMM and the horizontal dimension represents the frames of speech (i.e. time). Each large dot in the picture represents the log probability of observing that frame at that time and each arc between dots corresponds to a log transition probability. The log probability of any path is computed simply by summing the log transition probabilities and the log output probabilities along that path. The paths are grown from left-to-right column-by-column.

9.4 Experimental Verifications

9.4.1 Experiment Setup

The machine used in these experiments (shown as Fig. 9.9) is MAKINO V55 vertical milling machine driven by a 22kw spindle drive motor. The cutting force was measured with Kistler 9254 dynamometer. The work piece is clamped on the dynamometer on the feed table of the machining center with a holder. Tool wear is measured using the Olympus Toolmakers microscope, 213 times enlargement. The cutting force output was recorded on a Sony digital tape recorder. Cutting conditions: Feed-rate: 50–180 mm/min; spindle speed: 20,000/18,000 rpm/min; depth of cut: 30–300 μm. Cutter: C-CHES 2005-0150, diameter Φ500 μm; C-CHES 2005-0180, diameter Φ800 μm.

A total of 27 test experiments are conducted with different spindle speed, depth of cut, radial depth of cut, and feed rate for $\Phi 500\ \mu m$ and $\Phi\ 800 \mu m$ tools. The materials used are either copper or steel.

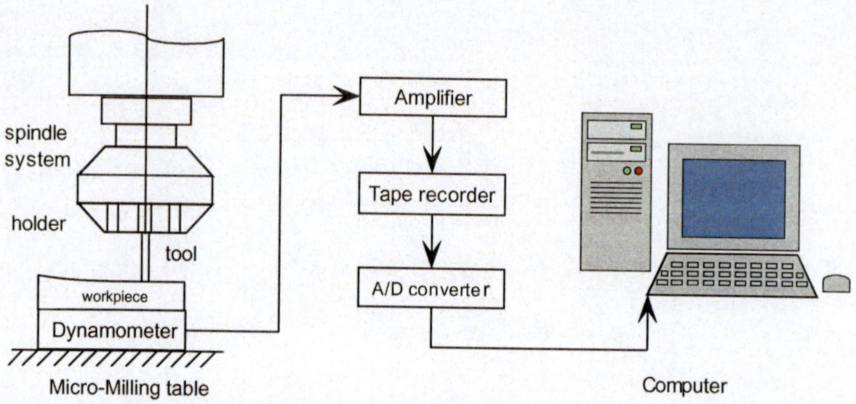

Fig. 9.9 Experiment setup

9.4.2 HMM Training for TCM

As mentioned above, we train each HMM with its own training sets. To train the HMM that represents the fresh tool state, we only use the feature vectors from that state, and similarly for the medium wear state and severe wear state.

We apply the EM algorithm to estimate the HMM parameter $\boldsymbol{\lambda}= (\boldsymbol{\pi}, \boldsymbol{A}, \boldsymbol{c}, \boldsymbol{\mu}, \boldsymbol{\Sigma})$. The algorithm is implemented to interactively adjust the model parameter to maximize the probability of the observation sequence. The EM method works as discussed in Sect. 7.3.

Figure 9.10 shows the training set for the HMM modeling, whose working conditions are listed on the right side. The selection of the training set is based on the distance to the mean working parameters. The EM algorithm converges quickly as illustrated in Fig. 9.10 from test 4.

Other working conditions are also trained to build HMMs in the same way. They test 10, test 13, test 20, and test 24, which will be discussed in [42].

9.4.3 HMM for Tool Wear State Estimation

Because there are 5 HMMs trained, we need to decide which HMMs suit best for the tool state estimation. Given the working condition, a similarity measure based on the Euclidian distance of the working condition is first determined as follows:

$$d_i = \left[(\frac{v - v_{ri}}{v_{ri}})^2 + (\frac{f - f_{ri}}{f_{ri}})^2 + (\frac{d_v - d_{vri}}{d_{vri}})^2 + (\frac{d_r - d_{rri}}{d_{rri}})^2 \right]^{1/2} \tag{9.28}$$

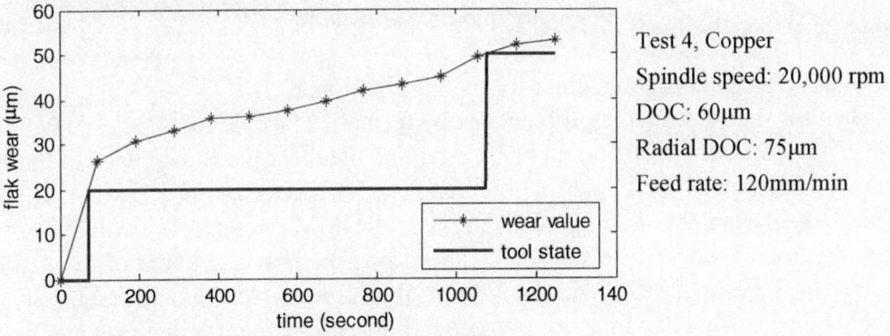

(a) Tool wear state for training the HMMs

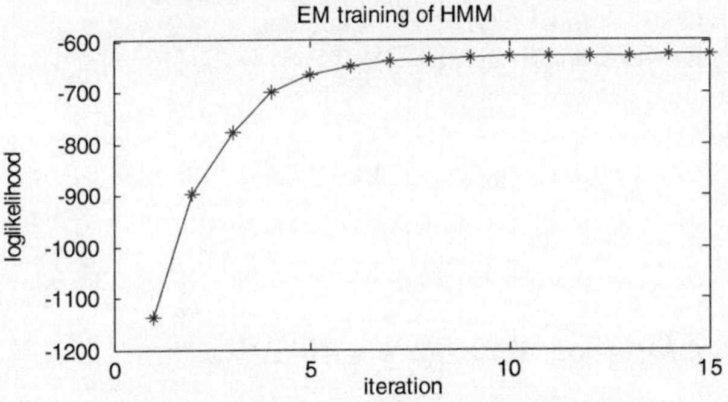

(b) Convergence of EM for HMMs training

Fig. 9.10 HMM training for test 4

where d_i is the distance, v_{ri} is the *i-th* reference spindle speed, f_{ri} is the *i-th* reference feed rate, d_{ri} is the *i-th* reference vertical depth of cut, and d_{rri} is the *i-th* reference radial depth of cut. We aim to find the one with the minimum distance, $\min d_i$. The other two variables, i.e. workpiece materials, and tool diameters are not included here as they are usually predetermined and assumed to be the same.

Wavelet features from various time-scale distributions are extracted and input to the trained HMM for recognition. We build an HMM for each tool state for classification, and each HMM is trained with its own feature set. To obtain the HMM for the fresh tool state, for example, we only select the features with tool wear from 0-20 μm, and likewise for the subsequent wear states. We then get $\boldsymbol{HMM_1 = \lambda_1 = (\pi_1, A_1, c_1, \mu_1, \Sigma_1)}$, $\boldsymbol{HMM_2 = \lambda_2 = (\pi_2, A_2, c_2, \mu_2, \Sigma_2)}$ …, for the corresponding tool wear states. When all the HMMs are trained, we obtain the HMM represented by the parameters: i.e. $\boldsymbol{HMMs = (\lambda_1, \lambda 2 \ldots)}$. The Viterbi algorithm is then implemented to estimate the most probable tool wear state sequence. For an observation feature set $\boldsymbol{Y}$, $p_w = p(Y|\lambda_i)$, $i = 1, 2, \ldots$ is calculated.

The unknown state is then uncovered with the model given the maximum probability of the observed features:

$$w^* = \arg\max p_w, w = 1, 2, ... \quad (9.29)$$

Thus, w* is the optimum class for the observation ***Y***.

The tool state estimation problem can be formalized as: *find* **λ**= *argmax P(Y*| **λ***).* Given a fixed HMM with parameters, determine the most likely sequence of hidden states $\{s_i\}$ for an observed set of features $\{y_i\}$. This decision process is made by the *Viterbi Algorithm.* Analogous to the use of HMMs in speech recognition, the classification of the wear level consists of finding the best alignment of the feature vectors to the HMM states via the Viterbi algorithm, which finds the most likely state sequence using dynamic programming. For each segment, a decision was made on the wear state based on the full history of observations $Y = \{Y_1, Y_2, ...Y_i\}$. The classification $\hat{W}_i$ corresponds to the most likely state sequence.

The best score accounts for the first *t* observations and ends in state S_i, which can be expressed as follows:

$$\delta_t(i) = \max P(q_1 q_2 ... q_t = i, Y_1 Y_2 ... Y_t | \lambda) \quad (9.30)$$

9.4.4 Moving Average for Tool Wear State Estimation Smoothing

In the multi-rate modeling, the classification of the tool states are also constrained to the left–right direction, that is, tool states change from initial to progressive and then to accelerated region but can't change back, the tool state may pre-enter into the next state because of the high noise level in the signal in micromachining. We eliminate this situation by introducing a 15-point moving medium filter. In the HMM recognition, we hold the consecutive 15-state estimations but not decide the state immediately, and then find the medium of these 15 estimates. This medium state is our final tool wear state. Figure 9.11 illustrates the improvement of this approach. It is observed that the low classification rate due to misclassifications of the neighboring states in Fig. 9.11a and the early state changing of the HMMs in Fig. 9.11b. After the left–right constraint and medium filter (Fig. 9.11c), the classification improves from 89.6 to 96.5% for test 12.

The classification rate is defined as:

$$\frac{Correctly\ classified\ samples}{Misclassifieds\ samples + Correctly\ classified\ samples} \times 100\% \quad (9.31)$$

Case studies are shown in Fig. 9.12 for the data set from tests 16 on copper and also test 27 on steel.

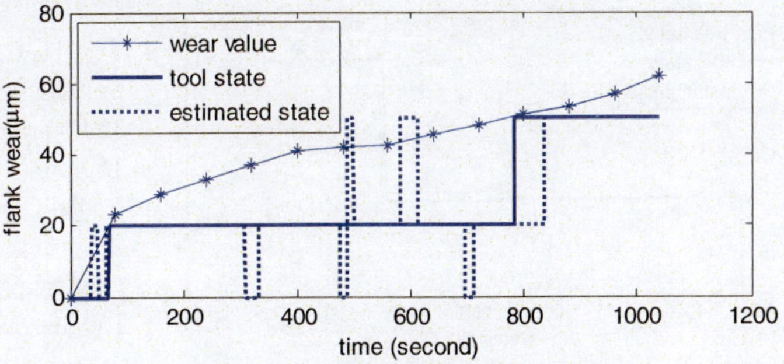

(a) (a) The general HMMs for TCM

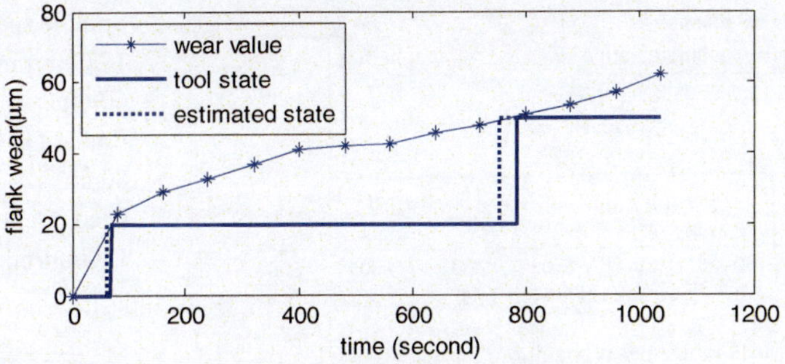

(b) The left-right HMMs for TCM

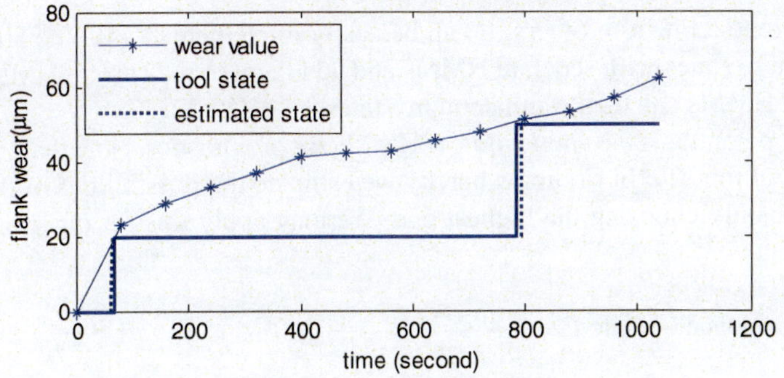

(c) The medium filtered left-right HMMs for TCM

Fig. 9.11 HMMs algorithm improvement

9.4.5 On the Generalization of the HMM-Based Algorithm for TCM

The above discussion assumes that the tool has three states: Slight wear, Medium wear and Severe wear. As a matter of fact, this choice is based on the similarity measurement of signal patterns of different wear values. It may be generalized to include more tool states, which emit features that represent corresponding states

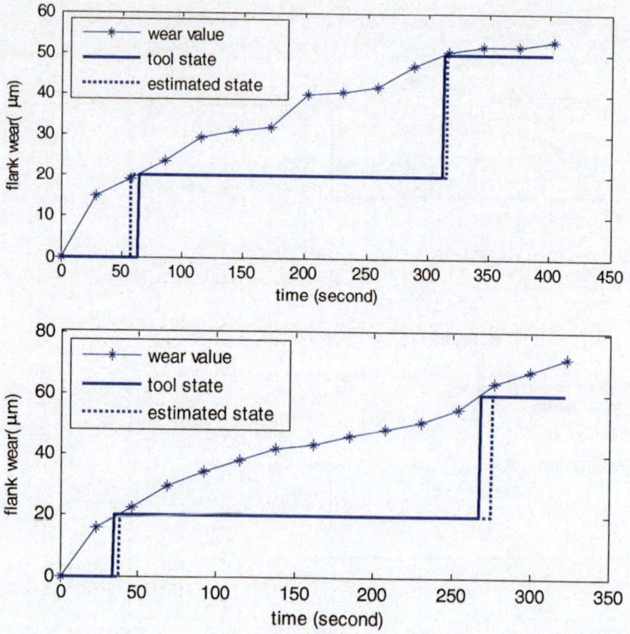

Test 16, Copper
Spindle speed: 18,000 rpm DOC: 150μm
Radial DOC: 225μm
Feed rate: 150mm/min

Test 27, Steel T4
Spindle speed: 18,000 rpm DOC: 60μm
Radial DOC: 80μm
Feed rate: 100mm/min

Fig. 9.12 HMMs tool states recognition

while differentiate it from others. It can be easily implemented with the HMMs—just train these states with separate HMMs and add them to the current HMMs. The augmented HMMs can then estimate more states.

Another potential is the application of HMMs for prognostics. Note that there is a by-product of the Viterbi Coding whereby we estimate the possibilities of the three states while only choosing the highest one. We may apply this for our confidence measurement.

$$\text{Confidence of state } C_i = \frac{p_i}{p_1 + p_2 + p_3} \times 100\%, \ (i = 1, 2, 3) \qquad (9.32)$$

For example, if our confidence of the severe wear state exceeds a certain level (e.g. 80%), we can be more certain predict that the tool will wear out and lose effect soon.

In this section, an approach based on noise-robust continuous HMM for tool wear multi-category classification in micro-milling is proposed. In the HMM, the observations are obtained from the most discriminant features according to their FDR scores to improve HMM's performance for TCM. To overcome the drawbacks of premature state-changing due to noisy signals in the left–right HMMs, the state sequence evolution is constrained and decisions of the states are made with a medium filter after Vertibi estimation. The effectiveness of the approach has been evaluated under different conditions, e.g. different working conditions, different workpiece materials, and variations of observation sequence length. The experimental results indicate that an average recognition rate of as high as 92.5% (ranging from 84.0% to 97.3%) and 90.5% ranging from: 84.6% to 95.6%) can be achieved for copper and steel respectively. Therefore, the approach based on continuous HMM proposed is highly effective for micro-milling tool wear monitoring. It can be generalized and implemented to include more states for TCM and can also be used for prognostics of tool life.

9.5 Diagnosis and Prognosis of Tool Life with Hidden Semi-Markov Model

Degradation of industrial equipment such as wearing and erosion is a common issue in the manufacturing process, and it restricts productivity greatly. The efficient approach that can monitor the degradation conditions and predict the remaining useful life of equipment with degradation process is necessary for improving the producing quantity and saving cost. According to the topology of HMM, existing HMM approaches could be divided into three categories: hidden Markov models (HMMs) [26, 27, 31, 46, 47], single HMM [48–53], and two-layer HMM [42, 54]. Approaches via HMMs assign one hidden Markov model for each health condition. The health condition which corresponds to one HMM with the maximum likelihood for observation sequence is seen as an estimated health condition. Athanasopoulou et.al [46] proposed an on-line recursive algorithm to choose the most likely HMM for given erroneous observations from multi HMMs corresponding to different failure modes. Approaches via single HMM build the entire varying process of health condition as a hidden Markov process. Given observations, the condition that corresponds to the state with maximum posterior probability is seen as estimated health condition. In [48], the fault propagation process was modeled as a high-order hidden Markov process. Then high-order particle filtering was utilized to obtain the posterior estimate of the state. Approaches via two-layer HMM not only include the global Markov transition of state but also include the Markov transition of sub-state.

Despite the wide applications of HMMs in machinery health monitoring, the time-invariant one-step transition probability and geometric distribution of duration make the HMM not match the real degradation process well. For instance, unlike the conventional HMM assumption, the probability that the wear state of a tool changes would mostly decrease as the cutting time increases. As a generalized HMM, the hidden semi-Markov model (HSMM) incorporates self-transition probability into the distribution of the duration of the state. Different forms of distribution of duration correspond to different forms of time-variant self-transition probability. To model the inhomogeneous Markov process, HSMM is more flexible and reasonable than HMM. For this reason, HSMM with explicit duration is widely used in the health monitoring domain [55–58]. Explicit duration HSMM assigns an explicit distribution of duration for each state and the distribution of duration is only determined by the corresponding state. The time-variant one-step transition probability of explicit duration HSMM is dependent on the current state and the lasted time of the current state. Geramifardet et al. [57] showed that the diagnostic result via a single explicit duration HSMM is more accurate than that via a single HMM in [51]. Another special form of HSMM commonly used in health monitoring is variable transition HSMM [59, 60]. Instead of assigning explicit distribution of durations, variable transition HSMM assigns explicit time-variant form for one-step transition probability. Like explicit duration HSMM, the one-step transition probability of variable transition HSMM is also dependent on the current state and corresponding lasted time. Of all the above studies focusing on applying HSMM to health monitoring, the dependence between durations of different states was not taken into account. However, there exists dependence between durations of different health states in the real case. And for the degradation process, the dependent relationship between durations of adjacent states is often remarkable. For example, long duration in fresh tool state may lead to long duration in the medium wear state. Therefore, including the dependent relationship between durations into the HSMM for the degradation process is more reasonable.

Based on such consideration, this section develops an improved HSMM model that includes the sequential transition of state and dependent durations of adjacent degradation states to describe the degradation process. In this model, the one-step transition probability of the degradation process is not only dependent on the current state and lasted time but also dependent on the duration of the preceded state. To avoid the underflow problem in the estimating of state and RUL, a modified forward–backward algorithm is proposed. And based on the modified forward–backward algorithm, on-line health monitoring and RUL estimation are then conducted conveniently. In this study, different from the prognosis methods via HSMM in [55, 56, 60], the distribution of RUL rather than limited statistics of RUL is derived to provide more comprehensive information.

9.5.1 *Hidden Semi-Markov Model for Degradation Process Modeling*

The hidden semi-Markov model [61] is a generalized hidden Markov model in which the transition from state and duration in a time segment to the state and duration in the next segment is considered. Different from the HMM, the transition in HSMM is a transition between 2-D vectors (state and duration) in adjacent time segments. State variable at time t is denoted by q_t and observation variable is o_t. A hidden semi-Markov model is described by the following three elements:

Transition probability:

$$a_{(i,d)(j,d')} \triangleq P\{q_{|t+1:t+d'|} = s_j | q_{|t-d+1:t|} = s_i\} \tag{9.33}$$

Observation probability:

$$c_i(v_j) \triangleq P\{o_t = v_j | q_t = s_i\} \tag{9.34}$$

Initial distribution:

$$\pi_{i,d} \triangleq P\{q_{|t-d+1:t|} = s_i\} \tag{9.35}$$

where s_i is ith state and v_j is jth observation. The set of model parameters is defined as:

$$\lambda = \{a_{(i,d)(j,d')}, c_i(v_j), \pi_{j,d}\} \tag{9.36}$$

To model the degradation process via HSMM, the degradation process is divided into N discrete states $\{s_i | i = 1...N\}$ according to the underlying degradation level. The state s_1 represents the initial state with slight degradation level and the state s_N represents the final state with severe degradation level. By assuming that the degradation state changes sequentially from the initial state to the final state and the durations of adjacent states are dependent, the transition probability of HSMM for the degradation process could be simplified as:

$$a_{d'(i,d)} \triangleq P\{q_{|t+1:t+d|} = s_i | q_{|t-d'+1:t|} = s_{i-1}\} \tag{9.37}$$

Namely, the transition probability represents the conditional probability of duration of the latter state on the condition that the duration of the preceded state is given. The degradation process is assumed to be irreversible and it will always stay at the final state as soon as it reaches the final state. So the duration of the final state is infinite and there is:

$$a_{d'(N,+\infty)} = 1 \ \forall d' \tag{9.38}$$

Initial distribution of HSMM for degradation process is defined as the distribution of duration of initial state:

$$p_1(d) \triangleq P\{q_{|1:d|} = s_1\} \tag{9.39}$$

Generally, the observation is continuous feature vector extracted from physical signals. To process the high-demission observation, two approaches are usually adopted. One approach is using continuous distribution with high demission such as multinormal distribution to descript the relationship between observation and state. The other approach is quantizing the observational feature via vector quantization (VQ) technique [62], and then the relationship between observation and state is modeled by discrete distribution function. In our paper, without assigning explicit distribution form for the feature vector, the second approach is adopted. The M

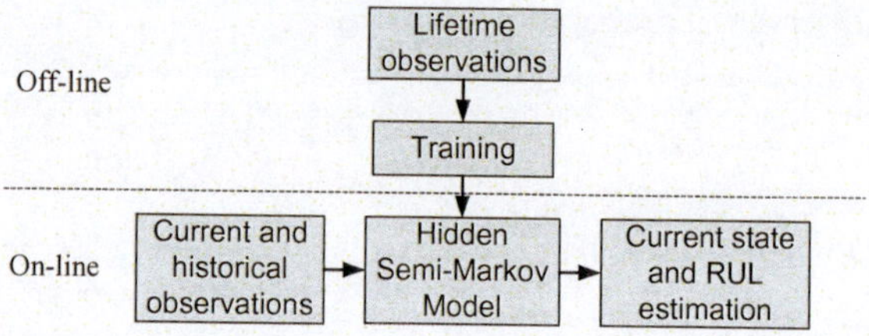

Fig. 9.13 Off-line training and on-line health monitoring via HSMM

possible discrete observation produced by the degradation process is denoted by $\{v_j | j = 1...M\}$ and observation probability is denoted by $c_i(v_j)$. A special HSMM for degradation process is defined as:

$$\lambda_d = \{a_{d'(j,d)}, c_i(v_j), p_1(d)\} \tag{9.40}$$

The parameters of HSMM could be estimated with experimental data. When the observations and states are both recorded completely in the experimental process, the maximum likelihood method could be used directly to estimate the parameters. When the state data cannot be recorded in experiments, with lifetime observation sequence of a degradation process only, the off-line training of HSMM for degradation process should resort to EM algorithm. Based on the trained HSMM, with current and historical observations, on-line estimation of the current degradation state and remaining useful life is then conducted. The idea of HSMM is given in Fig. 9.13 [58].

9.5.2 *On-Line Health Monitoring via HSMM*

In the theory of HSMM, forward and backward variables are two fundamental variables. Almost all of the posterior probabilities needed in inference could be represented by the two variables. Traditional forward–backward algorithm with forward (backward) variable as the joint probability of observation sequence encounters severe underflow problem. In this section, to conquer the underflow problem in HSMM for degradation process, we propose a modified forward–backward algorithm. Similar to the forward–backward algorithm which is proposed for explicit duration HMM in [63], the modified forward–backward algorithm defines forward variable as a posterior probability of observation sequence [58]. According to the modified forward–backward algorithm, the posterior probability of degradation state and RUL could be calculated conveniently.

(1) **Modified Forward–Backward Algorithm** [58]

The observation sequence is denoted by $o_1 o_2 ... o_{T_l}$. For $t \leq T_l$, the formal forward and backward variables [61] are defined as:

$$\alpha_t(i, d) \triangleq P\{q_{|t-d+1:t|} = s_i, o_{1:t}|\lambda_d\} \quad (9.41)$$

$$\beta_t(i, d) \triangleq P\{q_{|t-d+1:t|} = s_i, o_{t+1:T_l}|\lambda_d\} \quad (9.42)$$

where $q_{|t-d+1:t|} = s_i$ means that the state becomes s_i at time $t-d+1$ and changes to another state at time $t+1$. It could be found that the forward variable is generally very small when the time t lasts long and observation features are high. When the forward variables at the time t are less than the decimal of the computation system, they will be set as zero and the underflow phenomenon occurs. Then the forward variable cannot be used in the following inference or estimation process. The underflow phenomenon also occurs often in the computation of backward variables (10) when the time t is far from T_l. Like [61], reference [60] also defined the forward variable as a joint distribution of observation sequence and the underflow problem still existed as a result.

In this paper, to eliminate the underflow problem, the forward variable is instead modeled as the joint posterior probability of state at time t, corresponding sojourn time and the duration of last state on condition that the observation sequence $o_{1:t}$ is known:

$$\alpha_t(d^{'}; i, d) \triangleq P\{q_{|t-d+1:t} = s_i, q_{|t-d-d'+1:t-d|} = s_{i-1}|o_{1:t}, \lambda_d\} \quad (9.43)$$

where $q_{|t-d+1:t} = s_i$ means that the state becomes s_i at time $t - d + 1$ and the state s_i has lasted for d at time t. Different from $q_{|t-d+1:t|} = s_i$, the $q_{|t-d+1:t} = s_i$ does not require that the state at time $t+1$ must change. As it is not necessary to infer whether the current state will change at the next time in the on-line estimation process, the form $q_{|t-d+1:t} = s_i$ is more convenient for the on-line state estimation.

Auxiliary forward variable at time t is defined as the joint posterior probability on the condition that observation sequence $o_{1:t-1}$ is known:

$$J_t(d^{'}; i, d) \triangleq P\{q_{|t-d+1:t} = s_i, q_{|t-d-d'+1:t-d|} = s_{i-1}|o_{1:t-1}, \lambda_d\} \quad (9.44)$$

And there is:

$$J_t(d^{'}; i, d)c_i(o_t) = P\{q_{|t-d+1:t} = s_i, q_{|t-d-d'+1:t-d|} = s_{i-1}, o_t|o_{1:t-1}, \lambda_d\} \quad (9.45)$$

Then the one-step observation probability could be obtained as:

$$p\{o_t|o_{1:t-1}, \lambda_d\} = \sum_{d',i,d} J_t(d^{'}; i, d)c_i(o_t) \quad (9.46)$$

The forward variable $\alpha_t(d^{'}, i, d)$ could be seen as the normalization of $J_t(d^{'}, i, d)c_i(o_t)$ with one-step observation probability as normalization factor:

$$\alpha_t(d^{'};i,d) = J_t(d^{'};i,d)c_i(o_t)\Big/ p\{o_t|o_{1:t-1},\lambda_d\} \tag{9.47}$$

As depicted in the HSMM for the degradation process, the one-step transition is determined not only by the current state and corresponding lasted time but also the duration of the preceded state. So the one-step transition probability could be defined as:

$$\mu(d^{'},i,d;j) \triangleq P\{q_t = s_j|q_{|t-d:t-1} = s_i, q_{|t-d-d^{'}:t-d-1|} = s_{i-1},\lambda_d\} \tag{9.48}$$

When the state is $i > 1$, for the progressive characteristic of the degradation process, there is:

$$\mu(d^{'},i,d;j) = \begin{cases} \sum\limits_{k\geq d+1} a_{d^{'}(i,k)} \Big/ \sum\limits_{k\geq d} a_{d^{'}(i,k)} & j=i \\ a_{d^{'}(i,d)} \Big/ \sum\limits_{k\geq d} a_{d^{'}(i,k)} & j=i+1 \end{cases} \tag{9.49}$$

When the state is final degradation state $i = N$, according to Eqs. 9.38 and 9.49, the one-step transition probability is $\mu(d^{'},N,d;N) = 1$.

When the state is $i = 1$, the duration of preceded state does not exist. To ensure coincident mathematical description, an arbitrary value x is used to denote the non-existent duration before the initial degradation state. The one-step transition is:

$$\mu(x,i,d;j) = \begin{cases} \sum\limits_{k\geq d+1} p_1(k) \Big/ \sum\limits_{k\geq d} p_1(k) & j=1 \\ p_1(d) \Big/ \sum\limits_{k\geq d} p_1(k) & j=2 \end{cases} \tag{9.50}$$

According to the definition of auxiliary forward variable and forward variable, the recursive formula from forward variable at time $t-1$ to auxiliary forward variable at time t could be obtained:

$$J_t(d^{'};i,d) = \begin{cases} \alpha_{t-1}(d^{'};i,d-1)\mu(d^{'},i,d-1;i) & d\neq 1 \\ \sum\limits_{d^{''}} \alpha_{t-1}(d^{''};i-1,d)\mu(d^{''},i-1,d^{'};i) & d=1 \end{cases} \tag{9.51}$$

When the state at time t is the initial state, the lasted time must be t and the duration of preceded state is an arbitrary value. In this situation, the forward variable is denoted by $\alpha_t(x;1,t)$ and the auxiliary forward variable is denoted by $J_t(x;1,t)$. The initial condition of the forward variable could be set as $\alpha_1(x;1,1) = 1$.

Backward variable at time t is defined as the ratio between the joint posterior of state at time t, corresponding sojourn time and the duration of last state on condition that observation sequence $o_{1:T_l}$ is known and the forward variable at time t:

Table 9.3 Modified forward–backward algorithm

The forward recursion:
(1) For $t = 1$, set initial forward variable $\alpha_1(x; 1, 1) = 1$ (2) For $t > 1$, calculate the auxiliary forward variable by (9.51), the one-step observation probability by (9.48), and the forward variable by (9.47)
The backward recursion:
(1) For $t = T_l$, set initial backward variable $\beta_{T_l}(d'; i, d) = 1$ (2) For $t < T_l$, calculate the backward variable by (9.54)

$$\beta_t(d'; i, d) \triangleq \eta_t(d'; i, d) \Big/ \alpha_t(d'; i, d) \tag{9.52}$$

where:

$$\eta_t(d'; i, d) \triangleq P\left\{q_{|t-d+1:t} = s_i, q_{|t-d-d'+1:t-d|} = s_{i-1}|o_{1:T_l}, \lambda_d\right\} \tag{9.53}$$

The backward recursion formula is:

$$\begin{aligned} \beta_t(d'; i, d) = &\Big\{\mu(d', i, d; i)c_i(o_{t+1})\beta_{t+1}(d'; i, d+1) + \\ &\mu(d', i, d; i+1)c_{i+1}(o_{t+1})\beta_{t+1}(d; i+1, 1)\Big\} \times (p\{o_{t+1}|o_{1:t}, \lambda_d\})^{-1} \end{aligned} \tag{9.54}$$

The initial condition of backward recursion is:

$$\beta_{T_l}(d'; i, d) = 1 \tag{9.55}$$

The proof of backward recursion is given in the appendix. The modified forward–backward algorithm is concluded in Table 9.3.

(2) **Degradation State Estimation**

The two basic tasks of on-line health monitoring are the degradation state estimation and the remaining useful life estimation with the information inferred from the current and historical observations. The modified forward–backward algorithm is utilized for this purpose.

At current time T, the observation sequence is denoted by $o_{1:T}$. According to the definition of forward and backward variable, the posterior distribution of state at time $t \leq T$ could be written as:

$$\begin{aligned} P(q_t = s_i|o_{1:T}) &= \sum_{d',d} \eta_t(d'; i, d) \\ &= \sum_{d',d} \alpha_t(d'; i, d)\beta_t(d'; i, d) \end{aligned} \tag{9.56}$$

Because the initial condition of backward variable is $\beta_T(d'; i, d) = 1$, the distribution of current state could be represented only by the forward variables at the current time:

$$P(q_T = s_i | o_{1:T}) = \sum_{d,d'} \alpha_T(d'; i, d) \tag{9.57}$$

Forward variables at the current time could be calculated according to forward recursion listed in Table 9.3. In fact, only one step forward recursion needs to be conducted when the forward variable at preceded observation time is saved. Hence, the on-line estimation process equals the on-line forward recursion process.

To predict the state at future time $t > T$, the forward variables at time $t > T$ need to be calculated. Because the future observations have not been obtained at current time, the future observation could be set as an arbitrary value with uniform observation distribution. Then the auxiliary forward variable equals forward variable at future time and the forward recursion at future time could be simplified as:

$$\alpha_t(d'; i, d) = \begin{cases} \alpha_{t-1}(d'; i, d-1)\mu(d', i, d-1; i) & d \neq 1 \\ \sum_{d''} \alpha_{t-1}(d''; i-1, d')\mu(d'', i-1, d'; i) & d = 1 \end{cases} \tag{9.58}$$

With forward variables calculated at current time as initial condition, according to $t - T$ step forward recursion (9.58), the forward variables at future time $t > T$ could be obtained. Then the posterior distribution of state is obtained:

$$P(q_t = s_i | o_{1:T}) = \sum_{d,d'} \alpha_t(d'; i, d) \quad t > T \tag{9.59}$$

Given the distribution of estimated state, diagnostic analysis could be conducted. In the applications, if the degradation state is divided uniformly and each state corresponds to a specific degradation value, an exact degradation state will be obtained and decided. In most cases, each degradation state corresponds to a range of degradation values, and under this condition, the state with the maximum posterior probability is determined to be the final state.

(3) Remaining Useful Life (RUL) Estimation

It has been pointed out [64] that current HSMM (HMM)-based prognosis methods could only estimate limited information, such as mean and variance of RUL, and cannot provide more statistics of the RUL. Recent extended works [55, 56, 60] did not solve this limitation, however. In practice, it is important to compute the RUL distribution as it provides comprehensive information and confidence on the prognosis. In this section, the posterior distribution of RUL is derived via the proposed HSMM.

The remaining useful life is defined as the time interval between the current time and the time at which the state reaches to final degradation state. According to the

HSMM for the degradation process, the distribution of RUL is determined by these three variables: current state i, corresponding lasted time d and the duration of the last state $d^{'}$. Hence, these three variables are named as RUL factors in this paper. The remaining useful life is denoted by R and the duration of state k denoted by d_k. When the RUL factors are given, the RUL could be written as:

$$R = \sum_{k=i}^{N-1} d_k - d \tag{9.60}$$

The joint conditional distribution of variables $d_{N-1} \ldots d_{i+1}, d_i$ on the condition that the RUL factors are given could be written as:

$$P(d_i, d_{i+1}...d_{N-1}|d^{'}, i, d) = P(d_i|d_i \geq d, d_{i-1} = d^{'})P(d_{i+1}|d_i)...P(d_{N-1}|d_{N-2})$$
$$= \left(a_{d^{'}(i,d_i)} \Big/ \sum_{k \geq d} a_{d^{'}(i,k)}\right) \prod_{k=i}^{N-2} a_{d_k(k+1,d_{k+1})} \tag{9.61}$$

Then the conditional distribution of RUL $P(R|d^{'}, i, d)$ could be obtained by the joint distribution in (9.61):

$$P(R|d^{'}, i, d) = \sum_{d_i,..d_{N-1}} P(d_i, d_{i+1}...d_{N-1}|d^{'}, i, d) \tag{9.62}$$

where $d_i + d_{i+1} + ... + d_{N-1} - d = R$.

The conditional distribution of RUL is only related to the parameters of HSMM, so it could be calculated and saved off-line after the HSMM is trained. The forward variables at current time could be seen as the posterior distribution of RUL factors at the current time. With the conditional distribution of RUL calculated and saved off-line, the posterior distribution of RUL at current time is obtained:

$$P(R|o_{1:T}) = \sum_{d^{'},i,d} P(R|d^{'}, i, d)\alpha_T(d^{'}; i, d) \tag{9.63}$$

It is noted that the computation of the conditional distribution of RUL by (9.61) and (9.62) is very time-consuming and the numerical storage is huge especially when the maximum duration is large. While in many applications, only some posterior statistics rather than the posterior distribution of RUL need to be considered. The posterior statistics of RUL $T(R|o_{1:T})$ could be regarded as the weighted average of conditional statistics of RUL $T(R|d^{'}, i, d)$ with $\alpha_T(d^{'}; i, d)$ as weighted coefficient:

$$
\begin{aligned}
T(R|o_{1:T}) &= \sum_{R} T(R)\left\{\sum_{d',i,d} P(R|d',i,d)\alpha_T(d';i,d)\right\} \\
&= \sum_{d',i,d}\left\{\sum_{R} T(R)\sum_{d',i,d} P(R|d',i,d)\right\}\alpha_T(d';i,d) \\
&= \sum_{d',i,d} T(R|d',i,d)\alpha_T(d';i,d)
\end{aligned}
\tag{9.64}
$$

Hence, it just needs to calculate and save some conditional statistics (such as conditional expectation and variance) rather than the conditional distribution of RUL off-line. This will reduce the storage space greatly. The computation cost could also be reduced by computing the conditional statistics via the Monte Carlo simulation method [65].

9.6 Experimental Validation

9.6.1 Case Study

The case study is conducted on tool wear monitoring in a high-speed CNC milling center. The flank wear of carbide ball-nose cutters with three flutes and cutting force signal is recorded in experiments by a Kistler dynamometer mounted under the workpiece. The Inconel 718 is used as work-piece material, which is a typical hard-to-machine aerospace material. The spindle rotation frequency is 172 Hz and the sampling rate is 10 kHz. There are in total 23 tests carried out, with varying working parameters in feed rate and depth of cut. Every experiment lasts for 105 passes and the time interval of each cutting pass is 12 s.

Tool wear is defined as the mean flank wear of three flutes. Tool wear state is divided into 5 states according to the range of tool wear (shown in Table 9.4). The tool wear is measured during the recess of the tool holder. Thirteen tests are used as training samples and the remaining 10 tests are used as testing samples. In both of the training and testing tests, tool wear is measured and each cutting pass is labeled the wear state according to Table 9.4.

The number of cutting passes is utilized to represent the length of time for duration and RUL. For example, the duration of state 1 is 5 means state 1 will last for 5 cutting passes. Likewise, the RUL of a tool is 20 means the tool will reach the final state

Table 9.4 Tool wear state

State label	1	2	3	4	5
Wear range (μm)	0–60	60–90	90–120	120–150	>150

5 after 20 cutting passes. As the sequence of tool wear state of the testing test is recorded in experiments, the duration and the cutting pass at which the tool reaches state 5 could conveniently be determined. The real RUL at certain cutting pass is the number of cutting passes between the cutting pass and the cutting pass at which the tool becomes state 5. The sequence of real RUL of the testing test is also recorded and then used as a reference of RUL estimation in the testing process.

9.6.2 Feature Extraction and Quantization

Related researches show that the distribution of energy of cutting force signal in different frequency bands is capable to reflect the wear state. The cutting force signal could be approximately seen as a periodic signal with the spindle rotation frequency as fundamental characteristic frequency (172 Hz).

Wavelet packet decomposition (WPD) [66] is utilized to extract the features in different frequency bands of cutting force signals. In order to separate information inferred by different harmonic components of cutting force signal into different elements of the feature vector, the frequency band should be divided into at least 59 segments (10 kHz/172 Hz). To meet this need, the cutting force signal is decomposed into 64 frequency bands via 6-level db8 wavelet package decomposition of the cutting force signal. The mean energy of one frequency band is defined as the ratio between the sum of the square of wavelet packet coefficients $d^n_{j,k}$ and the number of wavelet packet coefficients in one frequency band:

$$E_j = \frac{1}{N_j}\sum_{k=1}^{N_j} |d^n_{j,k}|^2 \tag{9.65}$$

where $d^n_{j,k}$ is obtained by the wavelet packet transform of the signal localized at $2^j k$ in scale 2^j, and n is the oscillation parameter $n = 1, 2....$ Then the feature vector consists of mean energy in 64 frequency bands. The cutting force signal extracted from one cutting pass and its mean energy in each wavelet packet are shown in Fig. 9.14.

As introduced in Sect. 9.2, there exist two ways to build the generative model for the high-dimension observation feature. One way is to assign an explicit distribution for the features, and the other way is to quantize the features to obtain discrete observations. For the high-dimensional feature vectors in this study, it is difficult to assign an explicit distribution form for them. So the quantized approach is adopted. The discrete observation is used as the observation of HSMM and the probability distribution of discrete observation is used to descript the relationship between tool wear state and observation feature.

To obtain discrete observation, the k-means approach [67] is adopted to quantize the wavelet packet feature vector. The total 1365 wavelet packet feature vectors

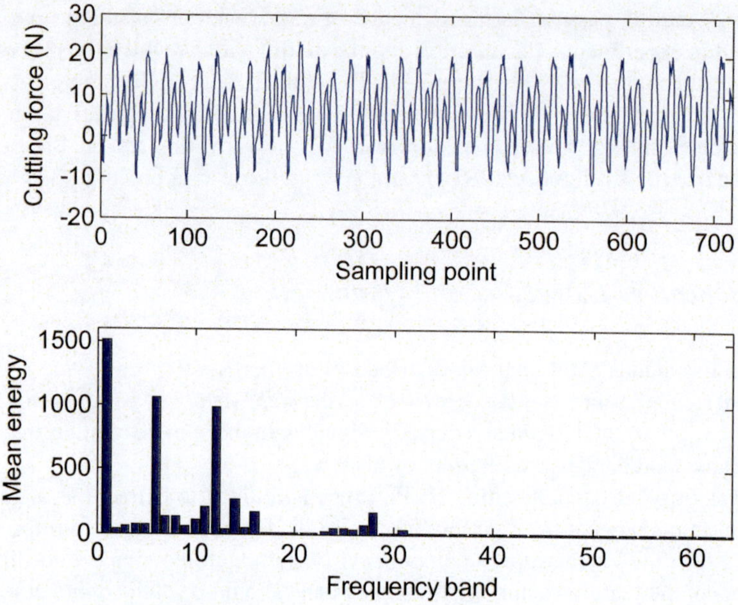

Fig. 9.14 Cutting force signal and its energy in wavelet packet

extracted from 13 training tests are divided into 20 clusters with the k-means method. The quantized result (discrete observation) is the cluster to which the wavelet packet feature vectors belongs to. For testing samples, the Euclidean distance from the wavelet packet features to each cluster's center are computed, and then the discrete observation is assigned to the cluster with the minimum distance.

Figure 9.15 shows the discrete observations of training test 1 and testing test 1. With all of the discrete observations and states in training set, the probability distribution of discrete observation in each state could be estimated, which are further studied in the next sections.

9.6.3 Training of HSMM for Tool Wear Monitoring

Apparently, the estimated probability $\hat{c}_i(v_j)$ is the frequency that discrete observation $j \in \{1, 2, ..., 20\}$ occurs when the wear state is $i \in \{1, 2, ..., 5\}$ in the training experiments. The estimated observation probability is given in Fig. 9.16.

The distribution of discrete observation could be used to estimate the tool wear state. For example, given observation 3, the state will be estimated as state 1 as the probability that observation 3 belongs to state 1 is bigger than the others (as Fig. 9.16 shows). Because the method does not utilize any prior knowledge of durations, it is named as distribution method without duration in this paper. Obviously, the method

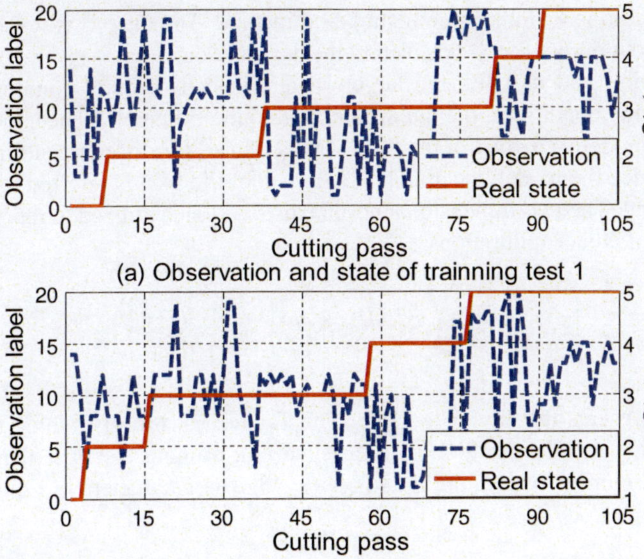

Fig. 9.15 Evolution of discrete observation and tool wear state

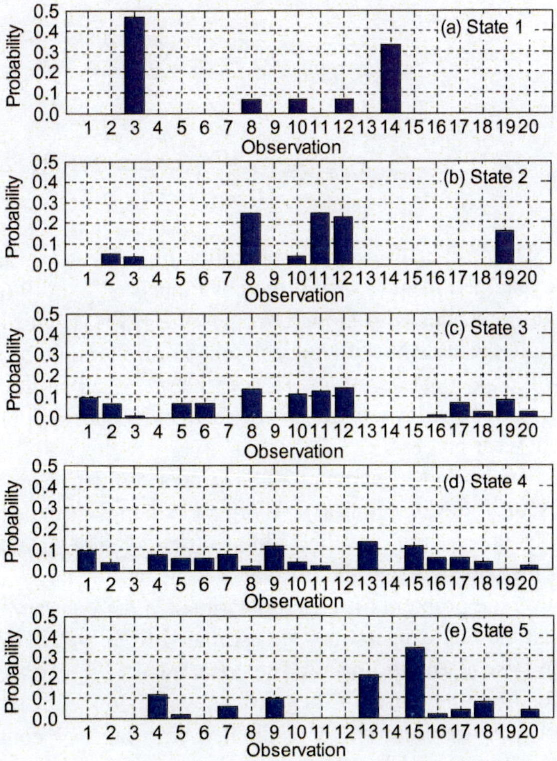

Fig. 9.16 Estimated discrete observation distribution

without duration is a kind of pure data-based method. The diagnosis result via the method without duration will be given in the next section.

The duration and tool life are largely determined by cutting conditions. For different cutting conditions, the duration may be quite different. Hence, there must exist a high modeling error if a fixed HSMM is trained for different cutting conditions. To reduce the model uncertainty, Taylor's tool life [42] is referred to the tool wear modeling. The two-dimensional normal distribution is utilized to represent the dependence of adjacent durations.

$$\begin{bmatrix} d_i \\ d_{i+1} \end{bmatrix} \sim N\left(T_e \begin{bmatrix} \mu_i \\ \mu_{i+1} \end{bmatrix}, T_e^2 \begin{bmatrix} \sigma_i^2 & \tau_{i,i+1} \\ \tau_{i,i+1} & \sigma_{i+1}^2 \end{bmatrix}\right) \tag{9.66}$$

where T_e represents the Taylor tool life. The Taylor tool life of training test p is denoted by T_e^p and the duration of wear state i in training test p is denoted by d_i^p. The total number of training tests is $K = 13$. The parameters in Eq. 9.66 are estimated as:

$$\hat{\mu}_i = \frac{1}{K} \sum_{p=1}^{K} \frac{d_i^p}{T_e^p} \tag{9.67}$$

$$\hat{\sigma}_i^2 = \frac{1}{K} \sum_{p=1}^{K} \left(\frac{d_i^p}{T_e^p} - \hat{\mu}_i \right)^2 \tag{9.68}$$

$$\hat{\tau}_{i,i+1} = \frac{1}{K} \sum_{p=1}^{K} \left(\frac{d_{i+1}^p}{T_e^p} - \hat{\mu}_{i+1} \right) \left(\frac{d_i^p}{T_e^p} - \hat{\mu}_i \right) \tag{9.69}$$

The discrete transition probability in (9.37) and initial probability in (9.39) could be obtained from (9.66). If the covariance $\tau_{i,i+1}$ in (9.69) is set as zero, the HSMM with dependent durations becomes the HSMM with independent durations. As a result, the HSMM with independent durations could be regarded as a special HSMM with dependent durations. Hence, the proposed on-line monitoring approach could also be applied when the HSMM with independent durations is utilized to model the tool wear process.

The estimated parameters have great influence on the diagnosis and prognosis results. If the expectation $\hat{\mu}_i$ is bigger than the real expectation, for most of the testing tests, the evolution of the estimated degradation state will lag behind the evolution of the real state and the estimated RUL will be larger than the real RUL. This will cause the problem of missing alarms. On the contrary, if the expectation is smaller than the real expectation, the estimated RUL will be smaller than the real one and the false alarm will occur. The covariance $\hat{\tau}_{i,i+1}$ represents the correlation between the durations. Positive (negative) covariance means positive (negative) correlation between durations. If the estimated covariance is contrary to the real covariance, the estimated RUL will seriously deviate from the real RUL. Variance $\hat{\sigma}_i^2$ reflects the confidence level of duration. Larger variance corresponds to smaller confidence. If the variance is set high, the information inferred from durations will have less influence on the diagnosis results and the diagnosis results will be determined by the observations more. If the variance is set small, the monitoring result will be more determined by the duration information and the estimated RUL will approximate to the prior RUL expectation.

As the HSMM is trained from the training set and the modeling accuracy is largely determined by the number of training samples, abundant training samples will be necessary and important for accurate modeling. In this study, the HSMM is trained from the 13 training tests, and it is sufficient to represent the 10 testing tests. This will further be verified by the monitoring results presented in the next section.

9.6.4 Diagnosis and Prognosis Results

(1) Diagnosis and Prognosis via HSMM Approach

In this section, the diagnosis and prognosis are shown for the on-line tool wear monitoring process via the HSMM method. At each cutting pass, based on the trained HSMM in the section above, the tool wear state and RUL are inferred from the discrete observations. The evolution of the diagnostic wear state is shown in Fig. 9.17. Figure 9.17 shows that utilizing duration information leads to more accurate diagnostic results and the HSMM with dependent durations has more accurate diagnostic results than HSMM with independent durations. This could also be verified by the recognition rate of all tests listed in Table 9.5.

Under the premise of correct estimation, higher confidence level shown as a more centralized distribution gives the decision-maker more confident diagnostic results. The evolution of the posterior distribution of state is shown in Fig. 9.18. From the figure, it could be found that the distribution of the estimated state is more centralized with dependent duration than that with independent durations when the wear state is estimated correctly (around pass 55). So it could be concluded that the fuller use of the dependent characteristic of adjacent durations leads to more confident diagnosis.

Figure 9.19 gives the evolution of the distribution of RUL at the passes in which

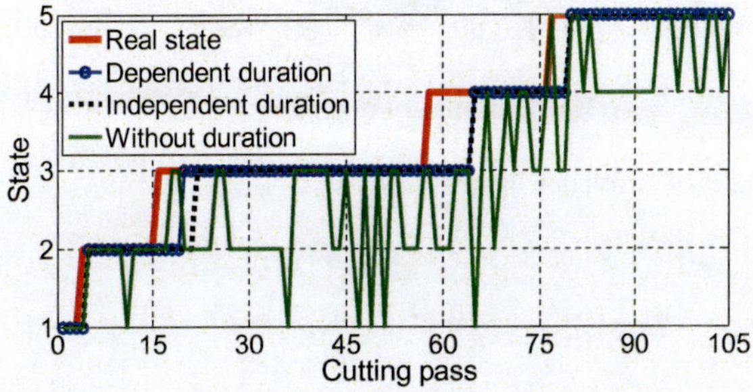

Fig. 9.17 Estimated wear state

Table 9.5 Recognition rate and ARER of all tests

Approach	State 1	State 2	State 3	State 4	State 5	Total	ARER
Dependent duration	0.8810	0.8773	0.9278	0.9270	0.9066	0.9143	0.1269
Independent duration	0.7619	0.7975	0.8763	0.7584	0.8132	0.8286	0.2986
Without duration	0.6429	0.6626	0.5711	0.5169	0.5659	0.5781	NA
SVM	0.5476	0.5890	0.8247	0.6067	0.5879	0.6990	NA
BPNN	0.4762	0.6196	0.8866	0.6348	0.6429	0.7431	NA

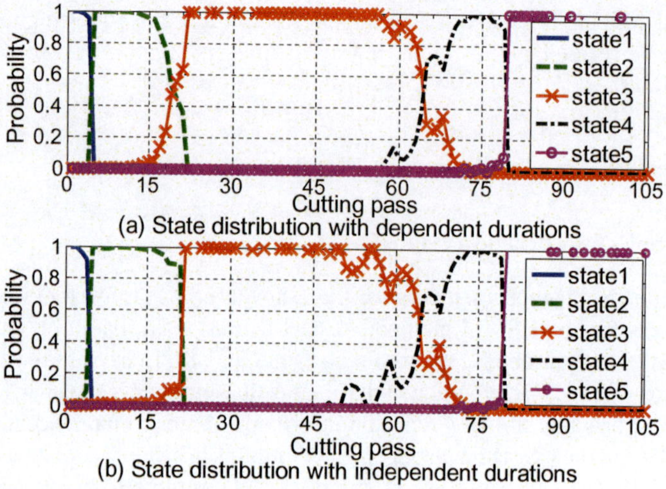

Fig. 9.18 Evolution of posterior probability distribution of wear state

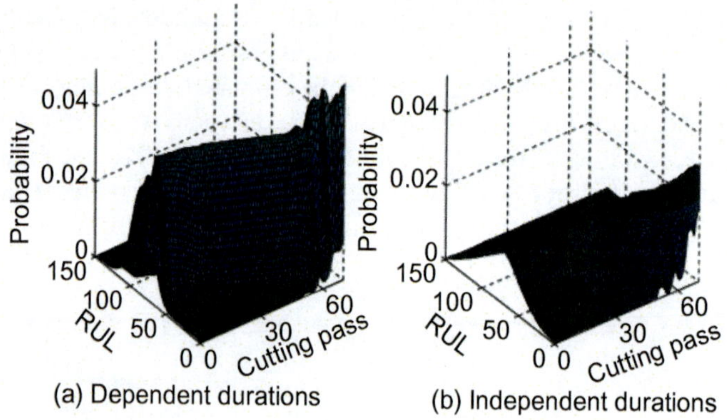

Fig. 9.19 Evolution of the distribution of RUL

the tool wear state is not estimated as the final state (when the estimated state is the final state, the probability that RUL is zero approximates to one). It shows that estimated RUL has a more centralized distribution with dependent duration than that with independent duration. Namely, it leads to a more confident RUL estimation with dependent durations. The estimated error of RUL which reflects the accuracy of RUL estimation is defined as the root of mean squared error. The evolutions of error and mean of estimated RUL are given in Fig. 9.20. It indicates that the RUL estimation with dependent durations is more accurate as the estimation error with dependent durations is less than that with independent durations. Figure 9.21 shows the evolution of optimal confidence interval at the 90% confidence level. It also shows that the RUL estimation with dependent durations is more confident as the width of confidence interval with dependent durations is narrower than that with independent durations. Therefore, we could say that RUL estimation becomes more accurate and more confident when the dependent relationship of durations of adjacent wear state is included in the HSMM for the tool wear process.

Figure 9.20 shows that the estimated RUL (mean RUL) is higher than the real RUL and missing alarms may occur. To eliminate this problem, the lower confidence

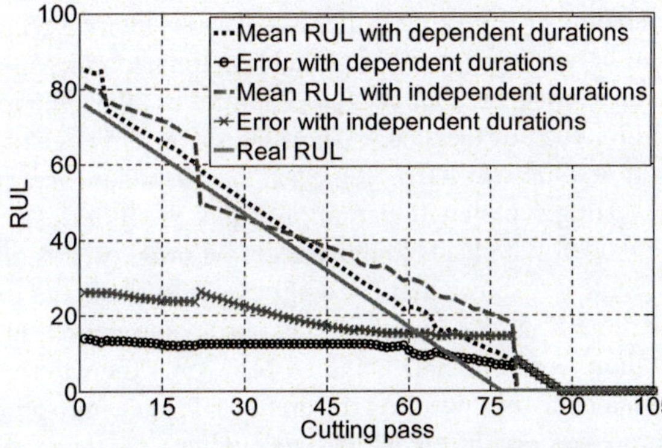

Fig. 9.20 The statistics of RUL

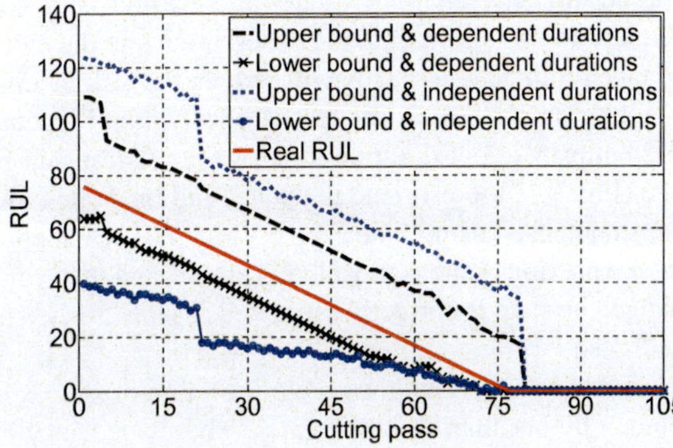

Fig. 9.21 Optimal confidence interval at 90% confidence level

bound of RUL is set as the estimated RUL after cutting pass 65 at which the estimated state becomes the last-to-second state (state 4). As the lower bound of RUL is smaller than the real RUL (Fig. 9.21 shows), the RUL of the tool will be estimated as zero before the tool actually reaches to the final wear state. In this way, the final tool wear state could be avoided and the process is relieved from loss due to worn tool.

(2) **Comparison to Pure Data-based Approach**

As classical pure-data monitoring methods, SVM and backpropagation neural network (BPNN) are applied to infer the tool wear state. The extracted wavelet packet features of cutting force are used as the input of SVM and BPNN. The training and testing sets used in SVM and BPNN approaches are the same as the sets used in the proposed HSMM approach. Generally, SVM is applied for binary classifications. To recognize 5 tool wear states, ten SVMs are trained, with each SVM separates two tool wear states. In the testing process, the wavelet feature is applied to all of the 10 SVMs, and each SVM outputs a state. Then the estimated state is set as the state which has maximum occurrence in the outputs of the 10 SVMs. With empirical studies, the trained BPNN has three hidden layers and the number of nodes at each layer is 32, 16, and 8.

As 10 tests are used as testing samples and each test has 105 cutting passes, the total number of testing samples is 1050. The total recognition rate is defined as the ratio between the number of states estimated correctly and the total number 1050. Table 9.5 lists the total recognition rates of all testing tests as well as the recognition rate for each of the states. The RUL error rate is defined as the ratio between RUL error and the real RUL. Table 9.5 also lists the average RUL error rate (ARER) of the testing samples. It also implies that the fuller use of the latent evolution law of the hidden state leads to more accurate prognosis results.

Table 9.5 shows that the diagnosis results via the data model-integrated method are more accurate than those via pure-data method. In pure data-based approaches (without duration, SVM, and BPNN), only the wavelet packet feature is utilized to estimate the tool wear. As different tool states may produce similar features and the feature space of different tool state overlaps, the recognition error would be high when overlapped wavelet packet feature occurs. HSMM approaches not only model the relationship between hidden states and features but also include the evolution model of states. In this study, both the observation and the evolution law of the hidden state are taken into account. This will reduce the risk of misclassification. For example, at cutting pass 90 of testing test 1, observation 9 occurs (shown in Fig. 9.15b), the distribution without duration utilizes the relationship between the observation and state (Fig. 9.16) to estimate the state, and the state is wrongly estimated as state 4. The same case occurs when the SVM or BPNN method is adopted. With the fact that the previous state is most likely state 5 and the evolution law that state 5 will not change, the state is correctly estimated as state 5 when the proposed HSMM method is adopted.

Table 9.5 shows that the recognition rate of state 3 in SVM and BPNN methods is much higher than in other states. This is because there are much more training samples of state 3 in the training set. On the contrary, the other four states could not be efficiently detected by the trained SVM and BPNN as there are much fewer training samples of these four states. While the generative model is used as the observation model and the duration of the state is included, the HSMM method is less influenced by the number of training samples in different states. This could be verified by the fact that the recognition rate for different states does not vary much when the HSMM method is adopted.

(3) **On the Adaptability of the Proposed HSMM Approac**

The experimental results imply that the proposed HSMM approach is applicable in progressive tool wear monitoring and tool life estimation. As the HSMM is built from the historical data set, the abundant historical data set is necessary for an accurate diagnosis and prognosis. For other applications with progressive degradation processes, the proposed HSMM approach is applicable provided with enough training data. It is noted that the degradation process is assumed to change progressively in the model and the proposed HSMM is not applicable in monitoring the degradation process with abrupt changes such as chipping of tools and the fracture of bearing.

In this study, an improved hidden semi-Markov process is proposed to model the degradation process with dependent durations. A modified forward–backward algorithm which can avoid underflow problem is also developed. And based on the modified forward–backward algorithm, on-line degradation state estimation and remaining useful prediction could be conducted conveniently. Case studies have shown that including the dependence of durations in the HSMM, it leads to more accurate and more confident estimation results, with the increase of both the state recognition rate and RUL estimation. And notably, the confidence of RUL has a great improvement. In further work, the jumping of HSMM states should be investigated in HSMM so that the abrupt changes of equipment conditions could also be modeled and predicted.

References

1. King RI (ed) (1985) Handbook of high-speed machining technology. Chapman and Hall, New York
2. Ikawa N, Shimada S, Tanaka H (1992) Minimum thickness of cut in micro-machining. Nanotechnology 3:6–9
3. Vogler MP, De Vor RE, Kapoor SG (2004) On the modeling and analysis of machining performance in micro-end milling, part II: cutting force prediction. J Manuf Sci Eng 126(4):694–704
4. Liu K (2005) Process modeling of micro-cutting including strain gradient effects. PhD thesis, Woodruff School of Mechanical Engineering, Georgia Institute of Technology
5. Jun MB (2005) Modeling and analysis of micro-end milling dynamics. PhD dissertation, University of Illinois Urbana-Champaign
6. Kim J-D, Kim DS (1995) Theoretical analysis of micro-cutting characteristics in ultra-precision machining. J Mater Process Technol 49:387–398
7. Tansel IN, Rodriguez O, Trujillo M, Paz E, Li W (1998) Micro-end-milling—I. wear and breakage. Int J Mach Tools Manuf 38:1419–1436

8. Tansel IN, Arkan TT, Bao WY, Mahendrakar N, Shisler B, Smith D, McCool M (2000) Tool wear estimation in micro-machining. Part I: tool usage–cutting force relationship. Int J Mach Tools Manuf 40:599–608
9. Zhu KP, Hong GS, Wong YS (2006) Cutting force denoising in micro-milling tool condition monitoring. Int Jo Prod Res. In Press
10. International Standard Organization (ISO 8688-2: 1989) Tool life testing in milling-Part 2: end milling
11. Altintas Y, Yellowley I (1989) The process detection of tool failure in milling using cutting force models. ASME J Eng Ind 111:149–157
12. Byrne G, Dornfeld D, Inasaki I, Konig W, Teti R (1995) Tool condition monitoring (TCM)—the status of research and industrial application. Ann CIRP 44(2):541–567
13. Wu Y, Escande P, Du R (2001) A new method for real-time tool condition monitoring in transfer machining stations. ASME J Manuf Sci Eng 123:339–347
14. Dimla DE Snr (2000) Sensor signal for tool-wear monitoring in metal cutting operations—a review of methods. Int J Mach Tools Manuf 40:1073–1098
15. Hong GS, Rahman M, Zhou Q (1996) Using neural network for tool condition monitoring based on wavelet decomposition. Int J Mach Tools Manuf 36(5):551–566
16. Saglam H, Unuvar A (2003) Tool condition monitoring in milling on cutting forces by a neural network. Int J Prod Res 41(7):1519–1532
17. Niu YM, Wong YS, Hong GS (1998) Multi-category classification of tool conditions using wavelet packets and ART2 network. ASME J Manuf Sci Technol 120:807–815
18. Obikawa T, Shinozuka J (2004) Monitoring of flank wear of coated tools in high speed machining with a neural network ART2. Int J Mach Tools Manuf 44:1311–1318
19. Sun J, Hong GS, Rahman M, Wong YS (2004) Identification of feature set for effective tool condition monitoring by acoustic emission sensing. Int J Prod Res 42(5):901–918
20. Wang WH, Hong GS, Wong YS, Zhu KP. Sensor fusion for on-line tool condition monitoring in milling. Int J Prod Res. In Press
21. Rabiner LR (1989) A tutorial on hidden Markov models and selected applications in speech recognition. Proc IEEE 77(2):257–286
22. Li D, Douglas OS (2003) Speech processing: a dynamic and object-oriented approach. Marcel Dekker, Inc
23. Bilmes J (2002) What HMMs can do? Technical report, University of Washington
24. Heck LP, McClellan JH (1991) Mechanical system monitoring using HMMs. In: IEEE international conference on acoustics, speech, and signal processing, vol 3, pp 1697–1700
25. Errtunc HM, Looparo KA, Ocak H (2001) Tool wear condition monitoring in drilling operations using hidden Marko modes. Int J Mach Tools Manuf 41:1363–1384
26. Baruah P, Chinnam RB (2005) HMMs for diagnostics and prognostics in machining processes. Int J Prod Res 43(6):1275–1293
27. Wang L, Mehrabi MG, Elijah KA (2002) Hidden Markov model-based tool wear monitoring in turning. ASME J Manuf Sci Eng 124:652–658
28. Scheffer C, Engelbrecht H, Heyns PS (2005) A comparative evaluation of neural networks and hidden Markov models for monitoring turning tool wear. Neural Comput Appl 14:325–336
29. Owsley LMD, Atlas LE, Bernard GD (1997) Self-organizing feature maps and Hidden Markov models for machine tool monitoring. IEEE Trans Signal Process 45(11):2787–2798
30. Atlas L, Ostendorf M, Bernard G (2000) Hidden Markov models for monitoring machining tool-wear. In: Proceedings of ICASSP, pp 3887--3890
31. Randall KF, Ostendorf M, Bernard GD, Castanon DA (2003) Multilevel classification of milling tool wear with confidence estimation. IEEE Trans Pattern Anal Mach Intell 25(1)
32. Cetin O, Ostendorf M (2004) Multi-rate Hidden Markov models and their application to machine tool-wear classification. In: The 2004 IEEE international conference on acoustics, speech, and signal processing (ICASSP 2004), pp 837–840
33. Vallejo Jr AG, Nolazco-Flores JA, Morales-Men´endez R, Enrique Sucar L, Rodr´ıguez CA (2005) Tool-wear monitoring based on continuous Hidden Markov models. In: Lazo M, Sanfeliu A (eds) CIARP 2005, LNCS 3773. Springer, Berlin, Heidelberg, pp 880–890

34. Kassim AA, Mian Z, Mannan MA (2006) Tool condition classification using Hidden Markov Model based on fractal analysis of machined surface textures. Mach Vis Appl 17:327–336
35. Konig W, Kutzner K, Schehl U (1992) Tool monitoring of small drills with acoustic emission. Int J Mach Tools Manuf 32:487493
36. Tansel IN, Mekdeci C, Rodriguez O, Uragun B (1993) Monitoring drill conditions with wavelet based encoding and neural network. Int J Mach Tools Manuf 33(4):559–575
37. Trujillo M, Li W, Fallerio B, Paz E, Tansel IN (1994) Inspection of Micro-tools at high rotational speeds. Int J Mach Tools Manuf 34:1059–1077
38. Tansel IN, Rodriguez O, Trujillo M, Paz E, Li W (1998) Micro-end-milling—III. wear estimation and tool breakage detection using acoustic emission signals. Int J Mach Tools Manuf 38:1449–1466
39. Bao WY, Tansel IN (2000) Modeling micro-end-milling operations. Part I: analytical cutting force model. Int J Mach Tools Manuf 40:2155–2173
40. Bao WY, Tansel IN. Modeling micro-end-milling operations. Part III: influence of tool wear. Int J Mach Tools Manuf 40:2193–2211
41. Tansel IN, Bao WY, Reen NS, Kropas-Hughes CV (2005) Genetic tool monitor (GTM) for micro-end-milling operations. Int J Mach Tools Manuf 45:293–299
42. Zhu KP, Wong YS, Hong GS (2009) Multi-category micro-milling tool wear classification with continuous Hidden Markov models. Mech Syst Signal Process 23:547–560
43. Bouchard G, Celeux G (2006) Selection of generative models in classification. IEEE Trans Pattern Anal Mach Intell 28(4):544–554
44. Theodoridis S, Koutroumbas K (2009) Pattern recognition. Academic Press
45. Fraley C, Raftery AE (1998) How many clusters? Which clustering method?—answers via model-based cluster analysis. Comput J 41:578–588
46. Athanasopoulou E, Li L, Hadjicostis CN (2009) Maximum likelihood failure diagnosis in finite state machines under unreliable observations. IEEE Trans Autom Control 55(3):579–593
47. Lee S, Li L, Ni J (2009) Online degradation assessment and adaptive fault detection using modified hidden Markov model. J Manuf Sci Eng Trans ASME 132(2):1–11
48. Chen C, Zhang B, Vachtsevanos G, Orchard M (2011) Machine condition prediction based on adaptive neuro–fuzzy and high-order particle filtering. IEEE Trans Ind Electron 58(9):4353–4364
49. Zhu JL, Ge Z, Song Z (2015) HMM-driven robust probabilistic principal component analyzer for dynamic process fault classification. IEEE Trans Ind Electron 62(6):3814–3821
50. Geramifard O, Xu JX, Zhou JH, Li X (2012) A physically segmented hidden Markov model approach for continuous tool condition monitoring: Diagnostics and prognostics. IEEE Trans Ind Inf 8(4):964–973
51. Geramifard O, Xu JX, Zhou JH, Li X (2014) Multimodal hidden Markov model-based approach for tool wear monitoring. IEEE Trans Ind Electron 61(6):2900–2911
52. Nakamura H, Chihara M, Inoki T, Higaki T, Okuda K, Mizuno Y (2009) Impulse testing for detection of insulation failure of motor winding and diagnosis based on Hidden Markov Model, IEEE Trns. Dielectr. Electr. Insul, vol. 17, no.5, pp.1619–1627. Oct. 209.
53. Ying J, Kirubarajan T, Pattipati KR, Patterson-Hine A (2002) A hidden Markov model-based algorithm for fault diagnosis with partial and imperfect tests. IEEE Trans Syst Man Cybern Part C Appl Rev 30(4):463–473
54. Camci F, Chinnam RB (2009) Health-state estimation and prognostics in machining processes. IEEE Trans Autom Sci Eng 7(3):581–591
55. Liu QM, Dong M, Lv WY, Geng XL, Li YP (2015) A novel method using adaptive hidden semi-Markov model for multi-sensor monitoring equipment health prognosis. Mech Syst Signal Process 64–65:217–232
56. Tobon-Mejia DA, Maderjaher K, Zerhouni N, Tripot G (2012) A data-driven failure prognostics method based on mixture of Gaussians hidden Markov models. IEEE Trans Reliab 61(2):491–503
57. Geramifard O, Xu JX et al (2011) Continuous health condition monitoring: a single hidden semi-Markov model approach. In: Proceedings of IEEE international conference on prognostics and health management (PHM), pp 1–9

58. Zhu KP, Liu T (2018) On-line high-speed milling tool wear monitoring via hidden semi-Markov model with dependent durations. IEEE Trans Ind Inf 14(1):69–78
59. Peng Y, Dong M (2011) A prognosis method using age-dependent hidden semi-Markov model for equipment health prediction. Mech Syst Signal Process 25(1):237–252
60. Wang N, Sun SD, Cai ZQ, Zhang S, Saygin C (2014) A hidden semi-Markov model with duration-dependent state transition probabilities for prognostics. Math Probl Eng 1–10
61. Yu SZ (2009) Hidden semi-Markov models. Artif Intell 174(2):215–243
62. Gersho A, Gray RM (2012) Vector quantization and signal compression. Springer
63. Yu SZ, Kobayashi H (2006) Practical implementation of an efficient forward-backward algorithm for an explicit-duration hidden semi-Markov model. IEEE Trans Signal Process 54(5):1947–1951
64. Si XS, Wang W, Hu CH, Zhou DH (2011) Remaining useful life estimation—a review on the statistical data driven approaches. Eur J Oper Res 213(1):1–14
65. Doucet A, Freitas ND, Gordon N (2001) An introduction to sequential Monte Carlo methods. Sequential Monte Carlo methods in practice. Springer, New York
66. Mallat SG (1998) A wavelet tour of signal processing. Academic
67. Jain AK. Data clustering: 50 years beyond K-means. Pattern Recogn Lett 31(8):651–666

Chapter 10
Sensor Fusion in Machining System Monitoring

10.1 Multi-sensor Information Fusion Principle

Multi-sensor information fusion utilizes multi-sensory sources and extract complementary information according to certain criteria, so that the information system obtained has a superior performance than its constituent components [1, 2]. It was first proposed by the Joint Directors of Laboratories (JDL) of United States in 1994 [1]. It defined sensor fusion as a multi-level and multi-faceted process, including detection, correlation, combination, and estimation of multi-source data, thereby improving the accuracy of state and characteristic estimation. It could carry out a timely and complete evaluation of the battlefield situation, threats, and important procedures. There is a fundamental difference between multi-sensor information fusion and classical signal processing methods. The key is that the information fusion is handled by the multi-sensor information in a more complex form, and can appear at different levels of information [2].

The basic strategy of fusion is to first integrate the information on the same level, to obtain higher level of information, and then import the corresponding level of information fusion. In general, the multi-sensor fusion is essentially a low-level to top-level integration of multiple information, layer-by-layer information processing process. It is generally a three-layer fusion structure, including data layer fusion, feature layer fusion, and decision layer fusion, as shown in Fig. 10.1.

(1) **Data layer fusion**: First, data fusion of all sensor observation data is carried out. After that, the feature vector is extracted from the fused data and identified. At this time, the sensors are required to be homogeneous (the sensor observes the same physical phenomenon). If multiple sensors are heterogeneous (the observation is not the same physical quantity), then the data can only be fused at the feature layer or decision layer. Data layer fusion does not have the problem of data loss, and the results obtained are the most accurate, but the requirements for system communication bandwidth are very high.

(2) **Feature layer fusion**: Each sensor provides representative features extracted from the observation data, fuse these features into a single feature vector, and

K. Zhu, *Smart Machining Systems*, Springer Series in Advanced Manufacturing,
https://doi.org/10.1007/978-3-030-87878-8_10

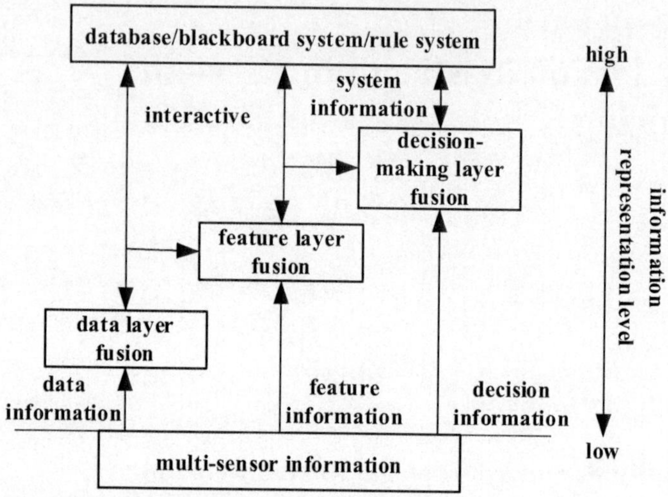

Fig. 10.1 Hierarchical model of fusion system [1]

then use the method of pattern recognition for processing. This method has lower requirements for communication bandwidth, but its accuracy is reduced due to data loss.

(3) **Decision layer fusion**: Decision layer fusion refers to the fusion of the recognition results of multiple sensors after each sensor recognizes the target. Due to the concentration of sensor data, the results obtained in this way are relatively inaccurate, but it requires the lowest communication bandwidth.

For the specific engineering application of the multi-sensor fusion system, factors such as the performance of the sensor, the computing power of the system, the communication bandwidth, the expected accuracy rate, and the financial capacity should be considered comprehensively to determine which level of fusion to choose. In addition, in a system, it is also possible to merge at different levels at the same time.

10.2 Multi-sensor Information Fusion with Neural Networks

In a multi-sensor information fusion system, the information provided by each sensor is generally incomplete, inaccurate, or even contradictory, and contains a lot of uncertainty. The information fusion should make inferences based on this uncertain information to achieve the purpose of process monitoring and state recognition. Therefore, uncertainty reasoning is the basis of state recognition and attribute information fusion.

Multi-sensor information fusion can be divided into numerical fusion and non-numerical fusion in the method. Numerical fusion solves the quantitative description of the system, that is, how to obtain a unified result from a set of related data to improve the accuracy of system measurement; non-numerical fusion gives the qualitative expression or decision of the system.

Currently, the commonly used information fusion methods are as follows.

(1) **Weighted average method**: The simplest and most intuitive method of fusing low-level data from multiple sensors. This method takes a weighted average of the redundant information provided by a group of sensor units and uses the result as the information fusion value.
(2) **Kalman filtering method**: For real-time fusion of dynamic low-level redundant multi-tactile group unit data, the method uses the statistical characteristics of the measurement model to recursively determine the optimal fusion data estimate in a statistical sense. If the system has a linear dynamic model, and the system noise and sensor noise are Gaussian distributed white noise models, then the Kalman filter fusion data provides the only statistically optimal estimate.
(3) **Bayesian estimation method**: A common method for fusing low-level information of multiple sensor units in a static environment. The information is described as a probability distribution, which is suitable for uncertain information with Gaussian noise.
(4) **Neural network method**: According to the similarity of the samples received by the current system, the classification standard can be determined, and the specific learning algorithm can be used to obtain knowledge, and the uncertainty reasoning mechanism can be obtained.

Neural networks use a large number of simple processing units (neurons) to process information. Neurons are organized in a hierarchical structure. Neurons on each layer are connected to neurons on other layers in a weighted manner, using the parallel structure and parallel processing mechanisms. Therefore, the network has strong fault tolerance and self-learning, self-organization and self-adapting capabilities, and can imitate complex nonlinear mapping. These characteristics and powerful nonlinear processing capabilities of neural networks happen to meet the requirements of multi-sensor information fusion. The signal processing and automatic reasoning functions of neural networks can be used to realize multi-sensor information fusion.

When a NN is applied for multi-sensor information, it has many steps. First of all, a suitable neural network model must be selected according to the requirements of the system and the characteristics of the sensor. Then, according to the existing multi-sensor information and system fusion knowledge, a certain learning method is adopted to perform offline learning on the established neural network system to determine the connection weight and connection structure of the network. Finally, the obtained network is used in actual information fusion. In the application of neural networks, the self-learning and self-organizing functions of the network should be used to continuously learn new knowledge of information fusion from practical applications, adjust its structure and weights to meet the real-time requirements of the constantly changing detection environment, and improve information fusion reliability.

Ghosh et al. [3] used BP neural network to analyze the multi-sensor signals collected by force sensors, current sensors, vibration sensors, sound sensors, and acoustic emission sensors to estimate the average wear of the tool flank; Rizal et al. [4] collected signals of cutting force, torque, vibration and cutting temperature in three directions, and optimized the prediction value of tool wear based on Taguchi method; Zhang et al. [5] proposed a multi-sensor tool condition monitoring method based on wireless interactive technology, which took power, vibration, actual machining parameters and cutting force as the collection objects, and realized tool wear recognition based on multi-source data.

In terms of multi-sensor information fusion at the decision-making level, many scholars have proposed different methods to meet their own problem needs. The more popular decision-making methods are Dempster-Shafer (DS), probability analysis method, subjective Bayesian reasoning method, probability analysis method, template matching method, and neural network methods. Ceng et al. [6] used a self-organizing feature mapping network combined with D-S evidence theory to diagnose gear multi-sensor fault signals based on extracting wavelet correlation scale entropy. Duro et al. proposed a multi-sensor fusion framework to distinguish the best sensor locations for monitoring cutting operations and identify the best signal of various sensors [7]. More studies are investigated in manufacturing process monitoring [8, 9].

Wang et al. [10] combined a direct sensor (vision) and an indirect sensor (force) to create an intelligent integrated tool condition monitoring (TCM) system for on-line monitoring of flank wear and breakage in milling, using the complementary strengths of the two types of sensors. For flank wear, images of the tool are captured and processed in-cycle using successive moving image analysis. Two features of the cutting force, which closely indicate flank wear, are extracted in-process and appropriately pre-processed. In-cycle refers to periodic, occurring between machining passes, tasks, or during part changeovers. An example of a machining task is a geometric feature to be machined, such as the machining of a region of a cavity to a certain depth. In-process, on the other hand, refers to machining that is in progress, and therefore, when the tool is engaging the workpiece.

A self-organizing map (SOM) network is trained in a batch mode after each cutting pass, using the two features derived from the cutting force, and measured wear values obtained by interpolating the vision-based measurement. The trained SOM network is applied to the succeeding machining pass to estimate the flank wear in-process. The in-cycle and in-process procedures are employed alternatively for the on-line monitoring of the flank wear. To detect breakage, two features in the time domain derived from cutting force are used and the thresholds for them are determined dynamically. Again, vision is used to verify any breakage identified in-process through the cutting force monitoring. Experimental results show that this sensor fusion scheme is feasible and effective for the implementation of on-line tool condition monitoring in milling, and is independent of the cutting conditions used. The overall scheme is shown in Fig. 10.2.

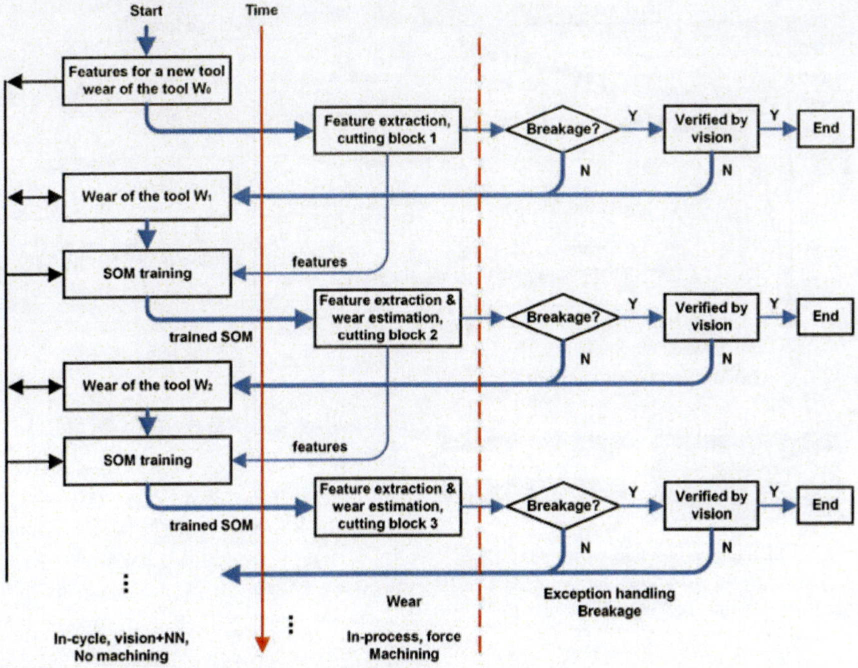

Fig. 10.2 Scheme of the sensor fusion method

Various cutting conditions and two types of inserts were tested [10], and one case study was shown in Fig. 10.3.

Figure 10.3a shows the wear estimation result throughout the entire test. (b) shows the breakage result inspected by vision, with the grey-level image of the insert on the left and the corresponding binary image on the upper right while the extracted breakage zone is shown on the lower right. (c) shows the force and force features, i.e. residual error and peak rate against time, superimposed with the thresholds (outlined by two horizontal lines) being equal to the maximum amplitudes of the features extracted in the previous pass. It has been concluded from the results [9] that the flank wear estimation result was good in association with the in-cycle measurement by vision, especially at the linear wear stage, although there was deviation at the initial wear stage. Tool breakage was successfully detected by vision, and the breakage can also be detected by force features using dynamic thresholding. Additionally, the residual error and peak rate are two complementary features to detect breakage.

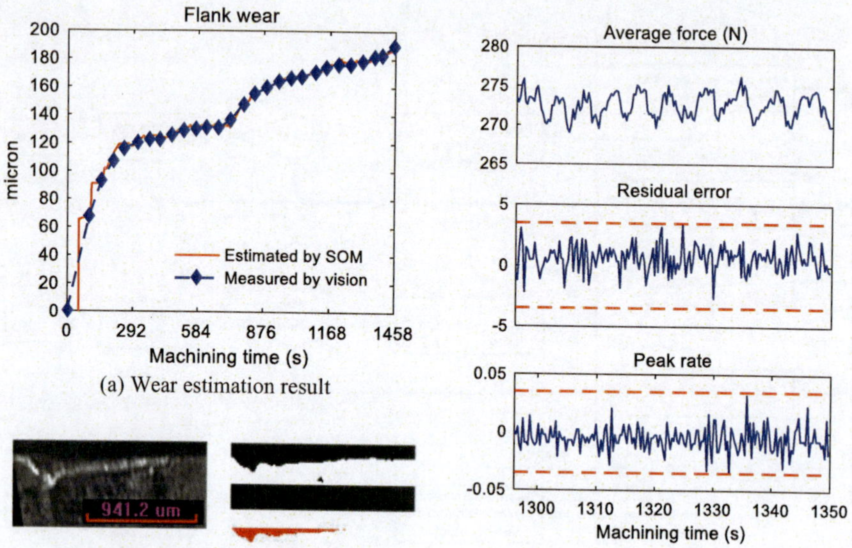

(a) Wear estimation result

(b) Breakage by vision after pass 27

(c) Force features in pass 27

Fig. 10.3 On-line TCM result for test 1 [10]

10.3 Sensor Fusion with Deep Learning

In the previous section, the SOM model combined a vision sensor and a force sensor to create an intelligent integrated TCM system for milling, using the complementary strengths of the two types of sensors. However, in this scheme, one needs to consider not only the relationship between various sensor signals and tool wear but also the relationship between different sensor signals. It is unavoidable to face the mapping difficulty, calculation load, and over-fitting risk which are similar to the problems caused by signal duration extension.

In recent years, with the rapid development of modern computing resources, deep learning has become a popular method in the prediction field due to its self-learning, complex non-linear mapping, and feature extraction ability, which is hopeful to solve the above problem. Deep learning generally uses multiple-layer neural networks to represent the data, in which the depth of layers corresponds to the levels of information extraction and feature representation. It has shown much better performances than traditional machine learning approaches in problem adaptability, network extension, and complex feature representation [11]. In the full scope of deep learning, deep belief network (DBN) [12], deep auto-encoder (DAE) [13], convolutional neural network (CNN) [14], and recurrent neural network (RNN) [15] constitute the fundamental models, and most follow-up studies are based on these or the modified models [16]. According to the training dependence on annotated data, DBN, DAE, and their derivatives are mainly unsupervised learning or semi-supervised learning, while CNN, RNN, and their derivatives are mainly supervised learning.

Of these deep learning methods, deep belief networks (DBN) [17] and stacked sparse auto encoders (SAE) [18] were applied to sensor fusion for tool wear monitoring, where the benefits of deep networks were not fully exploited due to their limitations in processing high nonstationary machining signals like cutting force and vibration, and prediction inadequacy of the long-term signal dependency. The long short-term memory network (LSTM) stands to be prominent in processing temporal series signals, and most suitable for cutting force and vibration signal in machining. The LSTM is a recurrent neural network (RNN) architecture originated for language processing [19]. Unlike conventional feedforward neural networks, LSTM has temporal connections. Li et al. trained an LSTM auto-encoder to preserve and hierarchically reconstruct multi-sentence paragraphs [20]. Chan et al. presented a pyramid LSTM sequence-to-sequence model to transcribe speech utterances directly to characters [21]. LSTM has been used to monitor tool wear in recent years. Zhang et al. proposed a bi-directional LSTM architecture specialized in discovering the underlying patterns embedded in temporal sequence to track the system degradation and consequently, to predict the RUL [22].

But LSTM can only analyze the current signal in each cycle, not as efficient as ANN in computing parallelism which makes the information of shallow layer difficult to transmit to the output. And the error also needs to transfer through several cycles from the output to the input during error back propagation. Capturing long-term trends by extending the length of the monitoring signal undoubtedly make the problem more serious. We need to consider not only the depth of the network but also the length of the signal sequence. It is easier to appear gradient explosion or disappear and difficult to form an information processing channel. To solve these problems, Zhao et al. designed a convolutional bi-directional LSTM which firstly extracts local features by convolutional neural network (CNN) and then encodes temporal information by bi-directional LSTM [23]. This method subtly reduces the computational pressure of LSTM by extracting local features from CNN, but it should be pointed out that CNN essentially treats signals as one-dimensional images and extracts spatial features rather than temporal features. Wang et al. developed an extended Kalman filter-based RNN training approach with a controllable training convergence to solve the problem of RNN convergence [24]. But in this method, the parameters need to adjust according to milling conditions artificially and the signal length is still limited.

In this study, a pyramid LSTM auto-encoder is proposed to extract features from long-term multi-sensor signals which can be generalized to unknown milling conditions as illustrated in Fig. 10.4. First, the periodic features are extracted layer by layer from the segmented multi-sensor signal by the pyramid encoder. Then the extracted features are sent to the predictor to map tool wear on the one hand and sent to the decoder symmetrical to the encoder to reconstruct the input on the other hand. The decoder based on unsupervised learning can improve the generality of the model under various machining conditions through regularization and prior knowledge introduction.

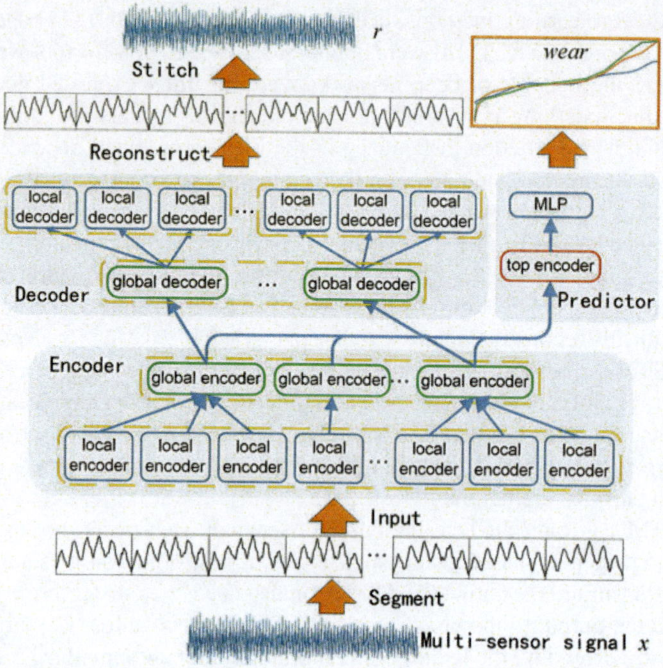

Fig. 10.4 Tool wears monitoring framework of 3-layer pyramid LSTM auto-encoder

10.3.1 Problem Formulation

The problem of tool wear monitoring can be regarded as a mapping from the monitoring signal space X to the tool wear space Ψ: $X \rightarrow \Psi$. In this approach, the temporal features y of the signal x is extracted by the pyramid LSTM layer by layer. In the k-th layer of the pyramid LSTM encoder, the long-term input $x(k) = [x_1(k), x_2(k), \ldots x_n(k)]$ with length n is first decomposed into several short-term signals $x_{(1)}(k), x_{(2)}(k), \ldots x_{(n/m)}(k)$ with length m according to the frequency spectrum of the input. The i-th short-term signal of the k-th layer can be expressed as

$$x_{(i)}(k) = [x_{(i,1)}(k), x_{(i,2)}(k), \ldots x_{(i,m)}(k)], 1 < i < n/m \tag{10.1}$$

The features extracted by the k-th layer from the i-th short-term signal through iteration at time j are

$$y_{(i,j)}(k) = f(y_{(i-1,j)}(k), x_{(i,j)}(k)), 1 < i < n/m, 1 < j < m \tag{10.2}$$

The features extracted from the i-th short-term signal are

$$y_{(i)}(k) = y_{(i,m)}(k) \tag{10.3}$$

The features extracted from the k-th layer are the combination of multiple features of the short-term signal, which can be expressed as

$$y(k) = [y_{(1)}(k), y_{(2)}(k), \ldots y_{(n/m)}(k)] \tag{10.4}$$

After each layer, the long-term signal with length n is compressed into the features with length n/m. The features y extracted from the last layer of encoder input to the predictor to map the tool wear:

$$s = \varphi(y) \tag{10.5}$$

In addition, the features y also input to the decoder to perform the inverse process symmetrical to the encoder G:

$$[r_1, r_2, \ldots r_n] = G(y) \tag{10.6}$$

The entire model is a multi-task system composed of tool wear monitoring and signal reconstruction.

$$[p, r] = \Psi(x) \tag{10.7}$$

where p represents the tool wear prediction and Ψ represents the mapping function of the entire model. The loss function L of the entire model consists of two parts: tool wear prediction error and reconstruction error,

$$L = \sum_i (s_i - p_i)^2 + \lambda \sum_i (r_i - x_i)^2 \tag{10.8}$$

where s represents the actual tool wear value and λ is used to adjust the weight of reconstruction error.

10.3.2 The Unit of Pyramid LSTM Auto-encoder

In the proposed model, the LSTM unit is a sub-network specially designed for temporal signal processing. The structure and symbolic meaning of traditional LSTM units as described in [25]. In order to match the model to the tool wear monitoring task, it is necessary to improve the ability of the LSTM unit in processing long-term signals. A reasonable method is to reduce the number of iterations required in long-term signal processing. In the pyramid LSTM encoder, the shallow layer tends to the periodic information compression. As the inverse operation of the encoder,

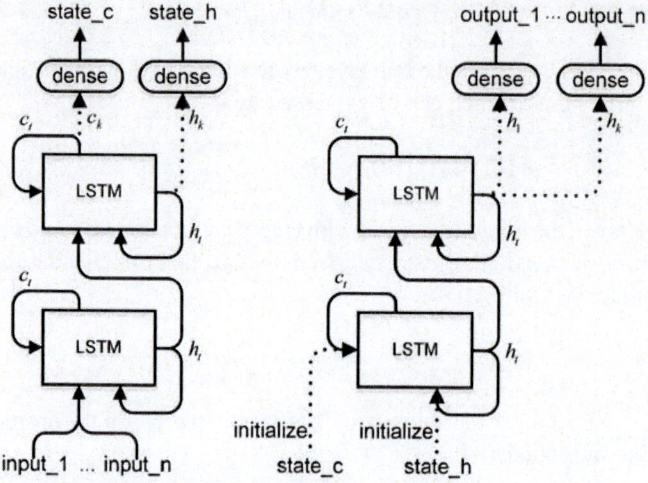

(a) Encoder unit (b) Decoder unit

Fig. 10.5 The structure diagrams of encoder unit and decoder unit

the decoder is more similar to the decompression process. Based on the traditional LSTM unit, the structure of the encoder and decoder unit are modified, which can be summarized as follows:

First, the encoder unit and decoder unit are both defined as a sub-model including a two-layer LSTM to analyze temporal information and a fully connected (FC) layer to make up for the shortcomings in non-sequential processing. The encoder unit is a many-to-one sub-model to extract temporal features from a period of signal or features, as shown in Fig. 10.5a. In contrast, the decoder unit is a one-to-many sub-model that takes the compressed features as input and spits out h_t in each cycle as shown in Fig. 10.5b.

Second, the encoder unit outputs h_k and c_k simultaneously. The traditional encoder only takes the last state h_k as the output [26, 27]. However, the c_k corresponding to long-term memory can better reflect the essence of information compression [28]. As a result, taking these two temporal features can better reflect the short-term and long-term features simultaneously.

Third, the input mode of the decoder unit is modified as shown in Fig. 10.6. In the classical LSTM decoder, the state h and c are initialized to 0 and the input is entered through the input channel as shown in Fig. 10.6a. In the first cycle under classical setting,

$$c_1 = g(W_z x_1 + b_z) \odot \sigma(W_i x_1 + b_i) \tag{10.9}$$

$$h_1 = g(c_1) \odot \sigma(W_o x_1 + b_o) \tag{10.10}$$

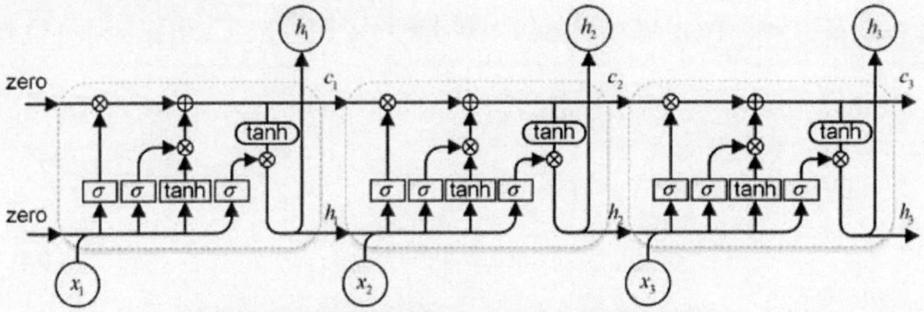

(a) The structure of decoder unit with input channel.

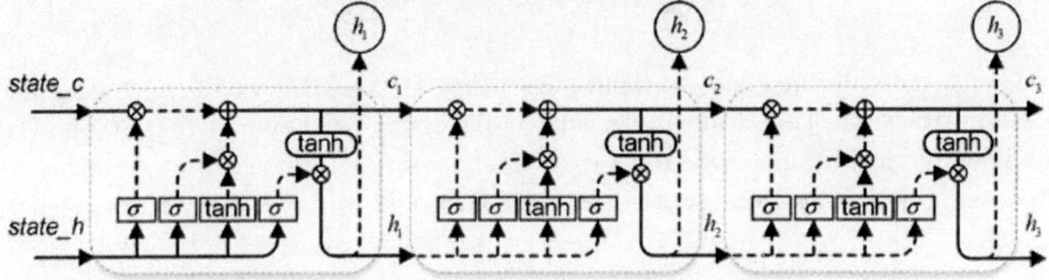

(b) The structure of decoder unit without input channel.

Fig. 10.6 Comparison of decoder units with different structures. The solid lines represent the variable flow that determines the output in each cycle. After the first cycle, no matter which mode is adopted, the information determining the output only

where W is the weight of LSTM unit, $\odot$ represents dot product, and $g(.)$ and $\sigma(.)$ are tanh and *sigmoid* activation functions respectively. In the following cycles, if we choose to input only in the first cycle x_1 and zero of the rest $x_2, \ldots x_m$, then:

$$c_t = g(R_z h_{t-1} + b_z) \odot \sigma(R_i h_{t-1} + b_i) + c_{t-1} \odot \sigma(R_f h_{t-1} + b_f) \tag{10.11}$$

$$h_t = g(c_t) \odot \sigma(R_o h_{t-1} + b_o) \tag{10.12}$$

Or, we choose to repeat the same input in each cycle:

$$\begin{aligned} c_t = g(W_z x_1 + R_z h_{t-1} + b_z) \odot \sigma(W_i x_1 + R_i h_{t-1} + b_i) \\ + c_{t-1} \odot \sigma(W_f x_1 + R_f h_{t-1} + b_f) \end{aligned} \tag{10.13}$$

$$h_t = g(c_t) \odot \sigma(W_o x_1 + R_o h_{t-1} + b_o) \tag{10.14}$$

As can be seen above, no matter which mode is used, the information comes from the input of the first cycle and transmit through h_t and c_t, as shown in Fig. 10.6a, x_t remains unchanged in the followed cycles. It involves in the model as an intermediate but a constant term, while it always needs to be re-calculated. Based on this observation and in order to improve the computing speed, we take the input to initialize h_t and c_t while discard the input channel. As a result, the first cycle is redefined as:

$$c_1 = g(b_z) \odot \sigma(b_i) + c_0 \odot \sigma(b_f) \tag{10.15}$$

$$h_1 = g(c_t) \odot \sigma(b_o) \tag{10.16}$$

And then in the following cycle, it gets:

$$c_t = g(R_z h_{t-1} + b_z) \odot i_t + c_{t-1} \odot f_t \tag{10.17}$$

$$h_t = g(c_t) \odot \sigma(R_o h_{t-1} + b_o) \tag{10.18}$$

Compared with the classical input model Eqs. 10.9–10.14, the decisive information transmission in the following cycles is the same, as shown in Fig. 10.6b. The modified model eliminates the redundant input operations which can speed up signal processing and guarantee the continuity of information flow from the encoder to decoder.

10.3.3 The Structure of the Pyramid LSTM Auto-encoder

In this study, the model is constructed in a modular way based on the frequency spectrum of the tool wear monitoring signal. There are three components contained in the model: encoder, decoder, and predictor. The framework of the model is shown in Fig. 10.4.

The encoder is a 2-layer pyramid network (4-layer LSTM) to extract features from the signal which match the harmonic characteristics of the signal shown by the frequency spectrum. The input length of each local encoder corresponds to the minimum period of the signal and the input length of each global encoder corresponds to the maximum period of the signal. During feature extraction, the long-term signal is first decomposed into multiple segments and then input into the local encoders in turn. Then the global encoders extract features from multiple local encoders. Conversely, the decoder is an inverse process to reconstruct the signal from features. The purpose of the decoder is to improve the generality of the model under various machining conditions by regularization and prior information introduction. In this sense, the structure of the decoder can be arbitrary. However, the symmetric structure of the decoder and encoder can more intuitively reflect the reconstruction. In addition, compared with the classic multi-layer LSTM, the computational cost of the pyramid LSTM decoder is much lower.

The predictor is composed of LSTM and MLP without corresponding decoder. The features extracted by the encoder from multiple periods of signals are sent to the LSTM of the predictor to improve the robustness of features and avoid the influence of instantaneous noise. The temporal signal has been converted into discrete features through 3-layer pyramid LSTM. The MLP with dropout, which is suitable for discrete data, can better map features to tool wear.

Compared with the classic LSTM, the structural characteristics of pyramid LSTM auto-encoder can be summarized as follows:

(1) **Multi-scale receptive**. The shallow layers with small receptive fields focus on the extraction of small period features and the deep layers with coarse information granularity focus on the extraction of large period features. The receptive field of each layer can be designed based on the frequency spectrum of the signal, as the spectrum of the monitoring signal is relatively clear and stable.

(2) **Weight sharing and translation invariance**. The kernels with shared weights scan the signal or features. The number of parameters that need to be learned is reduced through this setting.

10.3.4 The Training Method

Multi-task learning is adopted to alleviate the overfitting of sensor signals of known milling conditions. The pseudo code of the training method is shown in Table 10.1. It was attempted to use an auto-encoder to initialize the feature extraction network in [26]. However, the experiments show that it is difficult to determine the iteration numbers of pre-training and training. The multi-task architecture is used instead. By

Table 10.1 Multi-task training algorithm

Input: The milling dataset with label $\{X,T\}$ and without the label $\{X\}$.
Data pre-processing.
For each training iteration **do**:
Long signal splitting based on Fourier transform $x \to [x^1, x^2, \text{L}\ x^n]$
Feature extraction of encoder $y = F(x)$
Signal reconstruction of decoder $[r^1, r^2, \text{L}\ r^n] = G(y)$
Splicing of the reconstructed signal $[r^1, r^2, \text{L}\ r^n] \to r$
If $x \in$ *dataset with label* **do**:
Tool wear monitoring $wear = \varphi(y)$
Loss Function calculation and error back propagation $L = \sum_i \lVert wear - t \rVert^2 + \lambda \sum_i \lVert r - x \rVert^2$
Else do:
Loss Function calculation and error back propagation $L = \lambda \sum_i \lVert r - x \rVert^2$
End if
End for

adjusting the weight coefficients λ, the importance of tasks can be controlled. At the beginning of training, the optimization direction is not clear and the solution is easy to fall into the local optimum. Under the assumption that reducing information loss is helpful to avoid the model falling into the local optimal solution [29], auto-encoder is taken as an unsupervised learning method to introduce the prior knowledge. In the middle and later stages of training, tool wear prediction error becomes the main component of the loss function with a gradual decrease of λ. The reconstruction term plays the role of regularization to avoid the model falling into overfitting of known milling conditions.

10.3.5 Computational Efficiency

Computation complexity is an important index of the tool wear monitoring algorithm, which determines whether the algorithm has practical application value. LSTM network processes temporal signal by unit iteration. The computation of LSTM can be simply expressed as

$$time = unit\,number \times unit\,time \times feature + others \tag{10.19}$$

Item others is mainly the time required for the FC layer, which can be ignored compared with LSTM. The number of units in the classic multi-layer LSTM is

$$unit\,number = layer \times signal\,length \tag{10.20}$$

On the premise that the number of LSTM layers is proportional to the length of the processed signal, the complexity of the classic LSTM is $O(n^2)$, which is computationally expensive for processing long-term signal.

The complexity of the pyramid LSTM auto-encoder is $O(n)$, which is due to the optimization in the following aspects:

(1) The signal length is compressed layer by layer according to the period.
(2) The units of encoder and decoder are simplified.
(3) No connection between sub-modules in the same layer.

Whether in unit number, unit time, feature, or memory, the pyramid LSTM auto-encoder needs less, so it runs faster. It should be noted that since the decoder does not participate in the tool wear monitoring if online learning is not required.

10.3.6 Experimental Validation

10.3.6.1 Experimental Setup

The first dataset is an HSM dataset with different milling tools collected by multiple sensors [30]. The dataset contains six sub-datasets with different tools, three of which have the corresponding tool wear labels. These tools are all alignment tools with three flutes but different in geometry and coating. The material of the working piece is Inconel 718. In the milling experiment, the tool cuts from the upper edge to the lower edge in a zigzag pattern. The signal is collected by a 3-channel dynamometer, a 3-channel accelerometer, and an acoustic emission sensor, with a PCI 1200 board. After each transverse cutting, the tool wear in 3 directions is measured by microscope. These 7-channel sensor signals are used as input to predict the 3-direction tool wear. The experiment platform is shown in Fig. 10.7.

The second dataset is an HSM dataset with the same milling tool collected by a 3-dimensional force sensor under 9 different milling conditions [31]. The experiment is carried out on the 5-axis high-speed milling center Mikron HSM600U. All the tools are two-flute milling tools. The workpiece is made of AISI 4340 square mold with a hardness of 30–40 HRC which has been roughly milled to guarantee flatness before the experiment. The force sensor with three orthogonal directions is used for signal sampling at the sampling frequency of 24 kHz. The corresponding tool wear is predicted according to the 3-direction cutting force signal.

The experiments are all completed by i5-8400 8 GB processor on the platform of Python-Anaconda- tensorflow.

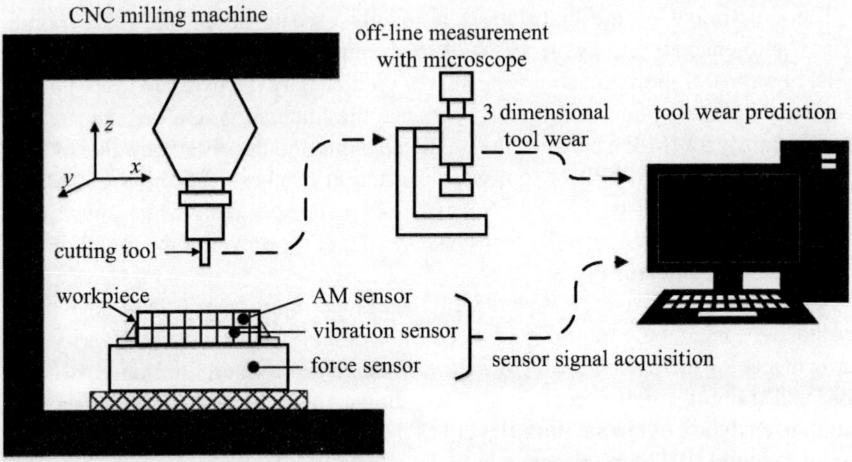

Fig. 10.7 The diagram of tool wears monitoring experiment platform

10.3.6.2 Data Pre-processing

The datasets are divided into the training set, test set, and reconstruction set, which contains independent samples of different tool types. The sub-datasets without labels are taken as reconstruction set to train the auto-encoder. The data pre-processing can be divided into the following three steps:

First, due to the difference of signal between uncut, cut-in, cut-out, and stable state, the signal monitored at the edge of the workpiece is eliminated.

Second, the signal is down-sampled. In the second dataset, the period of each milling condition is different, resulting in the mismatch between model size and signal period. Different down-sampling parameters are set based on the milling parameters.

Third, the labels are augmented artificially based on the premise that the prediction is to find the mapping from sample space to target space. What sample augmentation does is to sample more times in target space, to describe more comprehensively rather than change the mapping relationship. Considering the slow and continuous change of tool wear, linear interpolation is applied to augment the tool wear labels.

10.3.6.3 Results and Analysis on the First Dataset

The signal is first analyzed by Fourier transform. It is found that except for acoustic emission, the signal of other channels has harmonic characteristics, corresponding to the multiples of the tool rotation period, which is suitable for the structure of the pyramid LSTM auto-encoder. The parameters of auto-encoder are determined by the spectral characteristics of the signal.

The experiment on the first dataset is mainly used to prove that the pyramid LSTM auto-encoder can be generalized to the machining with unknown milling tools. Figure 10.8 shows the tool wear curves predicted by the pyramid LSTM auto-encoder. The final results are filtered to reduce the fluctuation. On the test set, the mean square error (MSE) between the predicted value and the observed value of tool wear is 0.0033, and MAPE of tool wear prediction reaches 9%. The average and max final prediction error (FPE) are 9 μm and 28 μm respectively which shows that the pyramid LSTM auto-encoder is insensitive to tool type, even if the tool type has never been encountered.

A lot of comparative experiments with the classic multi-layer LSTM have been carried out under the conditions of fixed and variable input length to verify the performance of the pyramid LSTM auto-encoder. The structure, training loss, test loss, and training time of each model are shown in Table 10.2. In the structure diagram, each box of classic multi-layer LSTM represents a 2-layer LSTM, and the box of pyramid LSTM represents a basic unit, both of which have almost the same complexity. The ellipse represents a 2-layer FC layer. The number of horizontal units only qualitatively represents the network width. According to the description of the basic unit, it can be inferred that the number of LSTM layers used for feature extraction in the same row is the same. And the length of the sensor signal processing

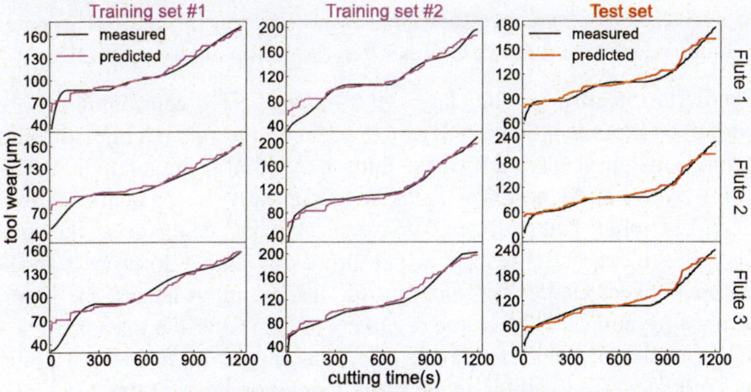

(a) The left two column: training result (b) Rightmost: test result

Fig. 10.8 The tool wear-time curves monitored by the 3-layer pyramid LSTM auto-encoder. The curves of **a** and **b** represent the tool wear prediction results in three orthogonal directions on the training set and the test set, respectively

Table 10.2 Performance of various LSTM networks on the first dataset

Network structure	Short-term signal			Medium-term signal			Long-term signal			Items
	Structure diagram	Fixed length	Variable length	Structure diagram	Fixed length	Variable length	Structure diagram	Fixed length	Variable length	
Classic 2-layer LSTM		0.0057 0.0062 0.0063	0.0083 0.0077 0.0080		0.0048 0.0049 0.0050	0.0111 0.0030 0.0029		0.0055 0.0056 0.0051	0.0115 0.0123 0.0114	Training loss
		0.0070 0.0070 0.0070	0.0095 0.0097 0.0081		0.0058 0.0057 0.0060	0.0131 0.0100 0.0105		0.0066 0.0060 **0.0058**	0.0145 0.0164 0.0133	Test loss
		129.7	102.6		1426.7	419.4		8553.5	7212.3	Training time(s)
Classic 4-layer LSTM		0.0061 0.0063 0.0061	0.0087 0.0086 0.0077		0.0050 0.0047 0.0048	0.0133 0.0145 0.0126		0.0042 0.0044 0.0045	0.0093 0.0105 0.0090	Training loss
		0.0078 **0.0068** 0.0074	0.0139 0.0242 0.0100		0.0055 0.0052 0.0053	0.0184 0.0194 0.0193		**0.0044** 0.0048 0.0048	0.0139 0.0138 0.0126	Test loss
		203.3	190.0		3413.0	622.7		28305.3	14690.2	Training time(s)
Classic 6-layer LSTM		0.0060 0.0063 0.0063	0.0077 0.0076 0.0072		0.0047 0.0047 0.0050	0.0074 0.0094 0.0078		0.0050 0.0048 0.0051	0.0082 0.0080 0.0085	Training loss
		0.0073 0.0073 0.0073	0.0100 0.0092 0.0091		**0.0049** 0.0049 0.0053	0.0094 0.0103 0.0085		0.0055 0.0054 0.0057	0.0115 0.0108 0.0120	Test loss
		334.5	235.2		4833.5	2622.3		40265.8	23849.7	Training time(s)
Pyramid LSTM					0.0037 0.0035 0.0036	0.0043 0.0043 0.0044		0.0030 0.0032 0.0029	0.0039 0.0038 0.0037	Training loss
					0.0050 **0.0044** 0.0045	0.0053 0.0054 0.0055		**0.0035** 0.0036 0.0037	0.0040 0.0041 0.0040	Test loss
					495.8	371.9		3529.2	3137.4	Training time(s)
Pyramid LSTM auto-encoder					0.0038 0.0039 0.0037	0.0046 0.0037 0.0041		0.0032 0.0034 0.0029	0.0036 0.0035 0.0032	Training loss
					0.0043 0.0043 0.0044	0.0050 0.0047 0.0046		0.0034 0.0036 **0.0033**	0.0037 0.0037 0.0034	Test loss
					824.9	793.6		5610.5	5433.2	Training time(s)

*The training loss represents the MSE loss on the training set, while the test loss represents the MSE loss on the test set. Training time represents the time required for the model to learn on the training set

in the same column is the same. The columns of fixed length correspond to the experimental results of the input signal with a fixed length, while the columns of Variable length correspond to the experimental results that the length of input signal varies from half-length to fixed length. The relatively small difference between the training and test error indicates that there is no underfitting or overfitting to the training data. Various models are compared in the following two aspects:

(1) **Prediction accuracy**. The classic multi-layer LSTM can monitor tool wear with fixed input length, but the gain brought by the signal length and network depth is insignificant. Under the condition of variable input length, the classic multi-layer LSTM performs weakly. In contrast, the precision of pyramid LSTM is much better. Extending signal and deepening networks can help models achieve better results, which proves its ability to process long-term sensor signals. Under the condition of variable input length, the prediction error of pyramid LSTM is also relatively small. After the introduction of the auto-encoder, the interference caused by the change of signal length almost can be ignored. Among all the models, the prediction error of the 3-layer LSTM auto-encoder is the lowest, no matter whether the input length is variable or not.
(2) **Computational complexity**. It can be seen from Table 10.2 that the training time of classic multi-layer LSTM increases significantly with the signal length and the number of layers. But the training time required for the 2-layer pyramid LSTM to process medium-term signal is about 15% of the classic two-layer LSTM. And the training time required for the 3-layer pyramid LSTM to process long-term signals is about 9% of the classic 3-layer LSTM. The introduction of the decoder increases the training time of pyramid LSTM by about 60%. Since the decoder is only used for model training, it does not increase the running time in the practical monitoring process.

To further understand the role of each layer in feature extraction, a visual analysis of the trained pyramid LSTM auto-encoder is carried out based on the test set. The output of each layer is arranged into feature curves in time order. Through the comparison of input and output signals, it is found that the original sensor signal has no clear trend with time, but the tool wear increases with time. Therefore, it can be assumed that the trend features have the potential to reflect tool wear, whereas features of other shapes tend to be auxiliary or non-convergent. Based on this assumption, the trend of the feature curves extracted by neural units is emphatically observed, as shown in Fig. 10.9.

The features of the local layer have no clear trend with time but vibrate irregularly. This is consistent with the conjecture: First, tool wear is unable to be extracted effectively by shallow LSTM layer. Secondly, limited by the receptive field, the shallow features fluctuate violently and are easily affected by the instantaneous noise. Compared with the local features, the fluctuation of global features is reduced. Some of the features have the potential to reflect the changing pattern of tool wear, such as features 2 and 3. However, the improvement of other features is not obvious, such as features 1 and 5. In contrast, the trend of top features is clearer and the fluctuation is smoother. Based on the top features, the stable tool wear curve can be obtained. It should be pointed out that the curve of global feature 4 is close to the curves of top features. This result shows that although more global features are not satisfactory, some of them already can be used to monitor tool wear preliminarily. What the

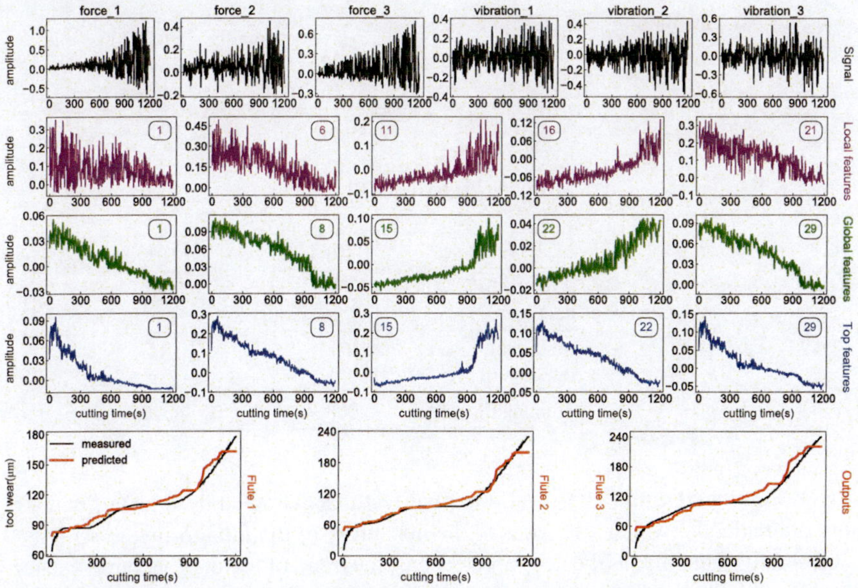

Fig. 10.9 The time-feature curves of neurons in each layer of 3-layer pyramid LSTM auto-encoder. From top to bottom, it represents the original signal, local features, global features, top features, and prediction results

top layer does is to improve the robustness of features, so that more features are associated with tool wear directly.

In order to avoid the influence of the top layer on local features and global features, the features of the 2-layer LSTM auto-encoder are also visualized. Figure 10.10 shows the feature-time curves of the 2-layer pyramid LSTM auto-encoder. It can be found that these features are not different from the corresponding features of the

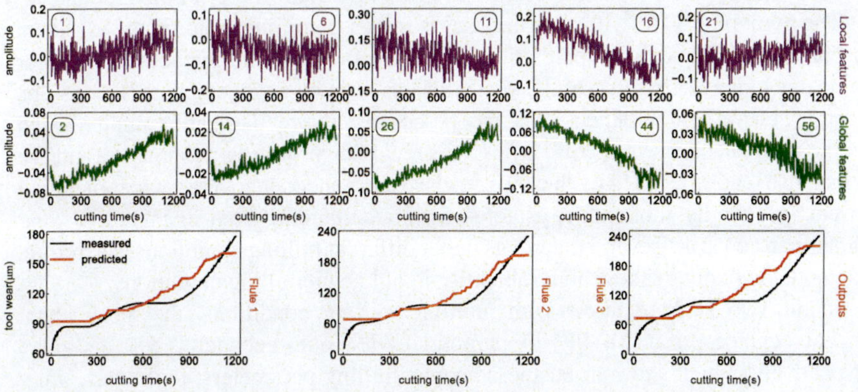

Fig. 10.10 Time-feature curves of neurons in each layer of 2-layer pyramid LSTM auto-encoder. The settings are the same as the 3-layer pyramid auto-encoder

Table 10.3 Tool monitoring results on the second dataset

Items	3-layer pyramid LSTM	3-layer pyramid LSTM auto-encoder
MAPE on the training set (%)	10.3	10.8
MAPE on the test set (%)	14.6	10.7
MSE on test set	0.0164	0.0133
Average FPE on test data (μm)	0.55	0.40
Maximum FPE on test data (μm)	1.66	1.14
Training time (s)	2008	3040

3-layer. Compared with the 3-layer, the final features extracted by 2-layer fluctuate more dramatically, which also leads to the instability of the monitoring.

Generally speaking, as the network deepens, the trend of features becomes clearer and the noise decreases gradually. The features of the deeper layer show excellent extraction ability for tool wear characteristics.

10.3.6.4 Results and Analysis on the Second Dataset

The experiment on the second dataset is mainly used to verify that the pyramid LSTM auto-encoder can be generalized to unknown milling conditions. The prediction results of the pyramid LSTM network are shown in Table 10.3. The pyramid LSTM without auto-encoder is not as good as the performance on the first dataset. This may be due to the greater difference in the mapping relationship from signals to tool wear under different milling conditions. The limited labeled signals could cover the entire mapping space, and decrease the generalization ability of the model.

The pyramid LSTM with auto-encoder is robust to milling conditions. As shown in Fig. 10.11, although the prediction results fluctuate around the actual tool wear value due to the sharp jump of the signal because of tool engagement variations, the trend is generally maintained. This may be because the regularization implemented by the auto-encoder avoids overfitting to a few known milling parameters, and the signal distribution learned by the unsupervised auto-encoder covers a wider mapping space. Specifically, it intends to establish a unified mapping from signal to tool wear rather than construct several mappings under different milling conditions. Therefore, the model is likely to show high reliability in unknown milling conditions.

In the two HSM datasets with multiple milling conditions, the mean absolute percentage error (MAPE) of pyramid LSTM auto-encoder is 9% and 11% respectively, which shows robustness against milling parameters, tool types, time-varying, and non-stationary signals. Compared with the classic multi-layer LSTM, the pyramid LSTM auto-encoder has advantages in precision, speed, and stability, regardless of the length of the sensor signal.

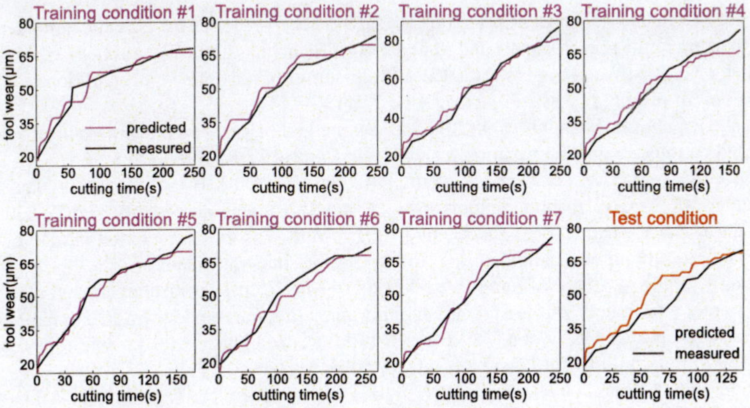

Fig. 10.11 Tool wear prediction results of pyramid LSTM on the second dataset

10.3.7 Conclusion

The hash working conditions, time-varying, and non-stationary signals are the important factors that affect the tool wear monitoring performance in HSM. The pyramid LSTM auto-encoder developed in this section is a network designed based on the harmonics of milling signal to monitor tool wear. Experimental on two HSM datasets have shown that the deep network model has the stronger capability and faster speed to process time-varying and non-stationary signal, and is robust to tool types and milling parameters. These advantages stem from the following reasons:

(1) The model is more likely to benefit from long-term multi-sensor signal and complex model, which can effectively avoid the instantaneous fluctuations and the limitations of the single sensor under a reasonable computational cost.
(2) Many unlabeled signals can be used by unsupervised training which not only makes full use of the data but also improves the robustness of the model under different milling conditions.
(3) The model can directly monitor tool wear with the arbitrary number of signal cycles, which is suitable for practical production.

References

1. Hall DL, Llinas J (2001) Handbook of multisensor data fusion: theory and practice, 1st edn. CRC Press
2. Khaleghi B, Khamis A, Karray FO, Razavi SN (2010) Multisensor data fusion: a review of the state-of-the-art
3. Ghosh N, Ravi YB, Patra A et al (2007) Estimation of tool wear during CNC milling using neural network-based sensor fusion. Mech Syst Signal Proces 21:466–479

4. Rizal M, Ghani JA, Nuawi MZ, Haron CHC (2017) Cutting tool wear classification and detection using multi-sensor signals and Mahalanobis-Taguchi system. Wear 376–377:1759–1765
5. Zhang X, Lu X, Wang S, Li WD (2017) A multi-sensor based online tool condition monitoring system for milling process. Procedia CIRP 72:1136–1141
6. Ceng G, Cen XH, Shan XL et al (2016) A new method of gear fault diagnosis in strong noise based on multi-sensor information fusion. J Vib Control 22(6):1504–1515
7. Duro JA, Padget JA, Bowen CR, Kim HA, Nassehi A (2016) Multi-sensor data fusion framework for CNC machining monitoring. Mech Syst Signal Process 66–67:505–520
8. Wang J, Xie J, Zhao R, Zhang L, Duan L (2017) Multisensory fusion based virtual tool wear sensing for ubiquitous manufacturing. Robot Comput Integrat Manuf 45
9. Kong K, Peng X, Chen Y, Wang P, Xu M (2020) Multi-sensor measurement and data fusion technology for manufacturing process monitoring: a literature review. Int J Extreme Manuf
10. Wang WH, Hong GS, Wong YS, Zhu KP (2007) Sensor fusion for on-line tool condition monitoring in milling. Int J Prod Res 45(21):5095–5116
11. Goodfellow I, Bengio Y, Courville A, Bengio Y (2016) Deep learning, 2nd edn. MIT Press
12. Salakhutdinov RR, Hinton GE (2009) Deep boltzmann machines. In: Artificial intelligence and statistics, pp 448–455
13. Hinton GE, Salakhutdinov RR (2006) Reducing the dimensionality of data with neural networks. Science 313(5786):504–507
14. LeCun Y, Bottou L, Bengio Y, Haffner P (1998) Gradient-based learning applied to document recognition. Proc IEEE 86(11):2278–2324
15. El Hihi S, Bengio Y (1996) Hierarchical recurrent neural networks for long-term dependencies. In: Advances in NIPS, pp 493–499
16. Goodfellow IJ, Pouget-Abadie J, Mirza M, Xu B, Warde-Farley D, Ozair S, Courville A, Bengio Y (2014) Generative adversarial networks. Adv NIPS 3:2672–2680
17. Chen Y, Jin Y, Jiri G (2018) Predicting tool wear with multi-sensor data using deep belief networks. Int J Adv Manuf Technol 99(5):1917–1926
18. Shi C, Luo B, He S, Li K, Liu H, Li B (2020) Tool wear prediction via multidimensional stacked sparse autoencoders with feature fusion. IEEE Trans Industr Inf 16(8):5150–5159
19. Hochreiter S, Schmidhuber J (1997) Long short-term memory. Neural Comput 9:1735–1780
20. Li JW, Luong MT, Jurafsky D (2015) A hierarchical neural autoencoder for paragraphs and documents. Comput Sci 1106–1115
21. Chan W, Jaitly N, Le Q, Vinyals O (2016) Listen, attend and spell: a neural network for large vocabulary conversational speech recognition. In: 2016 IEEE international conference on acoustics, speech and signal processing, Shanghai, pp 4960–4964, 20–25 March, 2016
22. Zhang JJ, Wang P, Yan RQ, Gao RX (2018) Long short-term memory for machine remaining life prediction. J Manuf Syst 48(part C):78–86
23. Zhao R, Yan RQ, Wang JJ, Mao KZ (2017) Learning to monitor machine health with convolutional bi-directional LSTM networks. Sensors 17:273
24. Wang X, Huang Y (2010) Convergence study in extended Kalman filter-based training of recurrent neural networks. IEEE Trans Neural Netw 22:588–600
25. Guo H, Zhu KP (2020) Pyramid LSTM auto-encoder for tool wear monitoring. In: CASE, Hong Kong
26. Sutskever I, Vinyals O, Le QV (2014) Sequence to sequence learning with neural networks. Adv Neural Inf Process Syst 3104–3112
27. Malhotra P, Ramakrishnan A, Anand G, Vig L, Agarwal P, Shroff G (2016) LSTM-based encoder-decoder for multi-sensor anomaly detection. arXiv:1607.00148
28. Yu Y, Si X, Hu C, Zhang J (2019) A review of recurrent neural networks: LSTM cells and network architectures. Neural Comput 31(7):1235–1270
29. Dong W, Yuan T, Yang K, Li C, Zhang S (2017) Autoencoder regularized network for driving style representation learning. IJCAI 1603–1609
30. 2010 PHM society conference data challenge. http://www.phmsociety.org/forum/583
31. Zhu KP, Mei T, Ye D (2015) Online condition monitoring in micromilling: a force waveform shape analysis approach. IEEE T Ind Electron 62(6):3806–3813

Chapter 11
Big Data Oriented Smart Tool Condition Monitoring System

11.1 The Big Data Issues in Manufacturing

In the era of rapid development of automation technology, computer technology, and information technology, CNC machine tools, data acquisition devices, intelligent sensors, and other intelligent devices with perception ability have been used more and more in the production system, and the production system has developed from automation and digitization to intelligence. Since the twentieth century, mechanical and electrical equipment has become more complex, with prolonged service life and significantly increased monitoring data. Based on the above characteristics, mechanical and electrical equipment has entered the "industrial big data" era [1, 2]. It is generally believed that the manufacturing big data includes the data generated in all aspects of workshop production and runs through the whole product life cycle. Big data can be divided into production and management data, equipment data, and external data [3, 4].

Manufacturing big data has the characteristics of general big data, namely volume, variety, and velocity [5, 6]. Volume refers to the type of data that can't be processed by the traditional enterprise; variety refers to the complex type of data, which includes both relational data from the enterprise, and un-relational data such as monitoring image and video. Velocity is reflected in two aspects: data acquisition and data analysis and processing. Manufacturing big data comes from the actual production and manufacturing and has high-speed requirements from data acquisition to analysis and processing. In addition, the industry generally believes that the manufacturing big data also has the forth "V", namely the value, which emphasizes the user value-driven and the usability of the data itself. What drives intelligent manufacturing isn't big data itself, but its analysis technology. Big data analysis technology is used to mine the potential relationship between data from complex data sets, find new patterns and rules, and feed them back to production, to realize the scientific management of workshops based on data-driven [7].

Manufacturing big data can manage workshop production in an all-around way, break through the bottleneck that can't be quantified in production, and eliminate the

K. Zhu, *Smart Machining Systems*, Springer Series in Advanced Manufacturing,
https://doi.org/10.1007/978-3-030-87878-8_11

uncertain factors in the decision-making process. Its application covers all aspects of workshop production [8]. The Predix cloud platform [9] released by General Electric can quickly obtain the production data of various equipment and conduct big data analysis, and optimize the production process by using the analysis results. The MindSphere platform [10] launched by Siemens can provide big data analysis services for enterprises. Through this platform, employees can detect the operation status of equipment.

From the above research, it can be found that although the application of manufacturing big data is becoming more and more extensive, the focus is mostly on mining relevant knowledge from massive historical data to optimize the existing machining process, and there is less research on how to deal with big data in real-time.

11.2 The Big Data Analytics in Smart Machining System

11.2.1 The Big Data Challenges and Motivation

The computer numerical control (CNC) machine tool is the foundation of the manufacturing industry. It is of great importance to monitor and control the machine tools in the manufacturing process to ensure product quality and realize intelligent manufacturing. For the high-end applications in aerospace, optics, and microelectronics industries, such as freeform surface optics, engine blades, gearbox, etc., the geometric accuracy and surface integrity are hard to guarantee by the traditional CNC machine tools due to the complex structure and poor material machinability. Under a specified CNC machine tool, the key techniques to solve these problems lie in the on-line monitoring of tool conditions and intelligent control of the machining process [11, 12]. It has been shown that the improvement of the next generation CNC machine performance depends greatly on the new tool condition monitoring (TCM) system [11].

Meanwhile, due to the development of modern sensing technology and the digitization of the CNC machining process, the machining process is producing a huge amount of heterogeneous "big data" such as process parameters, monitoring signals, and running historical records. These include the 1D signal such as force, vibration, acoustic emission, 2D image, 3D point cloud, and textual process data which enable digitized and networked manufacturing. The "big data" increases exponentially with the progression of the machining process, which makes the existing data acquisition and processing methods difficult to handle. It is hard to accurately judge the tool state and optimize the machining process as a result. At present, there has been extensive research in the field of CNC machining process monitoring [11]. These studies are mainly based on the traditional theory of signal processing and artificial intelligence methods with application to state monitoring and diagnosis of machine tools, which partially realized the intelligent manufacturing process [13, 14]. Others have also tried the STEP-compliant Numerical Control (STEP-NC) data models to include high-level machining process information for monitoring and optimization [15, 16]. While this brings much post-processing and data conversion and recognition, the machining efficiency and intelligence weren't improved as expected [16]. The key functions of tool condition monitoring are limited, and it lacks new perspectives of big data fusion for the monitoring system. The current machining system is limited by data acquisition methods and modeling approaches, and it is difficult to make full use of monitoring information to smartly assess and optimize the cutting conditions online.

To this end, and based on the characteristics of the CNC system, this chapter develops the big data-driven CNC machining process on-line monitoring system model and implementation method. It builds heterogeneous big data fusion and deep learning models, to enhance the data representation and learning capability, and prolong the useable tool life and improve CNC machining precision.

11.2.2 Related Works

The fundamental physical process of NC machining is cutting. In this process, the tool is constantly subjected to thermal-force coupling effect, gradually got worn and broken until failure. Tool wear is inevitable in the machining process. The tool states directly affect the machining accuracy and product quality, and serious tool wear will also cause tool breakage and failure, resulting in chatter, damage to the machine tool and workpiece. It is most important to monitor and control the tool wear condition and predict the remaining life and optimize the process, to improve the manufacturing intelligence and product precision [14]. Cutting force is most sensitive to tool conditions due to its direct reflection of the cutting process. The time and frequency characteristics of cutting force were investigated in [17, 18] and applied to model the dynamics of tool state with the stochastic process, which achieved good results for wear classification. The study [19] analyzed the force signal and established the correlation between the cutting force and the tool wear changes, then developed the sensor fusion model and the tool wear monitoring system based on a neural network. By tracking milling force model coefficients [20], presented a method of real-time tool wear monitoring, which showed that these coefficients could represent the wear state of the tool well and independent of processing conditions. The acoustic emission (AE) was studied for tool wear monitoring for its non-destructive inspection nature and its sensitivity to tool state variations. The AE signal and its spectrum characteristics were identified to meet the requirements of tool wear monitoring. Bhuiyan et al. [21] studied the input of the tool wear condition monitoring model by using the raw data, frequency characteristic, and RMS value of the AE signal respectively, and the experimental results showed that the tool wear status could be better detected by using the raw AE as the input of the monitoring model. The vibration signal was also studied on its mean power analysis for state monitoring and tool failure indication [22].

The image-based approaches of tool monitoring have attracted more attention lately due to its fast development of optical devices and their direct observation capabilities. The study [23] monitored the change of tool wear status by laser detecting the changes in the diameter of the tool. Due to the limitation of imaging equipment and the influence of the actual working environment, the tool state images obtained online are often blurred and the image resolution is difficult to recognize. Zhu and Yu [24] collected tool surface image and developed a morphological analysis approach to remove the background noise and enhance the tool wear region. It could identify tool conditions and quantify the wear levels.

The fusion of multi-sensor signals can fully reflect the change of tool state and improve the ability of recognition and fault tolerance. The performances of the tool condition monitoring system were improved by using the multi-sensor frame model [25]. Similar approaches were also developed in the area of machinery health monitoring. Xia et al. [26] incorporated both temporal and spatial information from multiple sensors which achieve higher and more robust diagnosis accuracy, and Sawo and Kempkens [27] developed information fusion for smart structural damage maintenance. Malekian et al. [28] applied the three different types of signals, AE, force, and vibration, and developed a fuzzy neural network model to predict the wear conditions, which achieved good forecast results. Due to the inherent defects of neural networks in training and learning, more researchers have turned to the probabilistic graph model methods that are easier to learn and expand, such as the Bayesian network [29], Hidden Markov model [30, 31].

From the above research, it can be seen the current research of tool condition monitoring mainly from the signal analysis and intelligent techniques, but the lack of the deep fusion of the process parameters, process model and monitoring data and other system changes. Without the inclusion of the system physical models, the current methods could detect the tool states but can't quantify the wear values or achieve online process optimization. At the same time, due to the complexity of processing conditions, the data collected by the monitoring system includes heterogeneous big data, such as system parameters, force, vibration, acoustic emission, 2D image, 3D point cloud, etc. How to effectively integrate these multimode big data is the basis of the monitoring system modeling and implementation. As the latest research achievement in artificial intelligence, deep learning has strong complex data expression capabilities and has a unique advantage in heterogeneous data fusion [32]. It is effective in training and learning the depth of data for multi-level expression, and is successfully applied in many fields [26, 33, 34]. Therefore, this study establishes a deep learning model for heterogeneous data fusion.

Based on the characteristics of the CNC machine tool, this chapter explores the theory and realization method of the smart tool condition monitoring and wear compensation with big data analytics. The system model and algorithms are developed for on-line condition monitoring, accurate estimation of tool wear, intelligent compensation, heterogeneous big data fusion, and deep learning, to improve CNC machining precision and extend tool life.

11.3 The Framework of Big Data Oriented Smart Machining Monitoring System

11.3.1 The Monitoring System Architecture

Taking the CNC machining center and tool monitoring unit as the research objects, the study constructs a new smart monitoring system model and architecture for big data-oriented TCM system [12]. It realizes the deep fusion of the physical model, process parameter, and monitor signal in the machining process. Under this framework, various intelligent monitoring function modules are modeled and built to process the big data. Due to the varied data structure of different machine components, there is a large number of heterogeneous data in the monitoring data, which is difficult to be collected and processed by a single communication protocol. Aiming at this problem, the data model, communication architecture, and access strategy of the cutting tool and machine tool based on OPC UA Technology [35] and MTConnect protocol [36] are adapted to solve the problem of multi-source heterogeneous data collection. The OPC UA technology is an important criterion to realize data interaction between NC machine tools, and the unified address space is designed so that it can describe highly complex data model.

In the data acquisition stage, the data obtained from the CNC system and the tool repository are uploaded to the front-end computer via OPC UA [12] as illustrated in Fig. 11.1. The data is then analyzed by machine learning methods for data fusion, feature extraction, and so on. These are stored in the database of the machine tool

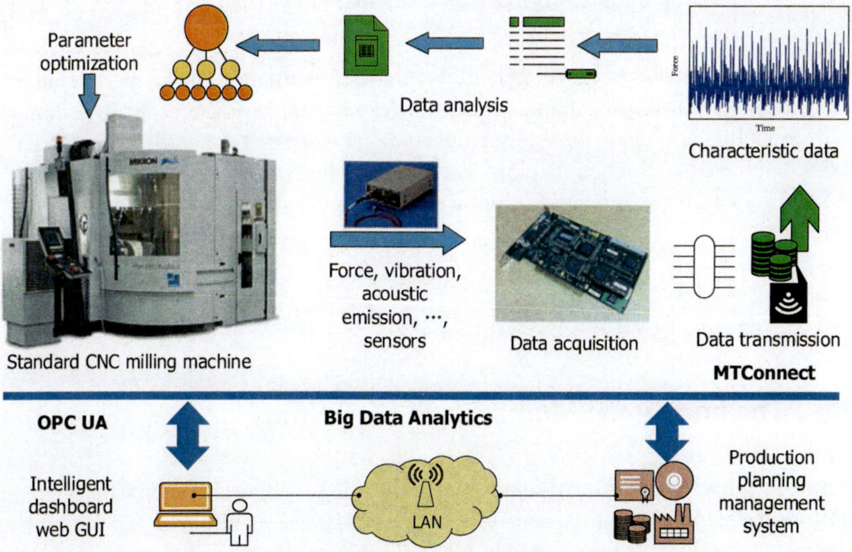

Fig. 11.1 The monitoring system architecture

in-situ measurement system and NC machining process information. The key of this monitoring system is to develop the theory and algorithms for on-line tool state monitoring, and the fusion and deep learning of multi-source heterogeneous data. Specifically, the system studies the feature extraction of weak signals, image super-resolution reconstruction algorithm, and the processing, feature representation, and fusion methods of multi-level heterogeneous big data. By verifying the mechanical model of the tool wear process and monitored signal patterns, the accuracy and reliability of the monitoring system are improved, and the machining parameters are adjusted by feedback control function to realize the wear compensation. The framework and four of the important functional modules are discussed with case studies in Sect. 3.

11.3.2 The Big Data-Oriented Formulation of TCM

From the big data analytics, the problem of TCM can be formulated as a machine learning problem, i.e., the objective of TCM is to search the most probable state $\theta_i(t)$ given the extracted signal feature $y(t)$. Depending on the monitoring implementation, the sensory signals are multimode and heterogeneous in nature, and the features $y(t)$ are time-varying and assumed to have a probability density function conditioned on the states. Specifically, let c be the number of states involved, the tool state estimation can be expressed as a feature vector $y(t)$ belonging to class $\theta_i(t)$ with maximum likelihood with the class-conditional probability function $p(y(t)|\theta_i(t))$,

$$\text{Determine: } p(y|\theta_i)p(\theta_c) = \arg\max_c\{p(y|\theta_c)p(\theta_c)\} \tag{11.1}$$

It is noted that, without loss of generality, the state estimation can be implemented by any machine learning techniques under this framework. In practice, and depending on the modeling approach, the tool state vector $\theta(t)$ can also be regarded as a system model parameter, and when it refers to tool wear, the value can be used for wear prediction with the increment of initial wear θ_0,

$$\theta(t) = \theta_0 + \Delta\theta(t) \tag{11.2}$$

11.4 The Functional Modules and Case Study

Compared with the traditional condition monitoring system, in which the signal acquisition, feature extraction, sensor fusion, and state recognition are divided into independent modules, while in this framework, these three functions are simplified and realized by the deep learning module as illustrated in Fig. 11.2. The TCM system

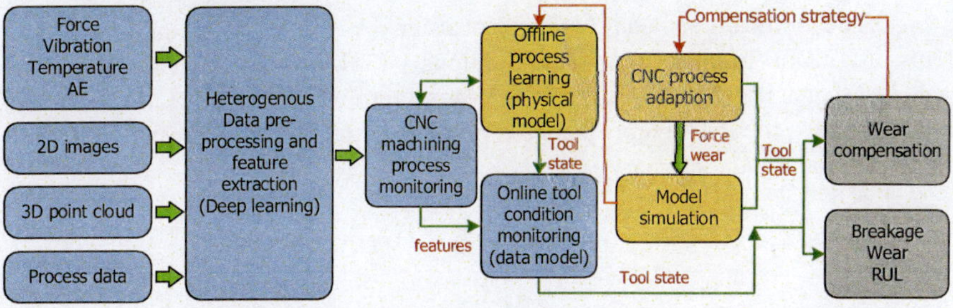

Fig. 11.2 Big data-oriented tool condition monitoring framework

firstly pre-processes the data and establishes the deep learning model for multi-source heterogeneous feature extraction to excavate the relationship between monitoring signal and tool state changes as represented in blue shaded boxes in Fig. 11.2. To improve the tool state estimation capability and online tool wear compensation, the machining process dynamics are modeled and physical model of tool wear is established, which are verified with the output of the data-driven model estimation for the decision as represented in yellow shaded boxes in Fig. 11.2. With these monitoring results, it finally realizes the on-line TCM and intelligent compensation of tool wear which is further represented with grey shaded boxes in Fig. 11.2. For this end and in the information flow, the following key modules are discussed, which start from the signal preprocessing to the final tool life prediction and wear compensation.

11.4.1 Sparse Coding Based Data Pre-processing

(1) Signal de-noising and feature extraction

In this module, the work studies the extraction of weak feature signal, image super-resolution reconstruction algorithm, and sparse decomposition theory and algorithm of the high-frequency wideband signal, and reveals the relationship between tool failure and cutting force, AE, vibration signal, and image change. Based on sparse decomposition, weak signal extraction, and high-resolution image reconstruction algorithm, the theoretical basis of milling monitoring signal and image enhancement is solved, and the complexity of the algorithm is reduced and the calculation speed is raised.

In the widely applied high-speed precision milling, the tool size and feed rate result in the low signal-to-noise ratio of monitoring signal, filtering noise, and improving signal-to-noise ratio are necessary to improve the reliability of on-line monitoring system. Based on sparse representation, a dictionary learning method can be used to learn the atoms of pure noise and noisy signals and to separate the noise by the correlation of atoms [37, 38].

Sparse decomposition means to represent a signal as a linear combination of a few atoms of a given (often overcomplete) dictionary [38]. In practice, a signal can be treated as sparse if most of its entries are close to zero while the others are relatively large. Formally, the sparse representation problem is to find the sparsest coefficients of signal y under the dictionary [38],

$$\min_{x} \|x\|_0 \text{ s.t. } \mathrm{y} = \mathrm{D}x, \tag{11.3}$$

where x_i are called the coefficients of y in the dictionary $D = [d_1, \ldots, d_T]$, and $\|x\|_0$ is the "l_0-norm" of the coefficient vector. It has been found that if the solution is sparse enough, the solution in Eq. 11.3 is equal to the solution of the convex l_1-minimization problem [36],

$$\min_{D,x} \|\mathrm{y} - \mathrm{D}x\|_2^2 + \lambda \|x\|_1 \tag{11.4}$$

where the l_1-norm is actually the sum of the magnitudes of elements of vector x. In the application, the dictionary is generally learned redundant to sparsify the coefficient x when the original data is dense.

With the idea of Fisher discriminant analysis, the constrained optimization can make the learned atoms of different tool states bear maximum discrimination. In this way, the sparse expression is more advantageous to the extraction of useful signals and the implementation of on-line monitoring in strong noise. With these considerations, the final objective function consists of three parts, with the optimization function,

$$\begin{aligned} J(D, X) &= \arg\min \left\{ \frac{1}{2} \frac{S_W(X)}{S_B(X)} + \lambda_1 \|X\|_1 + \frac{1}{2} \lambda_2 \left\| d^T x_i - W y_i \right\|_2^2 \right\} \\ s.t.\ \|d_i\|_2 &= 1,\ d_i \geq 0 \end{aligned} \tag{11.5}$$

where W refers to the wavelet packet decomposition (WPD) which intends to extract the multiscale properties of the signal for sparsification, λ_1 is the l_l-norm regularization parameter, and λ_2 is the l_2-norm regularization parameter, S_W is the within-basis scatter, and S_B is the between-basis scatter. The first term maximizes discrimination between different states, the second term maximizes the sparsity of estimated force, and the last term controls the reconstruction error. The $J(D, X)$ optimization criterion intends to learn the basis that good for tool state estimation when dictionaries under different states are learned [31]. In this study, the K-SVD approach [38] is adapted to learn the dictionary.

(2) **Tool wear image super-resolution**

An image Super-resolution method based on sparse decomposition is developed to solve the problem of low resolution and blur of tool wear image and to improve the detection accuracy of wear area. According to the imaging principle, the sparse decomposition algorithm is firstly adapted to separate the different source images, by which the noise and background are eliminated and filtered. Then, the redundant dictionary is optimized, and the dual target of de-blurring processing and super-resolution reconstruction is realized by exploring the multi-scale similarity nature of sparse expression by matching the low-resolution tile to the high-resolution tile. Without loss of generality, the low-resolution image is assumed to be the blurred and down-sampled version of the high-resolution image, then the super-resolution image could be reconstructed. According to the above ideas to establish the optimization equation:

$$\min_{\mathbf{D}_h, \mathbf{D}_l, x} \left\| \mathbf{D}_h x_i - \mathbf{p}_i^h \right\|_2^2 + \left\| \mathbf{D}_l x_i - \mathbf{p}_i^l \right\|_2^2 + \lambda \| x_i \|_1 \tag{11.6}$$

where x_i is the sparse coefficients, the $\mathbf{p}_i^h$ and $\mathbf{p}_i^l$ represent the high and low-resolution image patches, $\mathbf{D}_h$ and $\mathbf{D}_l$ are the high and low-resolution dictionaries respectively. In this study, the K-SVD approach [38] is applied to learn the overcomplete dictionaries. As the corresponding patches $\{\mathbf{p}_i^l\}$ and $\{\mathbf{p}_i^h\}$ have similar sparse coefficients $\{q_i\}$ according to their respective dictionaries, then,

$$\{\mathbf{D}_l, q^i\} = \arg\min_{\mathbf{D}_l, \{q_i\}} \sum_i \left\| \mathbf{p}_i^l - \mathbf{D}_l q_i \right\|_2^2 \tag{11.7}$$

$$\{\mathbf{D}_h, q_i\} = \arg\min_{\mathbf{D}_h, \{q_i\}} \sum_i \left\| \mathbf{p}_i^h - \mathbf{D}_h q_i \right\|_2^2 \tag{11.8}$$

11.4.2 In-process Workpiece Integrity Monitoring

In this module, it introduces a novel workpiece integrity monitoring and intelligent recognition method, by developing the diffusion map of 2D image and 3D point cloud recognition according to the characteristics of machining. It monitors the integrity of the workpiece surface and geometrical dimension by comparing the CAD model and the monitoring image (or point cloud). At the same time, it can provide the decision basis for the tool's remaining useful life (RUL) and working condition judgment.

The main idea is based on the spectral graph theory [39]. An undirected diffusion graph is constructed on the pixel (or cloud point), and the similarity between the data is represented by the diffusion distance. By transforming the cloud data U into a weighted non-direction graph G, the point cloud matrix contains plenty of information, such as contours, lines, and geometric dimensions, and tolerances. To build the graph G, the first is to calculate the degree d_i of vertex i, which represents the number of edges connected to the vertex, and then constructs the diagonal matrix Λ,

$$\Lambda_{N\times N} \stackrel{\text{def}}{=} \text{diag}(d_1, \ldots, d_N)$$
$$\text{where } d_i = \sum_{j=1}^{j=N} w_{ij}, \forall i, j \in \{1 \cdots N\}. \tag{11.9}$$

The edge weight w is non-negative, and normally a Gaussian describing the similarity connected vertexes u_i, u_j,

$$w_{ij} = \exp(-\|u_i - u_j\|^2/\sigma^2), (\sigma > 0) \tag{11.10}$$

Then, define the normalized Graph Laplacian L as:

$$L_{N\times N} \stackrel{\text{def}}{=} \Lambda^{-\frac{1}{2}} \times (\Lambda - P) \times \Lambda^{-\frac{1}{2}} \tag{11.11}$$

where $\Lambda^{-\frac{1}{2}} = \text{diag}\left(1/\sqrt{d_1}, \ldots, 1/\sqrt{d_N}\right)$.

The spectrum s of L is calculated regarding to eigen-vector $\mathbf{v}$:

$$L\mathbf{v} = s^*\mathbf{v} \tag{11.12}$$

The larger the dissimilar of the spectra, the lower the dimensional integrity. The graph spectrum L obtains fine information about the dimension integrity of the component, which can't be obtained from the traditional statistical analysis and surface marking methods.

For surface integrity monitoring, by constructing the weighted function, the background pixel and the anomaly pixel have different diffusion coordinates in the new mapping. Therefore, it is easy to remove the image texture and background noise and to overcome the disadvantage of the background influence on the workpiece surface recognition. For the dimensional integrity monitoring, the difference between the point cloud data and the original CAD model data is inferred by comparing the similarity of the graph spectrum.

11.4.3 Heterogeneous Data Fusion and Deep Learning

For the tool condition monitoring, how to effectively fuse the multimode monitoring data and process parameters is the key to successful implementation. As the newest research achievement in the field of artificial intelligence, deep learning can effectively train and learn the deep network to carry on the multi-level characteristic representation to the data and has the strong complex data expression ability. It has a unique superiority in heterogeneous data fusion [40, 41]. This system proposes a heterogeneous data fusion model based on deep learning, as shown in Fig. 11.6. For different monitoring signals, it designs different network structures for feature extraction and develops uninformed representations and fusion techniques of various data types.

Based on the idea of multimode and multi-level analysis, the nature of each sensor signal in the machining system is first analyzed, and the data fusion and feature fusion are then investigated. Different deep network learning models are established to eliminate the structure difference between different sources. For example, a recurrent neural model (RNN) model is constructed for one-dimensional time series, and convolution neural network (CNN) model is proposed for two-dimensional input of images. For the important monitored signal, the multi-view method is adopted, such as using wavelet transform to process time series, then using convolution neural network to process the time–frequency image, and using deep belief network (DBN) to process the raw time signal. In this way, different modal signals produce different characteristics and the unified feature is generated by feature fusion. In the feature fusion step, based on the study of the direct stitching method, the benefit of the attention model [40] is applied, and finally, it makes tool state estimation through the whole deep connection layer.

Figure 11.3 shows a structure of the end-to-end heterogeneous data fusion model based on deep learning, in which the network parameters involved in each modal feature extraction, the parameters of the model in the feature fusion, and the final model prediction parameters can be trained simultaneously. For example, the uppermost line of Fig. 11.3 illustrates a typical CNN for force modeling, whose structure consists of a stack of fully connected layers, convolutional layers, and pooling layers. The cutting force signal is transformed by wavelet analysis to get the two-dimensional

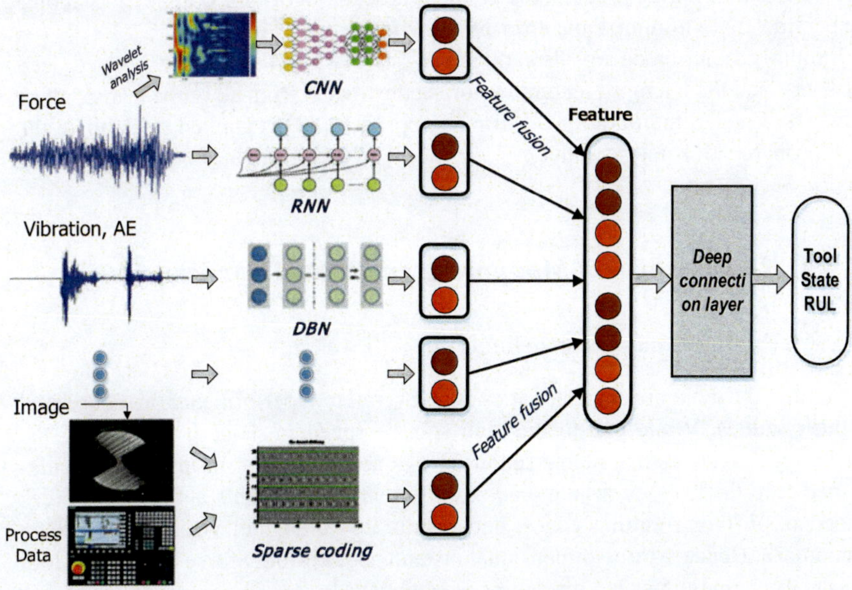

Fig. 11.3 Deep learned base heterogeneous data fusion for tool wear monitoring

time–frequency images. The input time–frequency image I is first convolved with kernel sets $W = \{w_1, w_2 \ldots, w_i\}$ with biases $\epsilon = \{e_1, e_2, \ldots e_i\}$, resulting a new feature map M_i. These features are subjected to a non-linear transform $f(\cdot)$ and this mapping is repeated for every convolutional layer l:

$$M_i^l = f(\mathbf{w}_i^{l-1} \otimes I^{l-1} + e_i^{l-1}) \tag{11.13}$$

The convolution operation greatly reduces the number of parameters to be learned and makes the network translation invariant [32]. The convolutional layers are then alternated with pooling layers to perform feature subsampling to decrease data dimension and over-fitting issues. In the training of CNN, the loss function is,

$$\begin{aligned} J(\mathbf{w}, e) &= \frac{1}{k}\sum_{i=1}^{k} L(\hat{m}^{(i)}, m^{(i)}) \\ &= -\frac{1}{k}\sum_{i=1}^{k} [(m^{(i)} \log(\hat{m}^{(i)})) + (1 - m^{(i)}) \log(1 - \hat{m}^{(i)})] \end{aligned} \tag{11.14}$$

To control the over-fitting weights, an L_2 norm is introduced with regularization term λ.

$$E(\mathbf{w}, e) = J(\mathbf{w}, e) + \frac{\lambda}{2} W^T W \tag{11.15}$$

In the CNN training process, the gradient descent algorithm is applied to update the parameters to minimize the error function is expressed in Eq. 11.15.

Similarly, other mode signals and features are extracted with corresponded deep networks, and then, a fused feature vector is constructed from all of the learned deep networks. Finally, the traditional Softmax approach [40] is applied to estimate the most probable tool states (values).

11.4.4 Intelligent Tool Monitoring and Wear Compensation

(1) **Machining dynamics modeling**

At present, the dynamic characteristics of an overall tool performance are concerned in most studies. While due to the high spindle speed in modern precision CNC machining center, the inevitable runout brings about unequal instantaneous undeformed chip thicknesses, which leads to different engagements and cutting forces loaded on different cutting edges, and results in varied tool wear and breakage phenomena. Hence, a three-dimensional instantaneous milling force model with tool run-out effect in high-speed machining is established.

Take the ball-nose end milling cutter as an example, within the tool runout effect, by integrating the elemental forces of each layer along the z-axis, the instantaneous milling force $\mathbf{F}(t)$ with the milling time t is obtained as [42],

$$\mathbf{F}(t) = \sum_{j=1}^{N} \left[\int_{z_{jd}}^{z_{ju}} \mathbf{TK}a_j(t, z(w))b(z(w)) \right] \quad (11.16)$$

where j the tool flute sequence number $j = 1, 2, \ldots, N_t$. The matrix $\mathbf{T}$ transforms the dynamic coordinate system into the fixed coordinate system O-xyz. The vector $\mathbf{K}$ is the milling force coefficients. The parameters z_{ju} and z_{jd} are the z-axial upper and lower boundaries of the instantaneous cutter/workpiece engagement. The b is the elemental chip width at the coordinate z, and a_j is the instantaneous undeformed chip thickness at milling time t and coordinate z, which are related to the real-time tool wear according to the geometric engagement. The model parameters $\{a_j, b, z\}$ are keys to accurate prediction of milling forces, and then the relationship between tool wear and machining dynamics is established.

(2) **Tool life prediction and compensation**

Currently, researches on tool wear are mainly focused on the physical and chemical mechanisms of wear. Modeling of an overall evolution process of tool wear and breakage is inadequate, of which the real-time, versatility, and predictability need to be strengthened. An exact tool wear process model makes a significant contribution to real-time tool monitoring and intelligent process optimization. According to the regular pattern and physical characteristics of tool wear, the entire tool wear process is divided into three stages: initial wear stage, steady-state stage, and accelerated/serious wear stage. A generic evolution model of tool wear and breakage is established to describe these phenomena, in which the wear rate and wear acceleration are approximated by the first- and second-order partial differential equations.

A typical tool flank wear curve is shown in Fig. 11.4. There are three wear stages divided by critical times, which are corresponding to different wear rates sequentially. In order to define the critical times, the value relation of transition functions' differentiations (w'_E and w''_L) is chosen as the evaluation indicator. Besides, the critical time is lower than the milling time when the tool wear VB reaches a critical limit VB^*, stipulated in the ISO standard (ISO 8688-2:1989).

To analyze the variation tendency of tool flank wear, the wear rate, and acceleration rate VB' and VB'' are calculated with the difference method as:

$$\begin{cases} VB'(t) = \Delta VB/\Delta t = [VB(t+\Delta t) - VB(t)]/\Delta t \\ VB''(t) = \Delta VB'/\Delta t = [VB'(t+\Delta t) - VB'(t)]/\Delta t \end{cases} \quad (11.17)$$

In order to describe wear conditions in different stages, two functions w'_E and w''_L in earlier and later wear zones are integrated. The tool flank wear model over the milling time is given as:

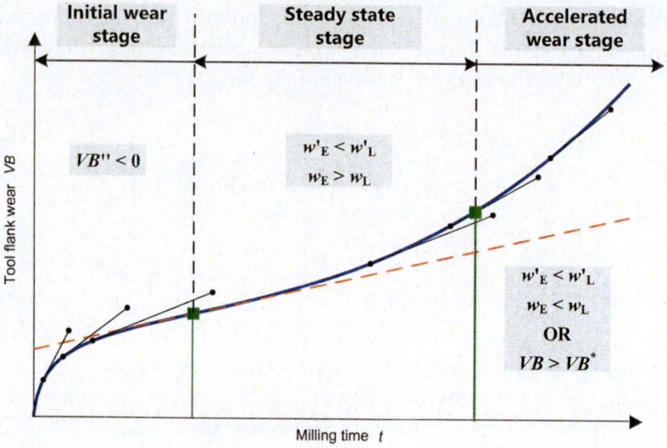

Fig. 11.4 Tool flank wear curve in different stages

$$w(t) = w_{\mathrm{E}}(t) + w_{\mathrm{L}}(t) = A\left[\ln(\nu_{\mathrm{E}}t + 1) + (\nu_{\mathrm{L}}t)^{H}\ln\left(\nu_{\mathrm{E}}/\nu_{\mathrm{L}} + 1\right)\right] \tag{11.18}$$

where A, ν_{E}, ν_{L}, and H are parameters which are characterized indirectly as the intrinsic amplitude of wear rate, the growth frequencies in earlier and later wear stages, and the growth index in the later wear stage, which can be obtained with milling experiments.

The critical average flank wear VB is set as the criterion to define an effective tool average life $\bar{t}$, which is obtained by the numerical method according to a time set of model solutions under the critical condition:

$$\bar{t} = \left\{ t \left| \left(\nu_{\mathrm{E}}/\nu_{\mathrm{L}} + 1\right)^{(\nu_{\mathrm{L}}t)^{x}} = \frac{\exp\left(VB^{*}/A\right)}{\nu_{\mathrm{E}}t + 1} \right. \right\} \tag{11.19}$$

Finally, the on-line error compensation method is designed. Through the tool wear model and the real-time state monitoring data fusion, it indirectly determines the milling tool wear state and the remaining useful life. The theoretical model is used to compensate the machining, and the intelligent process optimization is realized. For example, when the tool wear VB is estimated from the other module, the machining error compensation model is:

$$a_{ek} = a_0 + \mu_k \Delta_0 + \xi_k VB_k \tag{11.20}$$

where a_0 is the nominal deep of cut, Δ_0 is the prediction machining error, μ_k is the system theoretical error compensation coefficient, ξ_k is the compensation coefficient caused by tool wear [42]. The error compensation algorithm is shown in Table 11.1. In Table 11.1, the error is firstly evaluated, and if the machining requirements

Table 11.1 The algorithm for error compensation

Inputs:

Nominal depth of cut a_0
Current prediction error $\Delta(0)$
Machining requirements error accuracy $\Delta_{\lim}$
Initial tool wear $VB(0)$
Theoretical error compensation factor $\mu(0)$
Wear compensation coefficient $\xi(0)$

Procedure:

$k = 0$;
$a_e(0) = a_0 + \mu(0)\cdot\Delta(0) + \xi(0)\cdot VB(0)$;
Do {
    $k{+}{+}$;
    $\mu(k) = [a_0 - a_e(k-1)]/\Delta(k-1)$;
    $\xi(k) = [VB(k) - VB(k-1)]/VB(k)$;
    $a_e(k) = a_0 + \mu(k)\cdot\Delta(0) + \xi(k)\cdot VB(k)$;
    $\Delta(k) = a_0 - a_r(k)$;
} *While* $(\Delta(k) > \Delta_{\lim})$
return $\Delta(k)$, $VB(k)$;

Outputs:

Optimized prediction error $\Delta(k)$
Optimized tool wear $VB(k)$

are satisfied, that is, $\Delta_{rk} < \Delta_{\lim}$, the machining is finished. Otherwise, the related compensation coefficients μ_k and ξ_k need to be modified, and the error compensation value a_{ek} recalculated and updated. Thereupon the error compensation machining is reprocessed until the precision requirements are met.

11.5 Case Study

Experimental validation is conducted with a test from the micro-milling experiments [31]. The materials used is either copper. The tool is two-flute with diameter of 800 μm and works under spindle speed of 18,000 rpm, depth of cut 0.100 mm, feed rate of 0.150 mm/min. The tool wear was measured using the self-designed machine vision system. The monitored signals include multi-channel force, vibration, AE, CNC process data, as well as the tool wear image.

The first step is data pre-processing. Figure 11.5 shows a sample of cutting force with its sparse representation. The coefficient is very sparse, and only 85 of 1024 coefficients are non-zero. The coefficients are matching to the learned dictionary for state estimation. Some learned atoms are shown in Fig. 11.6. It is observed that very different atoms are found: from multiple localized atoms to impulse-like atoms.

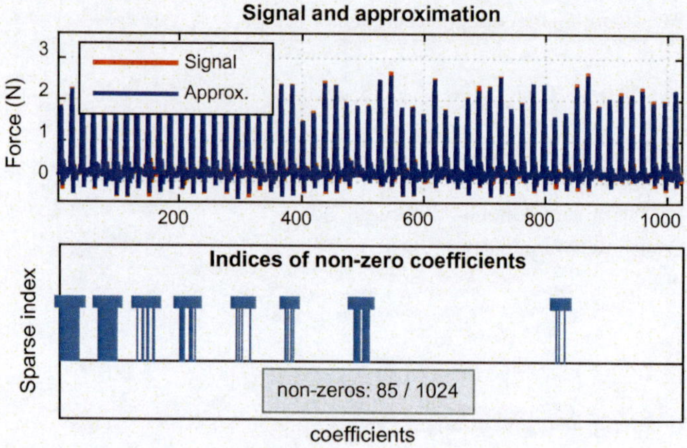

Fig. 11.5 Sparse representation of cutting force

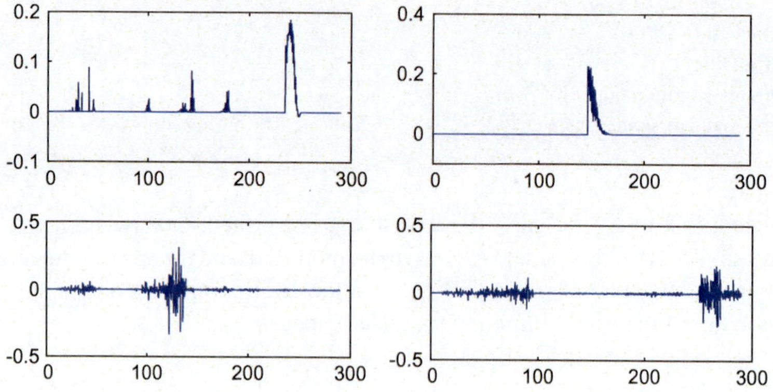

Fig. 11.6 The KSVD dictionary atoms after WPD: localized time and frequency properties

In addition, such a dictionary naturally obtains some favor of shift-invariance in Fig. 11.6, as similar patterns may appear in different locations in different atoms.

Simultaneously, the captured image is decomposed with sparse decomposition for tool wear image denoising and enhancement. Figure 11.7 shows an example of super-resolution. The captured low-resolution image and referred high-resolution image are learned with 256 patches, and then generate the corresponding high and low-resolution dictionaries (a), (b). In dictionary learning, the K-SVD approach is applied. In the super-resolution reconstruction, as in the bottom of Fig. 11.7, given the input original low resolution (c) and noisy images (e), their respective patches are learned and retrieved as in Eqs. 11.7 and 11.8. From the reconstructed image (d) (f), they are found that more tool flank wear appeared and geometry structures are clearer. This structural information is very important for the later tool wear estimation. The denoising effect is an important aspect of the super-resolution reconstruction, as

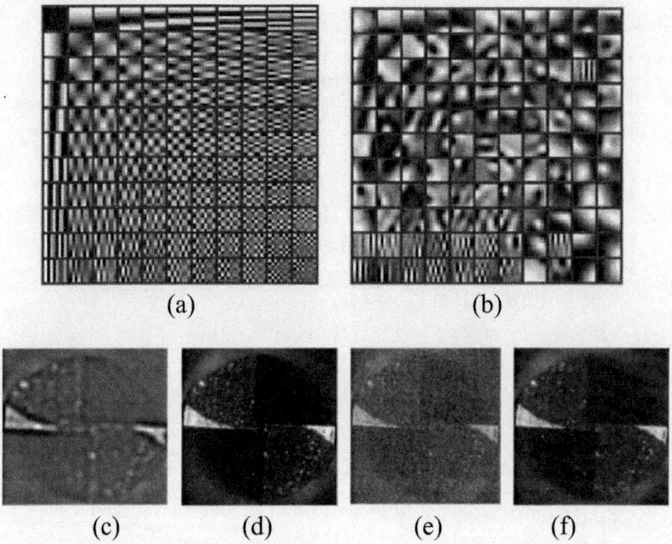

Fig. 11.7 Super-resolution reconstruction from low resolution and noisy images

the noise is relatively high in low-resolution images, and noise-robust features are important for accurate retrieval of the patches.

When the data are ready, the next step is to learn the features with the proposed deep network. Basically, based on the nature of each sensor signal in the machining system, different deep network learning models are established to eliminate the structure difference between different sources. For example, a long short-term memory (LSTM) recurrent neural model (RNN) model is constructed for one-dimensional time series, and convolution neural network (CNN) model is proposed for two-dimensional input of images. For the important monitored signal, the multi-view method is adopted, such as using wavelet transform to process time series, then using convolution neural network to process the time–frequency image, and using deep belief network (DBN) to process the raw time signal. In this way, different modal signals produce different characteristics and the unified feature is generated by feature fusion. Additionally, the different signals that are recorded at different frequencies can be synchronized by data segmentation and normalization. Table 11.2 lists the data modes for the different deep networks, and for the different structure deep networks, the parameters are empirically selected.

The tool wear is finally estimated with the network as in Fig. 11.8. It is observed the estimation is highly consistent with real measured values in the middle stage, while with larger errors at the early and later of tool wear stages. This is due to the nature of deep learning methods, which embedding the model with deep networks and robust to small variations. It could describe the steady tool wear progression well. However, the data-driven model lacks physical understanding of tool wearing process at the initial and accelerated stages, where the small changes of signals (i.e.

Table 11.2 The data types and CNN, DBN, RNN structures

Peered method	Network structure	Features
CNN	Three-convolutional-level CNN	Tool wear images, WT images
CNN	Four-convolutional-level CNN	
CNN	Five-convolutional-level CNN	
RNN	1 LSTM + 2Dense; 2 LSTM + 2Dense	Extracted WT, sparse features
DBN	Three-level DBN with 40 units for each hidden layer	
DBN	Four-level DBN with 20 units for each hidden layer	

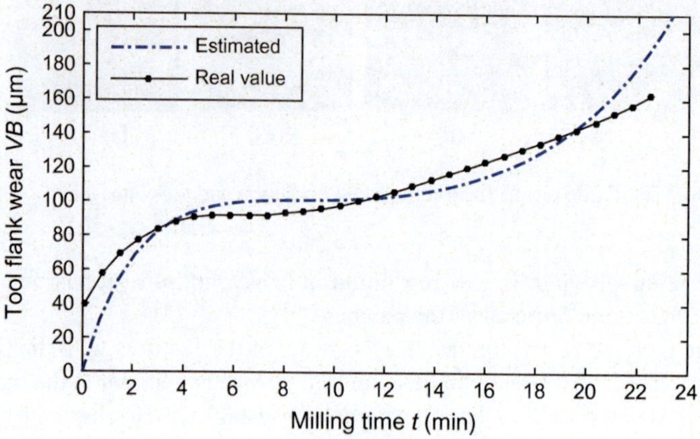

Fig. 11.8 Prediction of tool life with deep network

force) lead to the rapid increase of tool wear state, as illustrated in Fig. 11.8. As a result, this pure-data oriented approach could not meet the high precision demand of micro-milling alone, and the physical model should also be referred for decision.

The final step is for the tool remaining life prediction and wears compensation. With the combination of deep network estimation value and the physical model output as shown in Fig. 11.2, which models the tool condition with the variation of milling process parameters, the final tool state estimation is reached. In this process, the tool failure (breakage and chipping) is first checked, and in most conditions, the tool wear is found and estimated. The predicted tool life is shown in Fig. 11.9a. The milling time t starts with a fresh tool and ends with severe flank wear, where $\overline{t_1}$ and $\overline{t_2}$ are the estimations of tool life per-flute (two-flute micro-milling tool), and $\bar{t}$ is the average value. It is observed that the maximum error of predicted tool life after an initial milling period is less than 1–2 min, which means that the tool life can be predicted from tool flank wear in a short period of initial milling time. With the accurate estimated tool wear value, the wear compensation can be achieved with the algorithm implementation in Table 11.1, and also the tool RUL is estimated as

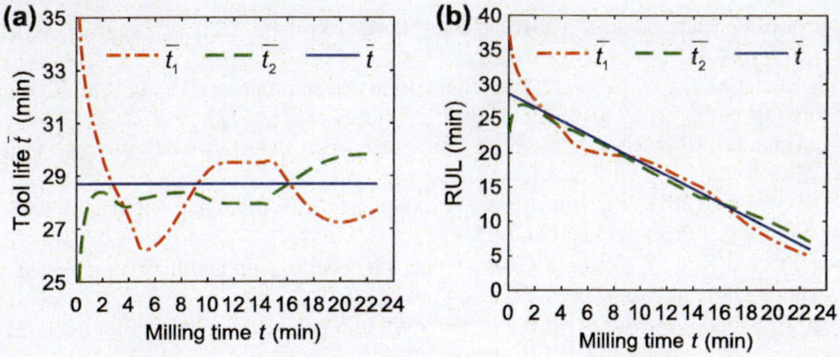

Fig. 11.9 Prediction of tool life over milling time

shown in Fig. 11.9b. The variation of tool life and RUL $\bar{t}$ with the milling time shows predictable changes of the total life under the machining conditions.

11.6 Summary

This chapter has developed a new idea and approach for smart tool condition monitoring, with tool wear and intelligent compensation model of machining process under big data analytics. By constructing a new type of intelligent monitoring system model and architecture for big data analytics, the deep fusion of machine tool and system parameters, theoretical model and monitored data, and the mutual verifications of on-line monitoring system and off-line physical models. Based on machine learning methods, it develops redundant dictionary learning methods for noise cancellation and weak signal extraction, extracts sensitive features, and increases discrimination. At the same time, the low-resolution image denoising and super-resolution reconstruction are realized, which can effectively improve the precision of the tool wear area recognition. The design of different structures of the deep learning network has enhanced the multimode heterogeneous data learning and fusion and extended the monitoring approach. The future study will establish the classification label of the historical operation mode to realize the on-line process mode identification, to carry out the efficient retrieval of the historical operation modal data, and to improve the process optimization function.

References

1. Lee J, Ardakani HD, Yang S, Bagheri B (2015) Industrial big data analytics and cyber-physical systems for future maintenance & service innovation. Procedia CIRP 38:3–7
2. Mourtzis D, Vlachou E, Milas N (2016) Industrial big data as a result of IoT adoption in manufacturing. Procedia CIRP 55:290–295

3. Monostori L, Kádár B et al (2016) Cyber-physical systems in manufacturing. CIRP Ann 65(2):621–641
4. Tao F, Zhang M, Liu Y, Nee A (2018) Digital twin driven prognostics and health management for complex equipment. CIRP Ann-Manuf Technol 67(1):169–172
5. Xu L, Duan L (2018) Big data for cyber physical systems in industry 4.0: a survey. Enterp Inf Syst 7575:1–22
6. He QP, Wang J (2018) Statistical process monitoring as a big data analytics tool for smart manufacturing. J Process Control 67:35–43
7. Liang YC, Lu X, Li W, Wang S (2018) Cyber physical system and big data enabled energy efficient machining optimization. J Clean Prod 187:46–62
8. Canizo M, Conde A et al (2019) Implementation of a large-scale platform for cyber-physical system real-time monitoring. IEEE Access 7:52455–52466
9. GE Predix Platform. https://www.ge.com/digital/iiot-platform
10. Siemens MindSphere. https://siemens.mindsphere.io
11. Teti R, Jemielniak K, O'Donnel G, Dornfeld D (2010) Advanced monitoring of machining operations. CIRP Ann-Manuf Techn 59(2):717–739
12. Zhu KP, Zhang Y (2018) A cyber-physical production system framework of smart CNC machining monitoring system. IEEE/ASME Trans Mechatron 23(6):2579–2586
13. Tapoglou N, Mehnen J, Vlachou A, Doukas M, Milas N, Mourtzis D (2015) Cloud-based platform for optimal machining parameter selection based on function blocks and real-time monitoring. J Manuf Sci E-T ASME 137(4):040909–040909–11
14. Morgan J, O'Donnell GE (2015) Cyber physical process monitoring systems. J Intell Manuf 26(6):1–12
15. Ridwan F, Xu X, Liu GY (2011) A framework for machining optimization based on STEP-NC. J Intel Manuf 23(3):423–441
16. Cha JM, Suh SH, Hascoet JY, Stroud I (2016) A roadmap for implementing new manufacturing technology based on STEP-NC. J Intel Manuf 27(5):959–973
17. Li X, Djordjevich A, Venuvinod PK (2000) Current-sensor-based feed cutting force intelligent estimation and tool wear condition monitoring. IEEE T Ind Electron 47(3):697–702
18. Zhu KP, Mei T, Ye DS (2015) Online condition monitoring in micro-milling: A force waveform shape analysis approach. IEEE T Ind Electron 62(6):3806–3813
19. Kious M, Ouahabi A, Boudraa M, Serra R, Cheknane A (2010) Detection process approach of tool wear in high-speed milling. Measurement 43(10):1439–1446
20. Nouri M, Fussell BK, Ziniti BL, Linder E (2015) Real-time tool wear monitoring in milling using a cutting condition independent method. Int J Mach Tool Manu 89:1–13
21. Bhuiyan MSH, Choudhury IA, Dahari M (2014) Monitoring the tool wear, surface roughness and chip formation occurrences using multiple sensors in turning. J Manuf Syst 33(4):476–487
22. Rmili W, Ouahabi A, Serra R, Leroy R (2016) An automatic system based on vibratory analysis for cutting tool wear monitoring. Measurement 77:117–123
23. Szydłowski M, Powałka B, Matuszak M, Kochmański P (2016) Machine vision micro-milling tool wear inspection by image reconstruction and light reflectance. Precis Eng 44:236–244
24. Zhu KP, Yu XL (2017) The monitoring of micro-milling tool wear conditions by wear area estimation. Mech Syst Signal Pr 93:80–91
25. Duro JA, Padget JA, Bowen CR, Kim HA, Nassehi A (2016) Multi-sensor data fusion framework for CNC machining monitoring. Mech Syst Signal Pr 66–67:505–520
26. Xia M, Li T, Xu L, Liu L, de Silva CW (2018) Fault diagnosis for rotating machinery using multiple sensors and convolutional neural networks. IEEE/ASME T Mech 23(1):101–110
27. Sawo F, Kempkens E (2016) Model-based approaches for sensor data monitoring for smart bridges. In: 2016 IEEE international conference on multisensor fusion and integration for intelligent systems (MFI 2016), Baden-Baden, Germany, September 19–21
28. Malekian M, Park SS, Jun MBG (2009) Tool wear monitoring of micro-milling operations. J Mater Process Tech 209(10):4903–4914
29. Tobon-Mejia DA, Medjaher K, Zerhouni N (2011) CNC machine tool's wear diagnostic and prognostic by using dynamic Bayesian networks. Mech Syst Signal Pr 28:167–182

30. Geramifard O, Xu JX, Zhou JH, Li X (2011) A physically segmented hidden Markov model approach for continuous tool condition monitoring: diagnostics and prognostics. IEEE T Ind Inform 8(4):964–973
31. Zhu KP, Lin X (2019) Tool condition monitoring with multiscale discriminant sparse decomposition. IEEE Trans Ind Inform 15(5):2819–2827
32. Droniou A, Ivaldi S, Sigaud O (2015) Deep unsupervised network for multimodal perception, representation and classification. Robot Auton Syst 71:83–98
33. Yang B, Liu R, Chen X (2017) Fault diagnosis for a wind turbine generator bearing via sparse representation and shift-invariant K-SVD. IEEE T Ind Inform 13(3):1321–1331
34. Sun W, Zhao R, Yan R, Shao S, Chen X (2017) Convolutional discriminative feature learning for induction motor fault diagnosis. IEEE T Ind Inform 13(3):1350–1359
35. MT (The Association for Manufacturing Technology) (2011) MTConnect standard Part 1—overview and protocol, The Association for Manufacturing Technology, McLean, VA
36. Niggemann O, Biswas G, Kinnebrew JS, Khorasgani H, Volgmann S, Bunte A (2015) Data-driven monitoring of cyber-physical systems leveraging on big data and the internet-of-things for diagnosis and control. In: International workshop on the principles of diagnosis
37. Zhu KP, Vogel-Heuser B (2014) Sparse decomposition in the time-frequency domain and its application to micro-milling monitoring. Int J Adv Manuf 68(1–2):1–17
38. Elad M (2010) Sparse and redundant representations: from theory to applications insignal and image processing. Springer
39. Mishne G, Cohen I (2013) Multiscale anomaly detection using diffusion maps. IEEE J-STSP 7(1):111–123
40. Jia F, Lei Y, Lin J, Zhou X, Lu N (2016) Deep neural networks: a promising tool for fault characteristic mining and intelligent diagnosis of rotating machinery with massive data. Mech Syst Signal Pr 72–73:303–315
41. Kiela D, Grave E, Joulin A, Mikolov T (2018) Efficient large-scale multi-modal classification In: Proceedings of AAAI 2018
42. Zhu KP, Zhang Y (2017) Modeling of the instantaneous milling force per tooth with tool run-out effect in high-speed ball-end milling. Int J Mach Tools Manuf 118–119:37–48

Chapter 12
The Cyber-Physical Production System of Smart Machining System

12.1 Introduction

In the year 2013, the German scientists introduced the concept of "Industry 4.0" [1]. They believed that in the next 10 years, the industrialization based on the cyber-physical system (CPS) will make the society enter the fourth revolution dominated by intelligent manufacturing. "Industry 4.0" will make the manufacturing process more flexible and strong, develop new business models, and promote the formation of a new cyber-physical system platform. The core of the "Industry 4.0" strategy is to realize the real-time connection, mutual recognition, and effective communication between people, equipment, and products through CPS network, to build a highly flexible personalized and digital intelligent manufacturing mode. In this mode, production changes from centralization to decentralization, and scale effect is no longer the key factor of industrial production; in the transformation from mass to individuality, future products will be produced completely according to personal demands; users will change from partial participation to full participation, and users will not only appear at both ends of the production process It also participates in the whole process of production and value creation extensively and in real-time. "Industry 4.0" can be regarded as a special application of CPS. The CPS application platform will provide comprehensive, fast, safe, and reliable services and application business processes, and support collaborative manufacturing, service, analysis, and prediction processes in mobile terminal devices and business networks.

12.2 The Cyber-Physical System in Manufacturing

12.2.1 The Definition

The concept of cyber-physical system was first proposed at an NSF workshop in the United States. There is no unified definition of CPS but several are popular: Lee

K. Zhu, *Smart Machining Systems*, Springer Series in Advanced Manufacturing,
https://doi.org/10.1007/978-3-030-87878-8_12

defined CPS from the perspective of informatics as "CPS is a close integration of a series of computing processes and physical process components, which monitors the operation of physical entities through the computing core, and the physical entities realize the perception and control of the environment through the network and computing components" [2]. According to Rajkumar and Lee [3], CPS is a physical and engineering system that realizes monitoring, coordination, control, and integration through computing and communication cores. Harrison et al. [4] defined the CPS from the perspective of automation, where the CPS is to achieve the detection and control of machine operation by embedding the cloud computing and communication chip into the mechanical system in the physical environment.

By deeply embedding computational intelligence, communication, and control capabilities into physical systems, and by utilizing pervasive sensing technology, the CPS enhances the adaptive function of physical systems through active and reconfigurable functional components. Based on the CPS, the manufacturing system significantly improves the informatization, automation, and intelligence, and intelligent manufacturing are realized to improve the flexibility of production and the utilization of resources. In the meantime, the product design, network communication technology, production, and process control technology are comprehensively applied to make the manufacturing equipment become a part of the CPS system.

12.2.2 The CPS Features

CPS can generally be regarded as an efficient networked intelligent information system based on embedded devices, which improves the system's ability in information processing, real-time communication, remote precise control, and component self-coordination through a series of highly integrated and interactive computing units and physical objects in the network environment. The CPS is composed of computing devices, network devices, and physical devices. All devices work together to determine their unique functions and characteristics. The main features of CPS can be summarized as follows.

(1) **Complexity and heterogeneity**: The CPS is composed of a variety of heterogeneous communication networks, computing systems, control systems, and heterogeneous physical devices. Through mutual integration, the physical devices have five functions: computing, communication, precise control, remote coordination, and autonomy. It is a multi-dimensional open system with high complexity and obvious heterogeneity.

(2) **Fusibility**: The CPS achieves deep fusion through the feedback loop of interaction between computing process and manufacturing process, and extends new functions through real-time interaction. Each physical device is deeply embedded with computing and communication functions. As a result, the computing object changes from digital to analog, from discrete to continuous, from static to dynamic. It is a system with many types of computing objects.

(3) **Autonomy and intelligence**: The CPS not only integrates computing, control, and manufacturing processes but also has powerful autonomous function and intelligent decision-making ability, which enables more flexible interaction and intelligent cooperation between computing components and the physical environment, which is mainly reflected in self-organization, self-adaptation, and self-management.

(4) **Real-time and magnanimity**: The CPS needs to know the current situation of physical devices in time and control and intervene the physical devices through network control. However, due to the random changes of equipment status caused by the access of mobile devices, it is necessary to conduct real-time dynamic reorganization of physical devices, which requires the time certainty and parallelism of the calculation process and the real-time performance of the network Very high. At the same time, CPS will produce a huge amount of data in real-time data acquisition and information interaction, so the demand for massive data processing will become very urgent.

The CPS mainly considers performance optimization in function and is a smart system integrating the "3C", i.e., computation, communication, and control technology [5]. The basic functional units of CPS include the drive execution unit, detection perception unit, and decision control unit. As shown in Fig. 12.1, the sensor and the actuator are the interfaces of the physical and computational worlds, and the decision control unit deploys the monitoring task according to the control rules; the sensor feeds back the sensing information to the decision control unit as the input of the control rule algorithm, and obtains the control instruction through calculation; the actuator controls the physical object according to the control command [6].

The CPS has the characteristics of high reliability, real-time, and autonomy. The stability and reliability of the system are the primary requirements for the construction

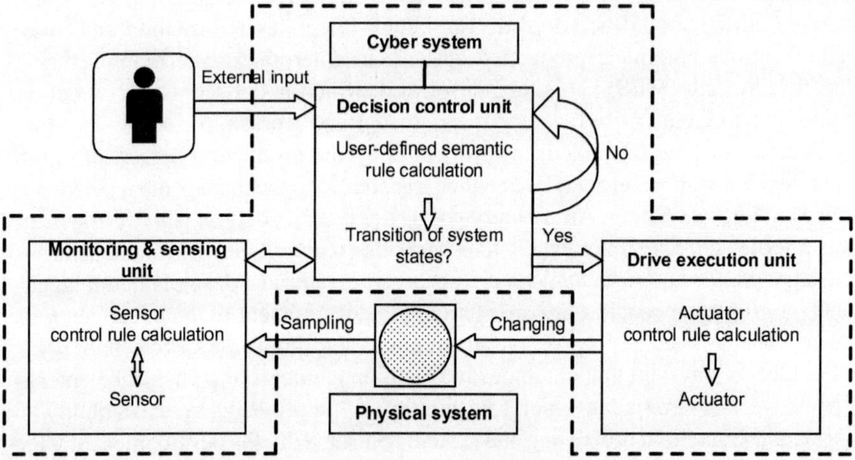

Fig. 12.1 The basic functional units of CPS

of complex large-scale manufacturing systems involving human–machine interaction. The CPS is a system with strict real-time requirements, which needs to complete pre-defined tasks in each specified period, and can process the input information of the system promptly. Meanwhile, there are a lot of uncertain and uncontrollable factors in the manufacturing environment of CPS, and unpredictable problems are likely to occur. Therefore, the CPS needs to have autonomy and achieve the established goals based on adaptive strategies.

The sensing and control method of the manufacturing process is key of CPS, which include sensor network, Internet of things network, smart devices, ubiquitous information acquisition and processing, and distributed intelligent control method of the manufacturing system. The CPS-based manufacturing system realizes the interconnection of manufacturing cells in the system through wired or wireless way to form a real-time distributed manufacturing system network. In this way, it integrates various types of terminals, devices, software, and human–computer interaction technologies with the ability of environment perception, and establishes a human–computer harmonious manufacturing system full of computing and communication capabilities.

12.3 The CPS of Machine Tool and Machining Process

12.3.1 The State-of-the-Art

As has been introduced in Chap. 1, the concept of manufacturing system was first proposed in the 1970s [7]. It is believed that each production link in an enterprise is an indivisible whole. The information existing in the enterprise manufacturing process must be integrated, and the computer integrated manufacturing systems (CIMS) should be adopted. The CIMS integrates information and functions of various automatic production systems in enterprises. The Manufacturing Execution System (MES) [7] is then developed, which is defined as the execution layer connecting enterprise resource management and workshop control. The data is collected from the manufacturing workshop by the production personnel to the system for processing and analysis and then transferred to the management personnel for production scheduling. After entering the twenty-first century, with the maturity of the Internet, cloud computing, big data, and other technologies, the manufacturing industry and information technology are further integrated and developed. Intelligent manufacturing has become a hot direction of the development of the manufacturing industry. The core technology of intelligent manufacturing is the CPS technology.

The CPS is a system that combines information elements and physical elements in the network environment, which can significantly improve the system's ability of information perception processing and system control. With the development of CPS technology, its concept has been extended to the field of manufacturing. Compared with the traditional manufacturing system, the CPS manufacturing system pays attention to the deep integration of the physical system and information space. Each module has the ability of intelligent perception. The modules are interconnected and highly autonomous. In the information space, the intelligent decision-making of the manufacturing process is realized through the collection and analysis of physical spatial data. Finally, the production efficiency, production flexibility, and cost reduction are improved. At present, the intelligent manufacturing system based on CPS has become a research hotspot in the field of intelligent manufacturing. Liu and Xu [8] proposed an information physical system architecture with three layers of physical connection layer, intermediate management layer, and computing layer in intelligent manufacturing workshop, aiming to solve the problem of CPS In the system, there are three major problems: equipment interconnection and assistance, multi-source heterogeneous data acquisition and processing analysis, and knowledge-based intelligent decision-making; Ghimire et al. [9] and others proposed a task management framework based on the Internet of things, using the Internet of things technology to obtain real-time events in the workshop, and according to the knowledge base to process and analyze the events, realized the intelligent perception and processing of workshop events; Zhong et al. [10] introduced the use of RFID And wireless network to capture the real-time status of the equipment status monitoring cloud platform, proposed the service-oriented cloud manufacturing workshop framework and the implementation scheme of the Internet of things in the workshop, and carried out an example verification; Zuo et al. [11] proposed a product energy consumption prediction analysis method, established a six-layer energy prediction and analysis system and studied the product energy consumption model. Cai et al. [12] proposed a sensor data integration method to establish the information model of CNC machine tools in the workshop, which combined sensor data with manufacturing data, and realized the prediction of machine tool processing quality by analyzing the characteristics of data; Zheng and Ming [13] and others established a general organization of five layers intelligent manufacturing workshop. A maturity model of enterprise manufacturing workshop is proposed to evaluate the current intelligent degree of the workshop, and it is verified in an automobile manufacturing enterprise.

It can be seen from the above studies that information fusion between the physical workshop and intelligent simulation is an effective way to eliminate information barriers. Although the specific concept and implementation form of intelligent manufacturing have not yet formed a unified definition and mode, its main components can be divided into the physical manufacturing, cyber-manufacturing system and data connecting physics and information system. Its main functions include process monitoring, simulation analysis, fault detection, process optimization, decision support, etc. As the core component of manufacturing system, the data acquisition and monitoring of CNC machine tools is an important part of an intelligent manufacturing.

12.3.2 The CPS of Machine Tool

Due to the significance of machine tools in modern manufacturing, Liu and Xu [8] introduced an initiative about cyber-physical machine tool (CPMT) and their application in a cyber-physical production system (CPPS) for intellectual, adaptivity, automaticity, and visuality. To illustrate the feasibility and functionalities of the proposed CPMT, a CPMT-centered CPPS is proposed as shown in Fig. 12.2 [8]. The CPPS comprises three layers: Physical Level, Cyber Space, and Service Cloud. In order to realize CPMT, four main components, which are CNC machine tools, data acquisition devices, machine tool cyber twin (MTCT), and smart human–machine interfaces (HMIs), are implemented in this model. The CNC machine tools are the whole machining equipment and system, accepting orders from controllers and taking machining. Data acquisition devices, which measure the real-time state data of manufacturing for record and analysis, are comprised of sensors like dynamometers, cameras, identification components, signal processing devices, and so on. MTCT, the kernel of CPMT, utilizes the real-time data from data acquisition devices to construct

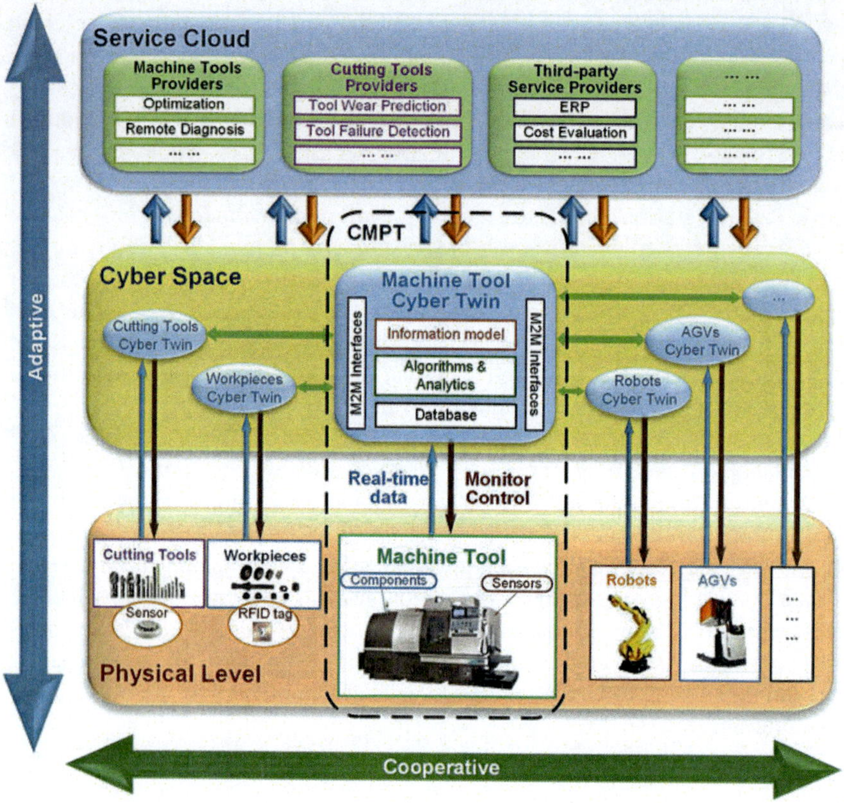

Fig. 12.2 System architecture of cyber-physical machine tool (CPMT) [8]

a specific description of machine tools for diagnosis and decision-making. Moreover, MTCT integrates the communication protocols and algorithms to ensure the interaction between machine tools and other field-level devices like robots for autonomy and cooperation. The service cloud creates a data port to the professional service providers for data analysis and decision suggestions of the particular component of CPPS. Meanwhile, smart HIMs can get access to the CPPS through the service cloud, making managers monitor and adjust the production process conveniently.

In general, the machine tool-centered CPPS not only realizes the horizontal integration, which means different components in physical level can cooperate well to monitor each other and reduce the human interference but also achieves vertical integration. This benefit means that the intelligent algorithms in the cyberspace and professional service from providers in the service cloud can utilize the real-time data of different components in the physical level for prognostic and decision-making [14, 15]. However, some fields need to be studied further for the implementation of the aforementioned CPPS. In the aspect of data, an appropriate description of data source and deployment of data acquisition devices are still challenging for CPPS. In the aspect of algorithms, more efficient, accurate, and intelligent monitoring and decision-making algorithms needed to be developed. In the aspect of interaction, communication methods between field-level equipment and human–machine interaction technologies are still in the early stage, needing to be promoted to match the machine tool-centered CPPS [16].

12.3.3 The CPS of Machining Process

Computer Numerical Control (CNC) machine tools are now moving towards high precision, high speed, and complex functional machining. The machining monitoring system is important to achieve intelligent control of the machining process to produce complex geometric features and precision parts. Traditional NC machining monitoring system is limited in data features, low in adaptability and slow in transmission efficiency, and hard to achieve intelligent online assessment.

To achieve the intelligent, modular, and reconfigurable monitoring system of the machining process, Morgan and O'Donnell [14] proposed a distributed cyber-physical machining monitoring system, which focuses on the signal processing chain to realize the aforementioned functions, as shown in Fig. 12.3. There are five kernel components, which are data acquisition, signal processing, fuzzy events, sequence identification, and sequence analysis, in this signal processing chain. The data acquisition is the beginning of the signal processing chain, which receives the raw measured state data from multiple sensors and transmits data to the Acquire Recognize Cluster (ARC) cloud in Binary Message Model (BMM) format for further analysis. Then the signal processing module acquires the raw signals for pre-processing in parallel to higher signal-to-noise ratio and some primitive features like mean, standard deviation (SD), and so on and transmits processed signals to the Shared Variable Engine (SVE) cloud. The fuzzy agent utilizes the pre-processed signals to form a Boolean

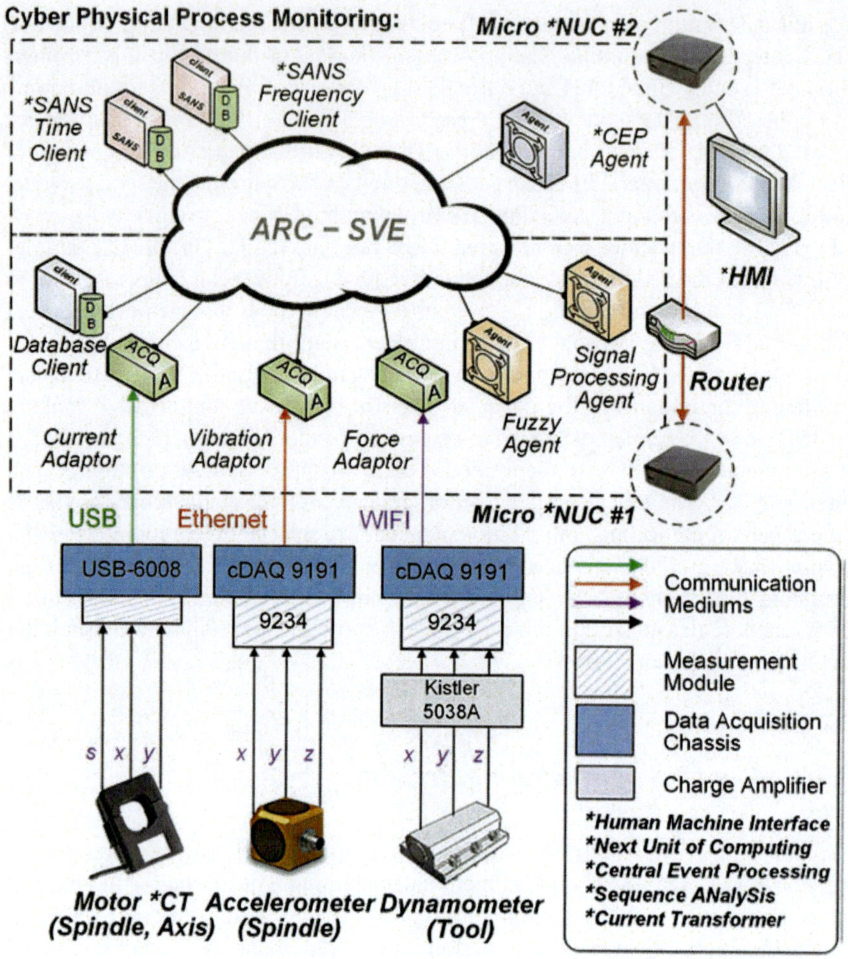

Fig. 12.3 Cyber-physical process monitoring topology [14]

state based on the fuzzy logic inference mechanism, whose function is equivalent to extracting the features and classifying the machining state. In the module of sequence identification, the complex event processing (CEP) agent identifies the machining process through the Boolean logic inference mechanism based on the state signals from the ARC data cloud. In the sequence analysis, clients correlate the further processed signals, which are in the time and frequency domain, with the corresponding machining operations for diagnosis and prognosis. The analysis is in parallel, improving the efficiency of the system.

For validation of the aforementioned cyber-physical system (CPS), an experiment, which is applied to a turning machine, depending on the signal processing chain has been done. In order to measure the state signals, a tri-axial dynamometer, measuring the force, a tri-axial accelerometer, measuring the vibration, and a Current Transformer (CT), measuring the spindle current are mounted to the turning machine. After acquiring the raw signals through multiple sensors, signals are sent to the corresponding adaptors, which are called Next Units of Computing (NUC), for decentralized pre-processing. The group of pre-processed NUC forms NUC#1. Processed signals are shared to an ARC-SVE cloud for interoperability through the internet. The CEP-agent utilizes the Root Mean Square (RMS) of spindle rotation and the SD of rapid movement in z-axial as the input to the fuzzy algorithm, which compares the inputs with the knowledge-based thresholds, to characterize the processing of machining, correlated with the sequence for further analysis. In sequence analysis in the frequency domain, it is found that the peak of the cutting force spectrum is related to cutting parameters, which can be used for monitoring along with the sequence characterization. In sequence analysis in the time domain, features like maximum, RMS, and SD in a particular window size of spindle vibration or cutting force to evaluate the performance of machining. Due to the parallel process of features, the condition can be evaluated from multiple dimensions for higher accuracy. All these analysis clients are formed as NUC#2. All the monitored results and characterization of the process are displayed on the Human–Machine Interferences (HMIs) for visualization along with a manual intervention port.

This model develops a decentralized machining monitoring CPS, which processes the signal in adaptors or agents and monitors the machining through the correlation between machining state and processed features. The aforementioned components constitute a complete signal processing chain, which is context-dependent and event-related, to improve the system accuracy and response speed. For further study, the working process specification can be more accurate by utilizing more events to characterize the process. Furthermore, events, which are triggered through processed signals in this system, need to be more specific through the collaboration between data and manual experience.

Caggiano et al. developed a cloud-based CPS architecture for smart machining process monitoring [15]. The CPS system combines the cloud service with the cyber-physical system to realize real-time catastrophic tool failure (CTF) and consumed tool life estimation along with the appropriate decision. The proposed cloud-based cyber-physical system has a three-layer architecture, which is consisted of physical resources, local server, and cloud, focusing on the signal monitoring and decision process (Fig. 12.4).

In the physical resources, multiple sensors are mounted on the CNC machine tools to acquire real-time machining state data for monitoring; CNC controllers monitor the on-going machining and pass the orders to CNC machine tool based on the decision; local terminals provide a human–machine interface to visualize the monitoring data and provide a manual intervention method. The local server receives the real-time monitoring data from physical resources for pre-processing including filtering, amplifying, and segmenting, and makes decisions. This would be transmitted to the physical resources and constructs a connection between physical resources and cloud, depending on the prognostic result from the cloud. Moreover, the local server has a partial function of storage, which records the processing specifications of machines,

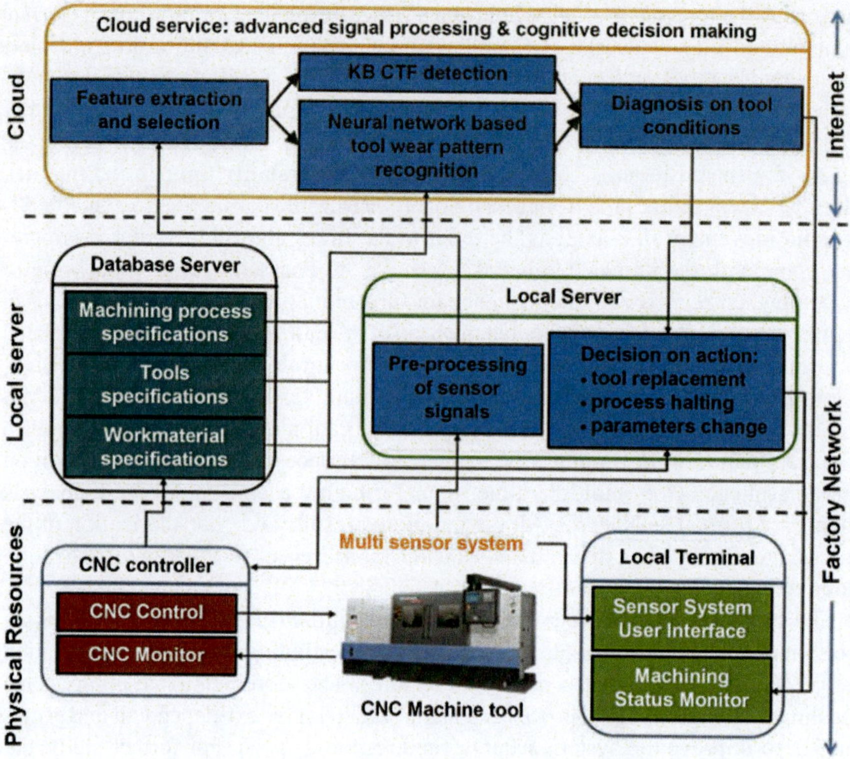

Fig. 12.4 Cyber-physical system for smart monitoring of machining process [15]

tools, and workpieces. The cloud server is the kernel of this architecture due to its functionality of feature extraction and selection, CTF detection, tool wear pattern recognition, and data storage.

The tool wear pattern recognition in this system is based on the Neural Network (NN). The NN uses this training set to train a consumed tool life model for online tool life diagnosis. The CTF and consumed tool life information are sent to the local server for decision making to avoid machining failure. Besides the function of diagnosis, the cloud server provides huge storage for all machining and monitoring data, which can be used for procedure improvement, fault retrospect, and training set enlargement.

The CTF based action time is less than 60 ms and the NN based is less than 2 s. The timely decisions avoid further damage to workpieces and machines. As a result, the enhanced computation and storage enable the cloud-based CPS to a real-time, quick-response, and robust for machining process monitoring. For further study, this cloud-based CPS system can be improved to support the cooperation between several machining systems and allow more smart portable devices to monitor the machining process.

12.4 A CPPS Framework of Smart Machining Monitoring System

12.4.1 Induction

Due to the rapid development of digital control technology, advanced measurement devices, high-performance driver, and controller, CNC machine tools have become more powerful and automated. The precision machine tools and centers have been widely used in the field of the high-end precision manufacturing industry, such as aero-engine blades, gears, crankshaft, etc. Because of their complex structure and poor material machinability, the geometrical precision and surface integrity are difficult to be ensured when the traditional CNC machine tools are used to process the parts [19, 20]. To improve the CNC machine's functionality and performance in the machining process, the intelligent on-line monitoring system needs to be developed under the established machine tool [16, 17]. The investigation has shown that equipped with such a system, the CNC machine tool can reduce downtime by 75%, improve productivity, and greatly improve the machining accuracy [18]. As stated in [16, 19], the improvement of the next generation of CNC machining performance depends largely on the new machining process monitoring system.

For most of the current CNC monitoring system, the data acquisition device performs single side function, either on the system parameters or on process measurements, and lacks interaction. The often custom-developed specific machine tools monitoring system is short of machining process adaptation and reconfiguration [19], and the one-way data acquisition and evaluation leads to the less smartness of the machining system [20, 21]. These have limited CNC machining monitoring capabilities in the applications. On the other hand, due to the development of modern CNC systems and sensing technology, the process is producing a large number of heterogeneous "big data" such as process parameters, measurements, and running history. The "big data" increases exponentially with the process progression, which makes the existing data processing algorithms difficult to handle, and cannot accurately predict the cutting state and optimize the machining process. Therefore, it is important to study the big data-driven machining monitoring system and utilize different sources of information to improve machining precision and enhance manufacturing intelligence. The cyber-physical systems (CPS) [22] integrates and interacts highly with computing units and physical objects in the network environment, and realizes the integration of information and physical systems, therefore, it has a unique advantage for the processing of big data and the intelligent process monitoring. The CPS approaches for condition monitoring were applied to many areas such as electric vehicles [23, 24] and manufacturing systems [18, 21]. It has been studied as the cloud-based approach for digital factory maintenance [25], and machining parameter selection in smart factories [26]. While these studies focused more on manufacturing system CPS modeling, the modeling specifications of the machining (cutting) process unit were not elaborated, and important part of tool condition monitoring was not discussed.

To build the CPS for machining process monitoring, the physical process of machining is first to be modeled. The fundamental physical process of CNC machine tools is the material cutting and relative movement of parts. In this process, the tool is continuously imposed with coupled heat-force interaction, which leads to tool wear. The tool state directly affects the machining accuracy and product quality, and serious tool wear will also cause cutting tool breakage and failure, resulting in chatter, damage to machine tools and workpieces. As a result, tool condition monitoring is one of the most important prerequisites to realize precision and intelligent machining [19]. Li et al. [27] modeled and classified the tool states through the time–frequency characteristics and dynamic characteristics of the cutting force, and achieved good results. Hung and Lu [28] showed the acoustic emission signal and its spectrum feature could meet the requirement of tool wear monitoring from different applications. Others researchers studied vision approaches [29, 30] and built-in sensors [31–33] for chatter and vibration detection so as to realize self-diagnosis and adaptive adjustment of process parameters.

The fusion of multi-sensor signals can reflect the change of machining conditions more comprehensively and improve the ability of state recognition and fault tolerance [34]. Malekian et al. [35] fused the three types of signal, i.e. force, AE, and vibration, and developed the adaptive Fuzzy Neural network model to predict the wear conditions. Wang et al. [36] presented a kernel method for feature selection and fusion and then monitoring the tool conditions with support vector regression. Because of the inherent defects of neural network and kernel methods in feature learning, more researchers turn to probabilistic models which are more convenient for learning and model extension, such as Kalman filter [37], Hidden Markov models [38, 39], and deep neural network [40], which can achieve better results in condition monitoring.

From the above analysis, in the current machining process monitoring studies, researchers generally consider only the CNC machine tool or monitoring system itself, and lack of deep fusion of the physical processes and versatile monitoring information [20]. This greatly limited the process monitoring and intelligent process optimization. How to effectively fuse these multimode heterogeneous big data is the basis of a successful monitoring system realization. In this aspect, deep learning could meet the need with powerful complex data expression ability by learning multi-level data features in the deep network, and it has the unique superiority in the heterogeneous data fusion [41, 42], with successful applications in state perception and fault classification [43, 44]. At the same time, the system needs to model and simulate the interaction to optimize the machining process, which requires that the system has the functions of physical modeling, on-line time-varying working conditions monitoring, big data learning, and analytics. To meet these requirements, this chapter proposes a smart monitoring system for CNC machining based on Cyber-Physical Production System (CPPS) framework. It is built on the CNC machine tool physical and virtual modeling, process monitoring, and big data analytics, and then synergized into a system through a distributed network. Under the CPPS framework, the smart monitoring system is divided into control layer, network layer, and decision layer. The function of each layer and the key technologies of the CPPS involved are discussed. Case studies of machining process monitoring are studied in the applications.

12.4.2 The Smart CNC Machining Monitoring CPPS Structure

12.4.2.1 The Technical Content of the Proposed CPPS

A CPPS framework is proposed for CNC machining monitoring system by synergizing the physical and virtual aspects of the machining system. The CPPS [45] is a specific form of CPS in the manufacturing areas. It relies on the development of information and communication technology and is also based on manufacturing equipment, measurements, and data processing technology breakthroughs [46]. On one hand, the CPPS is to make use of the physical system such as the machine tool entity, system components, and networked facilities to achieve interoperability and life-cycle management of manufacturing processes. On the other side, combined measurements and cyber system simulations to create expert analytics to optimize machining parameters, adjust its state, thereby enhances the adaptability and intelligence of the machining system.

According to the requirements of CNC machining process monitoring, the real-time status of the CNC machining process is estimated and intelligently adjusted based on the physical modeling, motion feature simulation, and data analytics, through a fused system formed in a distributed computing network. Referring to the theory of cyber-physical system [22] and combining it with the characteristics of the CNC machine tool, this paper develops a big data-driven smart machining monitoring CPPS architecture. The technical scheme and key parts are demonstrated in Fig. 12.5 [20].

12.4.2.2 The CPPS System Structure

The proposed CPPS system contains sensing and control layers, network communication layers, and decision and application layers, as shown in Fig. 12.6 [47]. The sensing and control layer describes the physical elements of the CPPS system, such as CNC machines, sensors, controllers, and numerical control physical components. The network communication layer is responsible for the transmission of large capacity, high reliability, and real-time data between virtual space and physical space. The decision and application layer is facing the application and the operator, through the high degree of autonomy in big data learning, real-time evaluation of the machining conditions, the prediction of tool states, and intelligent process optimization. Details of each layer are discussed in the following sections, with a view from bottom to top.

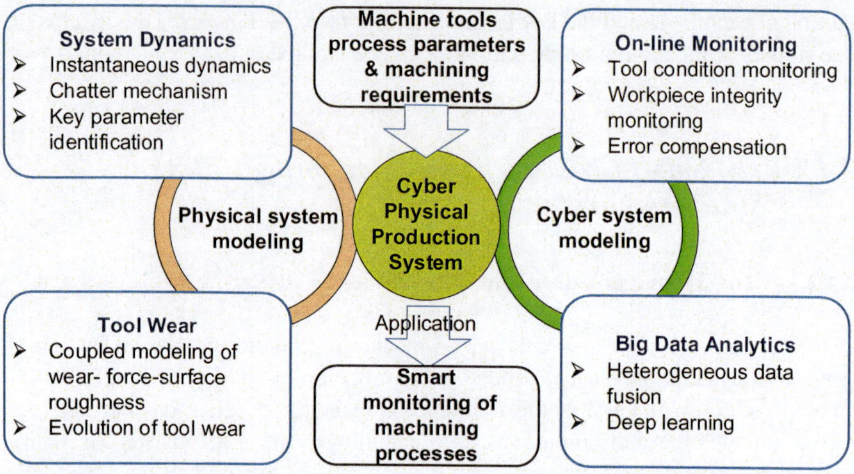

Fig. 12.5 The technical scope of smart machining monitoring CPPS

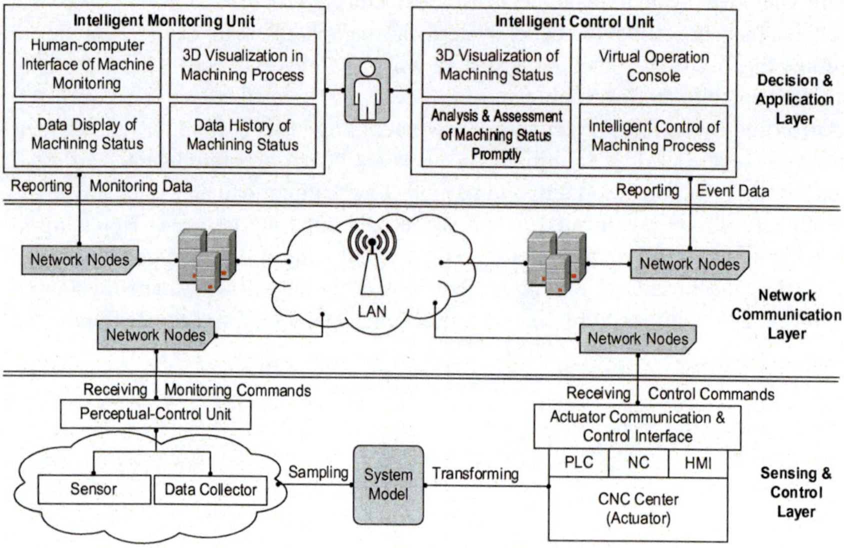

Fig. 12.6 The smart machining monitoring CPPS system structure

(1) The sensing and control layer

In the smart CNC machining monitoring CPPS, the sensing, and control layer describe the interaction of the CPPS and the physical process of the machine tool, which contains the physical elements of the CPPS, such as the machine entity, moving parts, sensor, all kinds of physical controller, driver and CNC physical units, such as cutting tools, tooling, etc. It mainly involves system modeling, numerical control, measurement system, perception, and communication technology. The core of the realization of the smart CPPS is the construction of the sensing function and the integration of the sensing network. It is implemented in the form of NC machine data inquisitor and control network. By mounting sensors and the corresponding data acquisition functional parts, and coupled with CNC machine tool entities (cutter, tooling, workpiece, etc.) coupling, this layer forms in the CPPS node with sensing, control, execution, and autonomous decision-making function.

(2) **The network communication layer**

The communication layer is responsible for data transfer between devices with different layers and includes topological features of packet routing and control system. In real smart machining monitoring CPPS, the network communication layer is constructed to ensure the large-capacity, high reliability, low delay transmission of real-time data between the virtual space and physical space. With the popularity of smart sensing devices and smart communication devices, the sites of network access will be increased exponentially on demand. In view of the existing facts on-site and network transmission reliability and performance, the distributed data acquisition system is designed based on the sensing and control layer to optimize data acquisition and processing strategy. With the sensing, data transmission, and machine-to-machine (M2M) communication mechanisms, it can effectively reduce the real-time data acquisition and processing units, improve network transmission efficiency, and meet real-time and high bandwidth requirements.

Due to the physical system modeling, perception, and simulation, the CPPS produces heterogeneous data sources and they form big data. These big data are to be stored, analyzed, and processed for decision making, but the sensing and control layer is limited in data storage and processing capacity. Therefore, another main function of the network layer is the acquisition and processing of real-time data from the massive data to extract useful information. It serves as a platform to support system operation in two folds. In the upper layer, it provides all kinds of data analysis, graphic operation, and data processing power to support the decision and application layer. In the lower layer, it provides mass data storage, data processing support to the sensing and control layer. At the same time, the sensing and control layer of the perception component and actuator realizes abstract modeling and forms a virtual space and physical space interaction service middleware to achieve status report, supervision instruction, machine tool operation commands, and control functions.

(3) **The decision and application layer**

The decision and application layer is connected to the application and faced to the operator, which is constructed to realize the visual monitoring of the cutting process of the machine tool and the intelligent control. As an enhanced facility, it provides real-time and more comprehensive reference of the NC machining process information, and provides intelligent evaluated data for the operator, to improve the operator's perception and control ability. On the other hand, the smart monitoring system has a high degree of autonomy, and the decision and application layer uses embedded computing and data analytics and assesses the actual machining process in real-time.

In this layer, the machining process can realize intelligent prediction, intelligent alarm, and error-proofing control, which makes possible the NC machining process further to intelligent, even unmanned machining.

12.4.3 Case Studies

This system developed a set of modular components of the CPPS monitoring system through the distributed sensor network. Based on the different functional modular, three cases are studied on the proposed CPPS according to physical system modeling, cyber-system modeling, and fusion of the cyber-physical systems.

12.4.3.1 Production System Modeling

The most widely applied 5-axis high-speed milling (HSM) is taken as an example in the modeling of cutting force. In high-speed milling, to improve the machining efficiency, a high spindle speed is required, and the runout generally exists and is no negligible under this condition, which would bring the tool wear, system vibration, and reducing of the machining accuracy. In most force models, the whole cutting cutter is studied while the variations of instantaneous cutting thicknesses caused by the tool runout are ignored. To describe the relationship between the real force and dynamic parameters more accurately, taking the ball nose end milling cutter, for instance, an instantaneous milling force model $\boldsymbol{F}(t)$ considering the tool runout is established [48]:

$$\mathbf{F}(t) = \sum_{j=1}^{N_t} \left[\int_{z_{jd}}^{z_{ju}} \mathbf{TK} h_j(t, z) db(z) \right] \tag{12.1}$$

where N_t is the number of cutter flutes. **T** is a transformation matrix, transforming the dynamic coordinate system into the fixed coordinate system *O-xyz*. **K** is the milling force coefficients matrix. *db* is the un-deformed chip width, and h_j is the un-deformed chip thickness considering the tool runout effect, as shown in Fig. 12.7. Based on the experimental data, the mechanism of tool runout and milling force is determined with this model, and the milling process can be optimized more accurately by force feedback online.

The exemplary results are illustrated with a high-speed milling test, with three-flute ball nose milling cutter Φ6 mm under spindle speed 10,360 rpm, feed rate 1.555 m/min, depth of cut is 0.25 mm, and width of cut 0.125 mm. Aluminum alloy 7075 was used as work-piece materials. The force was measured with Kistler 9119A dynamometer mounted under the workpiece. A typical example of milling forces component F_x (feed direction) with time t is shown in Fig. 12.8. The prediction

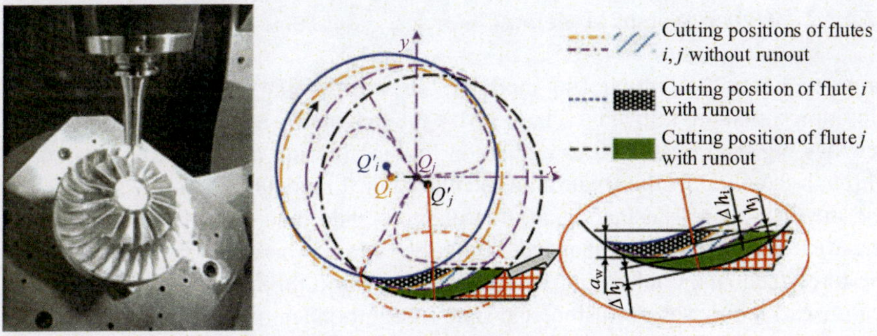

Fig. 12.7 System dynamics and instantaneous cutting force modeling

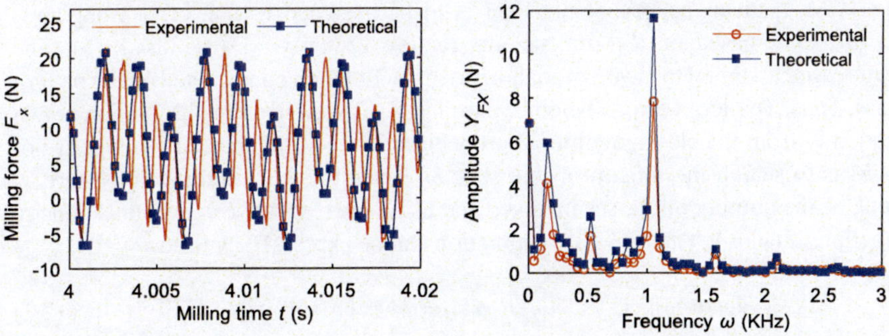

Fig. 12.8 The theoretical, experimental force and their spectrum

amplitudes and means of milling forces are in good agreement with the experimental data in the *x*-axial (feed) direction. The amplitude and frequency variations increase with the milling time, which is affected by the tool wear. As shown in the left part of Fig. 12.8, there are three main peaks in one period on the theoretical milling force curves, which correspond to the time when the respective three flutes are most engaging the workpiece. Amplitude frequency characteristic curves of milling forces obtained from both the experiments and the theoretical model are given in the right part of Fig. 12.8. The larger amplitudes are concentrated in the low-frequency region mainly in the initial milling stage, and the theoretical model can accurately describe the responses of milling force in the high amplitude and low-frequency region. Milling force spectrums of the theoretical model agree with the experimental data highly. When the modeled machining physical process is verified with the cyber system of sensory measurement, they are fused to determine the tool conditions and for further wear compensation strategy.

12.4.3.2 Cyber System Modeling

In the tool condition monitoring modular, the cutting force, acoustic emission, and vibration signal are utilized to characterize the tool states, and the fusion of heterogeneous big data is studied to realize on-line monitoring and tool life prediction. Through the mutual confirmation of the mechanical model of the tool wear process and the features of sensing signal, the reliability, and precision of tool monitoring and wear estimation are enhanced, which provide the basis for the realization of the force feedback control and wear compensation in the machining process. The feature extraction, sensor fusion, and state recognition are divided into independent modules in the traditional state monitoring system, while in this scheme illustrated in Fig. 12.9 these three functions are integrated through big data analytics.

The problem of TCM can be formulated as a big data analytics problem. For the TCM, a given pattern of big data is to be assigned to one of C categories, $\omega_1, \omega_2, \ldots\omega_c$, based on a d-dimensional feature vector $y = (y_1, y_2, \ldots, y_d)$. The features are assumed to have a probability density function conditioned on the pattern class. Thus, a pattern vector x belonging to class ω_i is viewed as an observation drawn randomly from the class-conditional probability function $p(y|\omega_i)$. The objective of TCM is to search the most probable state ω_i given the extracted signal feature y. Let k be the number of classes involved and using the respective conditional density functions $f(y|\omega_i)$. The tool state estimation can be expressed as follows,

$$\text{Determine}: f(y|\omega_i)P(\omega_k) = \underset{k=1,k}{Max}\{f(y|\omega_k)P(\omega_k)\} \tag{12.2}$$

Figure 12.9 shows a big data-oriented TCM system structure. It is implemented in the smart monitoring CPPS modular based on force, acceleration, image, and other types of sensor measurements. The sensor signals are sampled and then transmitted to the control unit for signal analysis. The data acquisition card can be directly installed in the controller, or may also be in separate external devices. Prior to use in the system, the system is trained with machine learning algorithms for the big data, and stores the process information. In the application, the real-time data are acquired and input to the system for decision making. When there is tool breakage, chipping

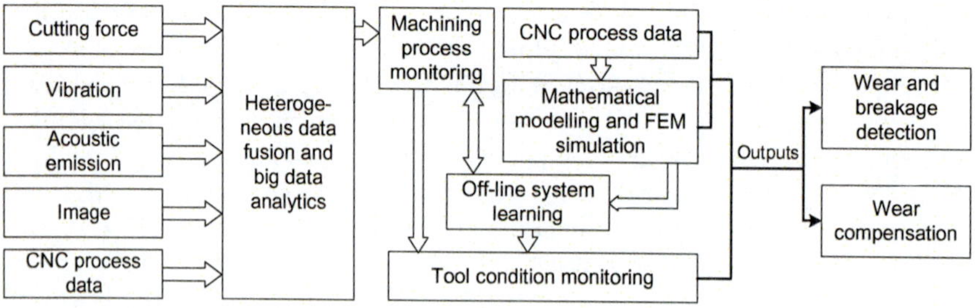

Fig. 12.9 Big data-oriented TCM

or severe wear, the system will automatically trigger the warning system and make decision analytics for further action when necessary.

Figure 12.10 presents an overview of the scheme for the online estimation of the tool state. The online estimation refers that at each time index t an estimate $\hat{\omega}_i$ of ω_i is produced based on all measurement feature $y(t)$ that is available at that time t. This state estimation keeps updating with new extracted features $y(t+1)$ at time index $t+1$, and recursively in this way, the online tool state estimation is achieved. It is noted that, without loss of generality, the state estimation can be implemented by any machine learning techniques in this framework.

Specifically, in this system, the tool state is modeled and estimated with variations of Hidden Semi-Markov Models (HSMM), which improves the current HMM-based models [38, 39] with the inclusion of state duration and dependency. Figure 12.11 shows that by modeling state duration it leads to more accurate estimate results, and the HSMM with tool state dependent durations has higher recognition rates than HSMM with independent durations. The recognition rate is defined as the ratio

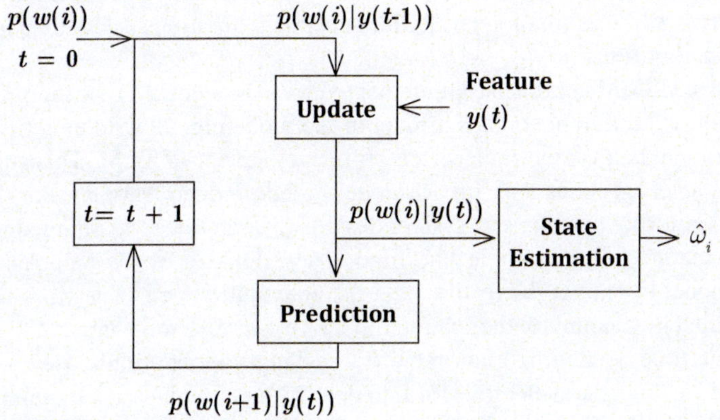

Fig. 12.10 Online tool state estimation with machine learning

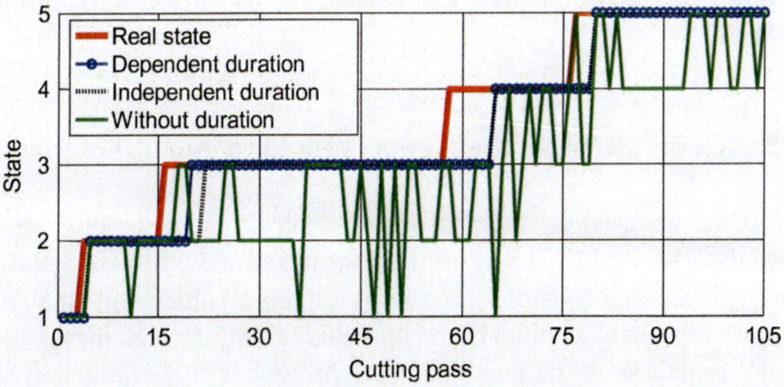

Fig. 12.11 Estimated wear state with HSMMs

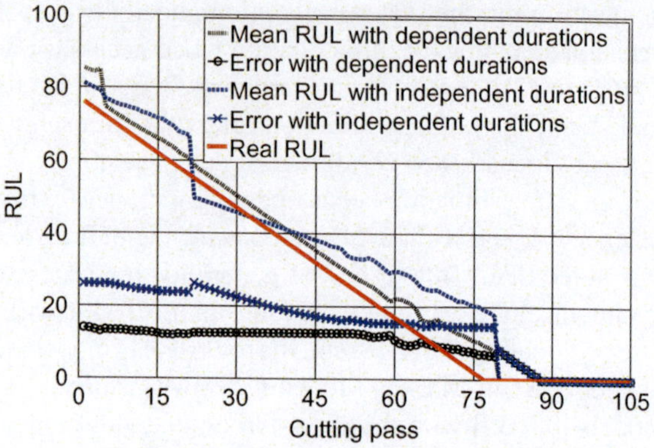

Fig. 12.12 The statistics of RUL with HSMMs

between the number of states estimated correctly and the total number of states. The evolutions of error and mean of estimated Remaining Useful Life (RUL) are given in Fig. 12.12. It indicates that the RUL estimation with dependent durations is more accurate as the estimation error with dependent durations is less than that with independent durations.

The above HSMM approach is implemented with the sole data type (cutting force) and hard to fuse different sources of data in the modeling. How to effectively fuse the multi-mode data is key to the successful development of TCM. This study also develops a heterogeneous data fusion model based on deep learning. For different types of monitoring signals, this system designs different deep learning algorithms to extract features. For example, the time-series signal uses the recurrent neural network model to extract the feature, and the convolution neural network model is used to extract the feature for the spatiotemporal two-dimensional input of the image. At the same time, some important monitoring signals adopt a multi-view learning, such that it can apply wavelet transform to deal with the time-series signal, then use convolution neural network to deal with the time–frequency image, and the recurrent model is used to deal with the original time-series signal. As a result, different modal signals produce different characteristics, and then a unified feature base is generated through feature fusion, and the state prediction is finally made through the whole connection layer.

12.4.3.3 The CPPS Synergy: The Fusion of Physical and Cyber System

The system fusion is based on the closed-loop control mechanism of "perception-simulation-analytics-control", and by utilizing the information from both the physical and cyberspace. In the cyber-space, the system contains all objects and their activities in the life cycle of the product models and knowledge, and through these models and knowledge, all functions in the manufacturing process will be simulated and optimized in cyberspace, so as to identify and avoid problems in the production process. This depends on the behavior of physical systems and its modeling approaches.

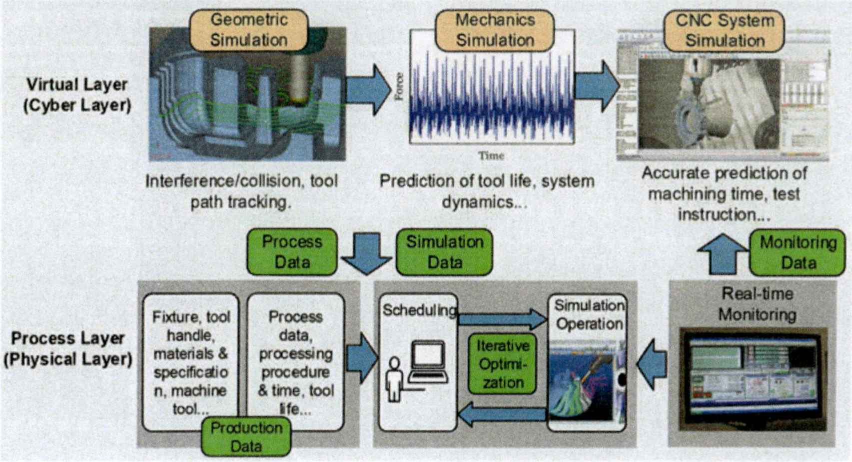

Fig. 12.13 The fusion of physical and virtual modeling of the machining process

Through the precise modeling of the moving parts of NC machine tools and multiple axis coupling dynamics, this study develops a three-dimensional simulation platform of the CNC machine tool, implements the dynamic loading and interactive control of static models and moving parts. Meanwhile, combined with the original CNC machine tool settings, tool geometry, workpiece, and tooling, the system automatically builds a virtual CNC machine tool simulation environment. With the real-time acquisition of CNC process data, it realizes the low delay, high precision simulation, and visualization of the three-dimensional machining process. The system interface is shown in Fig. 12.13.

Based on the big data analysis the fusion process is divided into three steps. The first step is the visualization of big data. Explicit and quantitative data is the basis for analysis and decision-making of the manufacturing process [49]. Most current control systems could provide data acquisition interfaces to facilitate the extraction of process parametric information. The second step is the realization visualization-aided machining process. After the manufacturing data is extracted, it is applied to build the mappings between the data and machining process, and mapping within the process instructions. Through the analysis of machining productivity or quality, improves and optimizes the NC programs. The final step is the implementation of big data analytics. It builds the knowledge base of manufacturing data associated with quality and productivity, and through the real-time analysis of machine operating parameters to predict product quality, machine breakdowns optimize them in real-time and finally reach the real smart machining.

For example, when the tool wear *VB* is estimated from the other module, the machining error compensation model is:

$$a_{ek} = a_0 + \lambda_k \Delta_0 + \xi_k V B_k \tag{12.3}$$

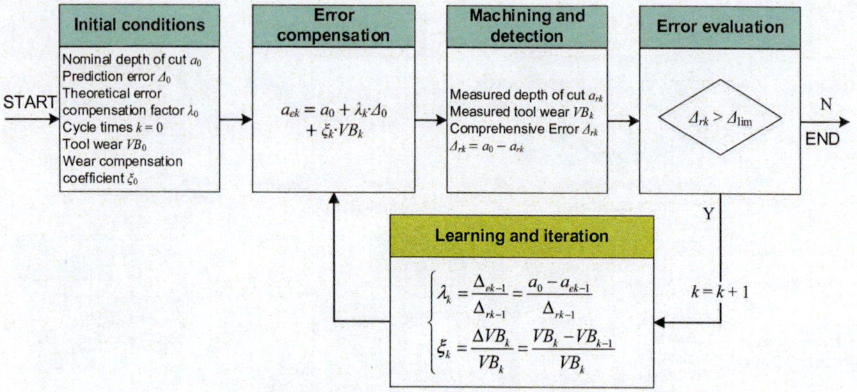

Fig. 12.14 The error compensation approach

where a_0 is the nominal depth of cut. Δ_0 is the prediction machining error, λ_k is the system theoretical error compensation coefficient. ξ_k is the compensation coefficient caused by tool wear. The error compensation algorithm is shown in Fig. 12.14.

In Fig. 12.14, λ_0 is a predictive simulation value of the error compensation coefficient, and ξ_0 is the position compensation coefficient caused by the continuous tool wear. $\Delta_{\lim}$ is an allowable maximum value of the machining error, which is the synthetical error caused by the system prediction and tool wear after the k-th machining process. According to the parameters Δ_0, λ_0, and ξ_0, the compensation value a_{ek} of the cutting depth error are calculated, and the initial error compensation machining is processed. After each process (or pass), the actual depth of cut a_{r0} is measured, and the system synthetical error Δ_{r0} is calculated. Then, the error is evaluated. If the machining requirements are satisfied, that is, $\Delta_{rk} < \Delta_{\lim}$, the machining is finished. Otherwise, the related compensation coefficients λ_k and ξ_k need to be modified, and the error compensation value a_{ek} of cutting depth is recalculated and updated. Thereupon the error compensation machining is reprocessed until the precision requirements are met.

12.5 Summary

To improve the machining system monitoring capabilities, this chapter develops a smart machining monitoring system in the CPPS framework. Based on the CPPS theories and machine learning techniques, the system parameters, CNC process data, and monitoring signals are synergized and communicated through the distributed network. Heterogeneous data fusion and big data analytics are applied to enhance the CPPS intelligence. The system implements a variety of monitoring modular based on the modeling of physical and cyber systems. Three case studies have demonstrated its capabilities in the physical and cyber system modeling of the machining system.

With this CPPS framework, the monitoring system would significantly increase the CNC machining intelligence and improve the machining precision and productivity. Future studies will be conducted on extending the heterogeneous data scope such as process textual data and the 3D CAD models for deep learning and machining process optimization.

References

1. Kagermann H, Wahlster W, Helbig J (2012) Securing the future of German manufacturing industry recommendations for implementing the strategic initiative INDUSTRIE 4.0. Federal Ministry of education and research Final report of the Industrial 4.0 Working Group, Germany
2. Lee EA (2006) Cyber-physical systems—are computing foundations adequate? Position paper for NSF workshop on cyber physical systems: research motivation, techniques and roadmap
3. Rajkumar R, Lee I (2010) Cyber-physical systems: the next computing revolution. Proceedings of the design automation conference. pp 731–736
4. Harrison R, Vera D, Ahmad B (2016) Engineering methods and tools for cyber–physical automation systems. IEEE Proc 104(5):973–985
5. Sztipanovits J, Karsai G, et al (2012) Toward a science of cyber–physical system integration. Proc IEEE 100(1)
6. Leitao P, Colombo AW, Karnouskos S (2016) Industrial automation based on cyber-physical systems technologies: prototype implementations and challenges. Comput Ind 81:11–25
7. Groover MP (2015) Automation, production systems, and computer-integrated manufacturing, 4th edn. Pearson
8. Liu C, Xu X (2017) Cyber-physical machine tool—the era of machine tool 4.0. Procedia CIRP 63:70–75
9. Ghimire S, Luis-Ferreira F, Nodehi T, et al (2017) IoT based situational awareness framework for real-time project management. Int J Comput Integr Manuf 30(1):74–83
10. Zhong RY, Wang L, Xu X (2017) An IoT-enabled real-time machine status monitoring approach for cloud manufacturing. Procedia CIRP 63:709–714
11. Zuo Y, Tao F, Nee AYC (2017) An internet of things and cloud-based approach for energy consumption evaluation and analysis for a product. Int J Comput Integr Manuf. 1–12
12. Cai Y, Starly B, Cohen P, et al (2017) Sensor data and information fusion to construct digital-twins virtual machine tools for cyber-physical manufacturing. 10:1031–1042
13. Zheng M, Ming X (2017) Construction of cyber-physical system–integrated smart manufacturing workshops: a case study in automobile industry. Adv Mech Eng 9(10):168781401773324
14. Zhang J, Deng C, et al (2021) Development of an edge computing-based cyber-physical machine tool. Robot Comput Integr Manuf 67:102042
15. Luo W, Hu T, Ye Y, Zhang C, Wei Y (2020) A hybrid predictive maintenance approach for CNC machine tool driven by digital twin. Robot Comput Integr Manuf 65:101974
16. Xie Y, Lian K (2021) Digital twin for cutting tool: modeling, application and service strategy. J Manuf Syst 58(B):305–312
17. Morgan J, O'Donnell GE (2015) Cyber physical process monitoring systems. J Int Manuf 26(6):1–12
18. Caggiano A, Segreto T, Teti R (2016) Cloud manufacturing framework for smart monitoring of machining. Procedia CIRP 55:248–253
19. Herwan J, Kano S, et al (2018) Cyber-physical system architecture for machining production line. In: IEEE industrial cyber-physical systems (ICPS). St. Petersburg, Russia
20. Zhu KP, Zhang Y (2018) A cyber-physical production system framework of smart CNC machining monitoring system. IEEE/ASME Trans Mechatron 23(6):2579–2586

21. Teti R, Jemielniak K, O'Donnel G, Dornfeld D (2010) Advanced monitoring of machining operations. CIRP Ann Manuf Technol 59(2):717–739
22. Deshayes L, Welsch L, Donmez A, Ivester R, Gilsinn D, Rhorer R, Whitenton E, Potra F (2005) Smart machining systems: issues and research trends. The 12th CIRP life cycle engineering seminar, Grenoble, France, pp 3–5
23. Gao R, Wang L, Teti R, Dornfeld D, Kumara S, Mori M, Helu M (2015) Cloud-enabled prognosis for manufacturing. CIRP Ann Manuf Technol 64(2):749–772
24. Altintas Y, Kersting P, Biermann D, Budak E, Denkena B, Lazoglu I (2014) Virtual process systems for part machining operations. CIRP Ann Manuf Technol 63(2):585–605
25. Lee J, Lapira E, Bagheri B, Kao HA (2012) Recent advances and trends in predictive manufacturing systems in big data environment. Manuf Lett 1(1):38–41
26. Monostori L, Kadar B, Bauernhansl T, Kondoh S, Kumara S, Reinhart G, Sauer O, Schuh G, Sihn W, Ueda K (2016) Cyber-physical systems in manufacturing. CIRP Ann Manuf Technol 65(2):621–641
27. Foundations for Innovation in Cyber-Physical Systems (2014) National Institute of Standards and Technology (NIST), Gaithersburg
28. Lv C, Liu Y, Hu X, et al (2018) Simultaneous observation of hybrid states for cyber-physical systems: a case study of electric vehicle powertrain. IEEE Trans Cybern 48(8):2357–2367
29. Lv C, Xing Y, Zhang J, Na X, et al (2018) Levenberg-Marquardt backpropagation training of multilayer neural networks for state estimation of a safety critical cyber-physical system. IEEE Trans Industr Inform 14(8):3436–3446
30. Mourtzis D, Vlachou E, Milas N, Xanthopoulos N (2016) A cloud-based approach for maintenance of machine tools and equipment based on shop-floor monitoring. Procedia CIRP 41:655–660
31. Tapoglou N, Mehnen J, Vlachou A, Doukas M, Milas N, Mourtzis D (2015) Cloud-based platform for optimal machining parameter selection based on function blocks and real-time monitoring. J Manuf Sci Eng 137(4):040909
32. Li X, Djordjevich A, Venuvinod PK (2000) Current-sensor-based feed cutting force intelligent estimation and tool wear condition monitoring. IEEE Trans Ind Electron 47(3):697–702
33. Hung CW, Lu MC (2012) Model development for tool wear effect on AE signal generation in micro-milling. Int J Adv Manuf Technol 66(9–12):1845–1858
34. Szydłowski M, Powałka B, Matuszak M, et al (2016) Machine vision micro-milling tool wear inspection by image reconstruction and light reflectance. Prec Eng 44:236–244
35. Zhu KP, Yu XL (2017) The monitoring of micro milling tool wear conditions by wear area estimation. Mech Syst Signal Process 93:80–91
36. Kuljanic E, Sortino M, Totis G (2008) Multisensor approaches for chatter detection in milling. J Sound Vib 312(4–5):672–693
37. Wang W, Jianu OA (2010) A smart sensing unit for vibration measurement and monitoring. IEEE/ASME Trans Mechatron 15(1):70–78
38. Rizal M, Ghani JA, Nuawi MZ, Haron CHC (2015) Development and testing of an integrated rotating dynamometer on tool holder for milling process. Mech Syst Signal Process 52–53:559–576
39. Mahajan A, Wang K, Ray PK (2001) Multisensor integration and fusion model that uses a fuzzy inference system. IEEE/ASME Trans Mech 6(2):188–196
40. Malekian M, Park SS, Jun MBG (2009) Tool wear monitoring of micro-milling operations. J Mat Process Technol 209(10):4903–4914
41. Wang J, Xie J, Zhao R, Mao K, Zhang L (2016) A new probabilistic kernel factor analysis for multisensory data fusion: application to tool condition monitoring. IEEE Trans Instrum Meas 65(11):2527–2537
42. Ompusunggu AP, Papy JM, Vandenplas S (2015) Kalman-filtering-based prognostics for automatic transmission clutches. IEEE/ASME Trans Mech 21(1):419–430
43. Zhu KP, Liu T (2018) Online tool wear monitoring via hidden Semi-Markov model with dependent durations. IEEE Trans Ind Inform 14(1):69–78

44. Geramifard O, Xu JX, Zhou JH, Li X (2012) Multimodal hidden Markov model-based approach for tool wear monitoring. IEEE Trans Ind Electron 61(6):2900–2911
45. Xia M, Li T, Xu L, Liu L, Silva CWD (2018) Fault diagnosis for rotating machinery using multiple sensors and convolutional neural networks. IEEE/ASME Trans Mech 23(1):101–110
46. Schmidhuber J (2015) Deep learning in neural networks: an overview. Neural Netw 61:85–117
47. Wang P, Yan AR, Gao R (2017) Visualization and deep recognition for system fault classification. J Manuf Syst 44:310–316
48. Droniou A, Ivaldi S, Sigaud O (2015) Deep unsupervised network for multimodal perception, representation and classification. Robot Auton Syst 71:83–98
49. Sun W, Shao S, Zhao R, Yan R, Zhang X, Chen X (2016) A sparse auto-encoder-based deep network approach for induction motor faults classification. Measurement 89:171–178
50. Monostori L (2014) Cyber-physical production systems: roots, expectations and R&D challenges. Procedia CIRP 17:9–13
51. Lee J, Bagheri B, Kao HA (2015) A cyber-physical systems architecture for industry 4.0-based manufacturing systems. Manuf Lett 3:18–23
52. Zhu KP, Zhang Y (2017) Modeling of the instantaneous milling force per tooth with tool run-out effect in high speed ball-end milling. Int J Mach Tools Manuf 118–119:37–48
53. Morgan J, O'Donnell GE (2017) Multi-sensor process analysis and performance characterization in CNC turning-a cyber physical system approach. Int J Adv Manuf Technol 92(1–4):855–868